Gender and Science

Culture and Environment of South Asia

This series aims at bridging the gap between the broadly defined cultural studies and research on various aspects of environmental history of South Asia. The word 'culture' is understood as a product of social practices, and therefore lies at the heart of society as well as the environment that shaped civilizations. Under the rubric of 'environment', the series proposes to look at not only ecology (air, water, forests, minerals, etc.) but also the social and cultural factors that shaped peoples' lives. Historically, it is the science and technology in a way that transformed culture and environment. Studies on science, technology and medicine are a crucial part of this series as they ideally demonstrate the mediation that articulates the mutually influencing relationship between culture and the living environment. The series brings into wider circulation the scholarly works that either analyses in detail any one aspect of the series or various aspects in terms of their broad interconnectedness and their influence on the history of South Asia.

Series Editor
Atlury Murali
University of Hyderabad

Books in the series

Becoming India: Western Himalayas under British Rule, *Aniket Alam*

Forest Ecology in India: Colonial Maharashtra 1850–1950, *Neena Ambre Rao*

Forest Policy and Ecological Change: Hyderabad State in Colonial India, *S. Abdul Thaha*

Himalayan Degradation: Colonial Forestry and Environmental Change in India, *Dhirendra Datt Dangwal*

Indigenous and Western Medicine in Colonial India, *Madhuri Sharma*

Gender and Science

Studies across Cultures

Edited by

Neelam Kumar

FOUNDATION®
BOOKS

Delhi • Bengaluru • Mumbai • Kolkata • Chennai • Hyderabad
Pune • Thiruvananthapuram

Published by
Cambridge University Press India Pvt. Ltd.
under the imprint of Foundation Books
Cambridge House, 4381/4 Ansari Road, Daryaganj, **New Delhi** 110 002

Cambridge University Press India Pvt. Ltd.
C-22, C-Block, Brigade M.M., K.R. Road, Jayanagar, **Bengaluru** 560 070
Plot No. 80, Service Industries, Shirvane, Sector-1, Nerul, **Navi Mumbai** 400 706
10 Raja Subodh Mullick Square, 2nd Floor, **Kolkata** 700 013
21/1 (New No. 49), 1st Floor, Model School Road, Thousand Lights, **Chennai** 600 006
House No. 3-5-874/6/4, (Near Apollo Hospital), Hyderguda, **Hyderabad** 500 029
Agarwal Pride, 'A' Wing, 1308 Kasba Peth, Near Surya Hospital, **Pune** 411 011
T.C. 25/2731 Lukes Lane, Ambujavilasam Road, **Thiruvananthapuram** 695 001

ISBN 978-81-7596-925-4

First Published 2012

Published by Manas Saikia for Cambridge University Press India Pvt. Ltd. and printed at Sanat Printers, Kundli, Haryana.

For

CSIR-NISTADS

Contents

List of Contributors

The Editor

Neelam Kumar graduated in psychology from Patna University and completed her PhD from Jawaharlal Nehru University. She is currently a Scientist at National Institute of Science, Technology and Development Studies, New Delhi. Her research interests lay in gender issues and organisational culture in scientific institutions. Her publications include an edited volume titled *Women and Science in India* (2009). Her forthcoming project would be on psychology of science in Indian context.

The Contributors

Luci Abrahams is the Director and Senior Lecturer at LINK (Learning, Information, Networking, Knowledge) Centre, Faculty of Humanities, University of the Witwatersrand, Johannesburg, South Africa. She writes on themes, institutions and economic sectors in the 'network, knowledge economy' and teaches on the Masters degree in ICT Policy and Regulations. Luci served as a council member on the National Advisory Council on Innovation from 1999–2009 and chaired the committee on SET for Women between 2003 and 2007. She is currently a member of the Financial and Fiscal Commission.

Mineke Bosch is Professor of Modern History at University of Groningen since 2008. She also held the Chair at the Centre for Gender and Diversity, University of Maastricht. Her specialisation includes gender history with a focus on the history of science, (international) feminism, and (auto)biography. She has published more than 150 papers and books, including *Curious Careers: An Unexpected History of Women in Science and Technology.*

Francesca Bray is Professor of Social Anthropology at the University of Edinburgh. She has held research and teaching posts at the Needham Research Institute in Cambridge, at the Centre National de la Recherche Scientifique (National Center for Scientific Research [CNRS]) in Paris, in the Department of Anthropology at the University of California, Los Angeles (UCLA), at the Centre for the History of Science, Technology and Medicine at the University of Manchester, and the Department of Anthropology at the University of California (UC), Santa Barbara. She has many publications to her credit including, the volume on *Agriculture* in the series *Science and Civilisation in China* (1984); *The Rice Economies: Technology and Development in Asian Societies* (1986, 1994); *Technology and Gender: Fabrics of Power in Late Imperial China* (1997); *Technology and Society in Ming China, 1368–1644* (2000) and *Technologies that Matter: Gender, Government and History in China* (forthcoming).

Françoise Cyrot-Lackmann is a former Director of Research at CNRS (France's National Centre of Scientific Research). Besides being an extremely successful physicist and a researcher, she has been devoted to fight for equality for women in science, and has been Director of the Mission for Gender Equality in Science and Technology, France. She has been member of various international advisory board and chief of MPST (Mission for Parity in Science and Technology). She has numerous publications to her credit.

Henry Etzkowitz is currently Senior Researcher at Stanford's Human Sciences and Technologies Advanced Reaearch Institute (H-STAR). Prior to that, he was a Fellow of the University's Clayman Institute for Gender Research, 2009–10. He is also Visiting Professor, Centre for Entrepreneurship Research, Edinburgh University Business School and Department of Management, Birkbeck, University of London. He is President of the Triple Helix Association and Editor of its forthcoming journal, *The Triple Helix Innovation Quarterly.*

He has held the Chair in Management of Innovation, Creativity and Enterprise at Newcastle University Business School and served as a Visiting Professor in the Department of Technolgy and Society, School of Engineering and Applied Sciences, Stony Brook University. He publishes regularly in Research Policy, Science and Public Policy and Social Science Information and is co-author of *Athena Unbound: The Advancement of Women in Science and Technology* (2000). His forthcoming book is 'The Gender Dynamics of Science and Technology: The 'Vanish Box' Phenomenon'.

Gregory J. Feist is Associate Professor of Psychology at the San Jose State University. He is founding President of the International Society for the Psychology of Science and Technology and editor-in-chief of its journal – *Journal of Psychology of Science* & *Technology*. Dr Feist has published widely on the psychology of creativity, the psychology of science, and the development of scientific talent.

Claudine Hermann is Honorary Professor of Physics at Ecole Polytechnique, France. She is the first woman ever appointed as Professor at Ecole Polytechnique (1992). She is an expert in the European Union European Technology Assessment Network (ETAN) group on Women and Science, and also is a French member of the Women in Science Helsinki group. A co-founder and the first president of the French Femmes & Sciences (Women and Science) association, she is now President of Honour and is Vice-President of the European Platform of Women Scientists.

Samira I. Islam is the first Saudi woman to obtain a PhD degree and to become a professor. She is the first Muslim and the first Arab to be nominated in the United Nations Educational, Scientific and Cultural Organization (UNESCO)/L'Oreal 'For Women in Science' Award and choosen as one of the 32 distinguished scientists of the world for the year 2000.

J. Scott Long is Distinguished Professor and Chancellor's Professor of Sociology and Statistics at the Indiana University, Bloomington. His research focuses on gender differences in the scientific career, aging and labour force participation, human sexuality, and statistical methods. He is the former editor of *Sociological Methods and Research*. Recently, the National Academy of Sciences has published his major work 'From Scarcity to Visibility'.

Mariko Ogawa was Executive Advisor to the President, Director of the Support Office for Women Researchers and Professor of History of Science and Science Studies, Mie University, Japan. She is also an Associate Member of the Science Council of Japan, a Visiting Professor of the Institute for Gender Studies, Ochanomizu University. Her specialty is scientific history and philosophy of science. She is the author of several books research papers including 'Uneasy Bedfellows', *Bulletin on History of Medicine* (2000), 'The Mysterious Mr. Collins', *Journal of History of Biology* (2001) and 'The Languages of Science: The Circulation of Knowledge in Translation' (Ogawa and Endo) Archives Internationales d' Historie des Sciences (2006).

Jeanne Peiffer is a historian of mathematics at the French Centre National de la Recherche Scientifique (French Centre of National Research [CNRS]). She is head of the Centre Alexandre Koyré in Paris. As a feminist in France, she has participated, with Hélène Rouch, in the first reflections on gender and science, and has been active in organising research in the field.

Hélène Rouch is a biologist at Centre d'enseignement, de documentation et de recherche féministes (CEDREF). She is one of the pioneers of the reflection on gender and science in France, and is very active in establishing feminist research at French academic institutions. She acted as scientific editor of two important book series, one called *Recherches* with the publisher Côté-femmes and another entitled *Bibliothèque du féminisme* with L'Harmattan.

Londa Schiebinger is the John L. Hinds Professor of History of Science at the Stanford University and Director of the Eu/US Gendered Innovations in Science, Health & Medicine, and Engineering She is author of numerous papers and books including *The Mind Has No Sex? Women in the Origins of Modern Science* (1989); *Has Feminism Changed Science?* (1999). She has edited *Feminism and the Body* (2000) and has co-edited several volumes including *Feminism in Twentieth-century Science, Technology, and Medicine* (2001) and *Gendered Innovations in Science and Engineering* (2008).

Arpita Subhash worked as Assistant Director, Science and Technology/ Innovation Division at Federation of Indian Chambers of Commerce and Industry (FICCI). Her M. Phil dissertation was on gender and science in the Indian context and she is working for Ph.D. on 'Renewable Energy Technologies and Women in Selected Agro-Climatic Region in India' at Jawaharlal Nehru University, Delhi.

Acknowledgements

THE preparation of this volume has been a rewarding and challenging experience for both, the editor as well as the contributors. Though the gender and science issues depict some universal characteristics, bringing together chapters based on diverse conceptual and methodological approaches was a daunting task. Despite recent changes, science remains gendered and women a minority in the scientific disciplines all over the world. The effort was to present a picture of all possible regions however the volume is restricted to data and findings from South Africa, US, Europe, Asia, and the Middle East. Delays have been inevitable in the production of such a complex text, and although some efforts have been made to update the findings as well as the references in the chapters, as an editor I do accept my limitations in getting the latest data and findings in various countries, which I wish should have happened. Here one can only be optimistic about other parts of the world as even the latest report by the European Union, *She Figures, 2009* makes clear that there are still too few women in science in Europe and there is an imbalance in the number and seniority of women and men in scientific professions.

I wish to register my gratitude to the editorial team of Cambridge University Press India Pvt. Ltd. in Delhi for their care and meticulous work on the manuscript. Sincere thanks to my colleagues at the National Institute of Science Technology and Development Studies (NISTADS) for their support and encouragement. Finally, I am grateful to Cambridge University Press for the permission to reprint Mineke Bosch's paper titled 'Science in the Netherlands: A Dutch Case?' published in the *Science in Context*, 15(4): 483–527 (2002). The press permitted me to include a revised and enlarged version of Claudine Hermann and Françoise Cyrot-Lackmann's paper entitled 'Women in Science in France' published in the *Science in Context*, 15(4): 529–56 (2002). I am thankful to the *Harvard Journal of Law and Gender* for the permission to reprint Londa Schiebinger's important and illuminating essay 'Getting More Women into Science: Knowledge Issues'. I express my sincere gratitude towards *Annual Reviews* for granting me the permission to reprint Francesca Bray's review 'Gender and Technology' published in the *Annual Review of Anthropology*, 36: 37–53 (2007). Finally, I wish to thank all those who have contributed to making this collection an addition to the literature on gender and science.

Neelam Kumar

Introduction

Reflections and Realities across Cultures

Neelam Kumar

> Whatever we may say, we don't in reality regard women as suitable for science careers
>
> *C.P. Snow (1993: 103)*

GENDER inequality and segregation have characterised science for centuries. Gender biases has been shown in science in terms of its nature and style, content and practices. Various ideological constructions of gender through different eras have served as barriers to women's access and progress in the sciences. In the mid-nineteenth century, for example, social Darwinists invoked evolutionary biology to argue that a woman was a man whose evolution – both physical and mental – had been arrested in a primitive stage. Women's intellectual development, it was argued, would proceed only at great cost to reproductive development. There was a prevalent myth, which claimed that, *as the brain developed, the ovaries shrivelled* (Schiebinger, 1989). Women were thus perceived intrinsically unsuited to natural philosophy and those who did show any aptitude were made the butt of savage satires. This attitude however continued, with few revisions, throughout the history of Western science. In the twentieth century, scientists gave new interpretations to the prejudices on women in science. For decades, whether boys' mathematical skills are superior to girls' has been a controversial topic among social scientists. Scientific and mathematical abilities were thought to be written into our genetics.

From the end of the twelfth century, university was the institutional home of science where women were denied entry. Universities were found in Europe between the twelfth and fifteenth centuries. The influence of Aristotelian philosophy in the universities' curricula provided a scientific basis for this exclusion, since Aristotle and his followers viewed women as intellectually inferior. Women did not have access to these institutions until the late nineteenth century (even until the twentieth century in a few countries), with an exception of Italy. In Italy, a few exceptional women did study and teach at universities as early as the thirteenth century and that too in such fields as physics and mathematics, which are thought to be especially resistant to women even today! During the seventeenth and the eighteenth century, a few outstanding women practiced and even lectured on natural philosophy (Fara, 2002). The professionalisation of science in the seventeenth century also had a tremendous impact on the exclusion of women from this discipline (Whaley, 2003). As science became institutionalised in academies, women were formally banned from study and research.[1] The new scientific societies, like the university, were closed to women. In this regard, Italy was again an exception among European countries. The first academy of scientists which admitted women was Italian. Britain's Royal Society, although established in 1662, did not admit women until 1945. Similarly, the Académie Royale des Sciences in Paris refused to admit women. Women were also excluded from the meetings of the scientific societies and were largely absent from the pages of scientific journals until the early twentieth century.[2] Women botanists were marginalised from the meetings during nineteenth-century England. Though duly qualified, women were not permitted to register as delegates in international medical congresses (such as the one in London in 1881). It was feared that the presence of female delegates might have hampered a resolution stressing the import of animals for medical research (Bynum, 1994).

[1] In the sixteenth century, science was carried out in domestic setting and females did play some role in it. Once science was withdrawn from the domestic sphere, women participated less as throughout the seventeenth century, the universities of Europe remained closed for them.

[2] The appearance of the scientific journals started in the 1660s. Harmon and Gross make a convincing case that Marie Curie is the first woman to publish scientific articles of major significance, thrusting her into the front ranks of important scientists worldwide. In the year 1898, aged 30, she along with her husband published two important articles. Earlier a few did work as technical assistants (for the most part, unpaid and unacknowledged), while still others played an important role in communicating scientific knowledge through books published in the seventeenth through nineteenth centuries. In the late eighteenth century, Caroline Herschel published a *Catalogue of Stars* under the auspices of the Royal Society of London. In fact, Herschel sighted eight new comets in her long life, three of which she reported in very brief *Philosophical Transactions* articles.

The eighteenth century constituted a distinct era in the organisational and institutional history of science. This century witnessed the transition from natural philosophy to the beginning of an array of scientific disciplines and largely proscribed women.[3] But the world of organised science in the eighteenth century was almost exclusively male (McClellan, 2003). The eighteenth century, however, also saw intense debates over the different natures of females and males – whether women were mentally and socially inferior to men or were they equal but different or at least potentially equal! Women were perceived weak physically and intellectually, and also unstable emotionally to endure, for example, the rigours of medical education and the fierce competition of practice. In the nineteenth century, science took a new meaning – science became devoid of its philosophical and theological concepts. In the late nineteenth century, the doors of universities, scientific societies, and research laboratories gradually opened for women. Women gained access to institutions of higher education in USA in 1833; Germany in 1908; and Japan in 1913. In India, the first graduate degrees were granted to women in 1883. One of the first university professions that women gained access into was medicine, in the face of tremendous resistance in most European countries. Women began to be admitted in recognised medical schools by the mid-nineteenth century. In Philadelphia, the Women's Medical College was founded and a complete medical course for women was offered in 1850. But around 1920 there was a reaction towards the growth of female education saying that it 'disadvantages every student' (Bonner, 2000: 339). Segregated education remained a hallmark of the medical practice (Bynum, 1994). Only a few top schools opened their doors to women medical students and it proved, however, easier to establish separate schools for women. In Germany, medical schools remained closed to women until 1900. The leading British and American universities with a few exceptions barred women until the World War I (Harvard Medical School did not have a woman student until 1945!).

The progress and development of science over the centuries had its own effects on the women's access to it. The institutionalisation of science, for example, resulted in the women's marginalisation. Universities became the home of

[3] The organisational process of science at the various stages had implications for the exclusion of women. Some historical work on masculinity and science are suggestive. Nye (1997) has pointed out that even as science and medicine became more professional at the end of the nineteenth century, men encoded masculine displays of 'honour' by requiring certain rituals for admission that encouraged a degree of gentlemanly debate and contention. Oldenziel (1997, 1999) has studied just how in the twentieth century engineers worked to make technology masculine.

science at the end of twelfth century where women were denied entry earlier. Later, the introduction of professional academies in the seventeenth century further reinforced the exclusion of women from science. The new scientific societies, like the universities, were closed to women. Britain's Royal Society, established in 1662, for example, did not admit women until 1945. Similarly, the Académie Royale des Sciences (1666) in Paris refused to admit women till 1979. Even the illustrious Marie Curie (1867–1934) was turned away (Schiebinger, 2003). The appearance of the scientific journals in the 1660s marked a final step of the structure and process of the institutionalisation of science, but women have been largely absent from the pages of scientific journals until the early twentieth century. The writings on science have also undermined the roles of women. The history of science has been overwhelmingly dominated by the contributions of male scientists. A book entitled *English Men of Science: Their Nature and Nurture* (Galton, 1874) was not a strange idea. Another illustration of ignoring women is the editions of *American Men of Science* (Cattell, 1906). There were women, in hundreds and later thousands in each edition but the name chosen entirely minimised them till 1971 (Rossiter, 1993). The history and contributions of female scientists, with some notable exceptions, have remained largely unknown.

The twentieth century has been described as a historic turnaround for women. Women were not only admitted to universities, but they also found their place as professors. They became part, though in limited ways, of nurturing scientific subculture. Women began to enter into all fields of science.[4] Did science, ultimately, become free from discrimination on the basis of gender?[5] Unfortunately, not! But, during the twentieth century, inequalities between female–male participation and recognition in the sciences began being questioned. Feminist theorists started emphasising the biased nature of science. Feminist debates about science, however, experienced a fundamental

[4] Women though particularly remained concentrated in the fields like botany, sociology, and psychology. For example, in the United States, it was not until the rise of the women's movement in the 1970s, which grounded its arguments in the importance of the individual, the equality of men and women, and the rejection of gender stereotyping, that the proportions of women began to increase in all of the sciences, including psychology (National Science Foundation, 1982).

[5] Gender and science first made its appearance in a Psychoanalytic journal (Keller 1978). Like the concepts of class, race, and ethnicity; gender is an analytical tool for understanding social process. Gender is a culturally specific set of characteristics that identifies the social behaviour of women and men, and the relationship between them. Gender, therefore, refers not simply to women or men, but to the relationship between them, and the way it is socially constructed.

change in the way science came to be understood and practiced. The concept of gender and science thus came to the fore (Keller, 1978). Yet gender blindness in science has been prevalent till 1987 (Delamont, 1987). The turn to gender as an analytic category (Connell, 1987; Harding, 1986; Scott 1986) later found new avenues as the ways of bringing gender into the discussions of power, culture, and the politics that seemed to have obvious implications for the study of science.[6] By the 1980s, some feminist literature on science (such as *Science and Gender* by Ruth Bleier) appeared along with the works on the history of women in science (for example, Margaret Rossiter's book. *Women Scientists in America: Struggles and Strategies to 1940*. Baltimore, MD: Johns Hopkins University Press, 1982). It was declared that science is 'masculine' not only in person of its practitioners but also in its ethos and substance (Keller, 1985).

The idea that the social structure and processes are gendered has slowly emerged in diverse areas of feminist discourse. Feminists have elaborated gender as a concept to mean more than a socially constructed binary identity and image. Scientific institutions somehow continue to discriminate against women. Race and ethnicity have erected additional barriers to the full participation of women in science. While overt discrimination, observed in the earlier decades is out of style, covert and subtle forms of discrimination still prevail today. Not only historical but also contemporary sociological researches show marked disparity in the status of women. Women face barriers at entry and at all stages of the academic ladder in science. Under-representations at higher echelons and slower advancement in the academic ladder constitute two persistent problems, apart from the scarcity of women in science. Women are also under-represented on editorial boards (White, 1985). Recent study provides evidence of massive gender bias in the peer review of research grant applications to Sweden's Medical Research Council. During the past fifteen years, there has been increasing international concern about the more effective integration of women into S&T (Science and Technology). Numerous international conferences have tried to determine the causes of this under-representation which is an issue not just for a single nation, but across all cultures.

[6] Scott, for example, defines gender as follows: 'The core of the definition rests on an integral connection between two propositions; gender is a constitutive element of social relationships based on perceived differences between the sexes, and gender is a primary way of signifying relationships of power' (1986: 1067).

Today, the vital importance of gender equity has often been proclaimed universally, but the world of science is still a different one for women than it is for men. Sex-segregation is a feature of scientific careers – women tend to be more concentrated in certain disciplines or fields of studies. The issue of gender discrimination crosses national borders. Universally, women are just a tiny minority of people in top scientific jobs. Ironically, the universal socio-psychological conditions related to gender and science is aggravated by some local circumstances. Thus, the problems that women scientists face is to be tackled best at a variety of levels from local and regional to national and international. Questions of gender and science have now come into the foreground in sociological theory, feminist research, human resource policy, and so on. The European Union, for example, has recently set up a 'Women and Science' unit to gather statistics (of which few are currently available) and has created a network for women in these fields, actions that are in accord with its general principles regarding equal opportunities.

Gender as a concept has developed further in recent years in order to contest the naturalisation of sexual differences (Haraway, 1991). New, elusive, and explosive questions have emerged (Schiebinger, 1999). It has been noted that gender has moulded the very content of science (Schiebinger, 2003). Biological myths and assumptions have also been questioned and during the 1990s, feminist reconsideration of the sex/gender problem moved into a full swing. Fausto-Sterling, for example, examined numerous scientific claims about biologically based sex differences between men and women. She argued that the sex-gender or nature-nurture accounts of difference fail to appreciate the degree to which culture is a partner in producing body systems commonly referred to as biology – something apart from the social. Our bodies physically imbibe culture (Fausto-Sterling, 2005). In other words, works have started examining and developing feminist perspectives on the epistemology and methodology, history, philosophy and sociology of science and its processes. In dialogue with the history and sociology of technology, feminist technology studies have also emerged. Interesting questions have been raised by the feminists on gender and technology relations too. It became clear that not only gender influenced technology, but one fundamental way in which gender is expressed in any society is through technology. Women kept away from technology as it was considered 'masculine' and at the same time 'masculinity' was being defined in terms of man's use of technology and its tools.

Thus, in the past thirty years, feminists have produced major critiques of science. Feminists elaborate gender as a concept to mean more than a socially constructed binary identity and image. Yet, at the beginning of the twenty-first century, one of the unresolved problems is equality in terms of gender.[7] Will getting more women bring any change or open science to new perspectives, new questions, and new missions? Londa Schiebinger (chapter 1) discusses how gender analysis, when turned to the sciences, can profoundly affect human knowledge. She argues that the removal of gender bias can open science to new perspectives, new questions, and new missions. Her essay also analyses: (a) the participation of *women* in science; and (b) gender in the cultures of science. Using several examples her essay investigates how understanding of gender functions in science and society can spark creativity in particular fields of science.

Women still remain in subordinate positions. The number of women who are awarded degrees in science and engineering has increased but their representation in university and college faculties are still inadequate. Women do not receive recognition in the same degree as men do for similar contributions (Cole, 1987; Scott Long, 1992). Women and minorities face various institutional and organisational barriers to science. Henry Etzkowitz (Chapter 5) clearly demonstrates this. Women and minorities often experience 'social capital gaps', exclusion from social networks. Discrimination due to gender biases and non-recognition at the highest levels is also faced by women scientists. This, in turn, reduces future career chances for them. An in-depth interview with 400 female graduate students and faculty members at US research universities forms the basis of his conclusions. The contradiction between the ideology and practice of science is exemplified by inflexible rules that uniformly reduce women's chances for the scientific achievement. Etzkowitz believes that these vary with societal contexts. He argues that negative effects of ingrained cultural practices on women in science are hidden behind a normative structure that proclaims universality. According to him, women in science produce a 'surplus of anxiety' by consequences of the triple burden of gendered discrimination, 'role overload', and social capital depletion that interferes with scientific production and recognition of accomplishment.

[7] Gender and science first made its appearance in a Psychoanalytic journal (Keller 1978). Like the concepts of class, race and ethnicity, gender is an analytical tool for understanding social process. Gender is a culturally specific set of characteristics that identifies the social behaviour of women and men, and the relationship between them. Gender, therefore, refers not simply to women or men, but to the relationship between them, and the way it is socially constructed.

The concept of gender has been used as one aspect of a larger intellectual approach, a form of comparative and social history (Jordanova, 1993). Concerns of gender are thus essential to a social theory of science. Gender does constitute an important theme for history, philosophy, and sociology of science. Gregory J. Feist (chapter 4) introduces the study of gender and science from the 'psychology of science' perspective. He argues that psychology is conspicuously absent – at least until the mid 1980s – from the studies of science, also called metascience. He finds this particularly puzzling, given the fact that philosophers, historians, and even sociologists of science often touch on inherently psychological processes in their writings on science and scientists. According to him, the role that gender plays in science in general, and in scientific and mathematical ability, and achievement in particular, constitutes one of the more contentious and polemical questions in the psychology of science. Feist also discusses how the psychology of science can contribute to an understanding of the gender role in scientific and mathematical interest and behaviour. He believes that interest-attrition, mathematical ability, and productivity are the three main areas, which have accumulated enough literature in psychology to warrant attention on gender and science.

Gender binaries such as masculine–feminine have been not only associated with science, these have deeper connotations for technology. One fundamental way in which gender is expressed in any society is through technology – as a result technology is gendered to its core! Bray in her essay (Chapter 3) outlines feminist technology studies (FTS), one of the most vibrant and coherent school of gender and technology studies, its concepts and methods. She then turns to the anthropology of technology, which also offers useful conceptual frameworks and methods for exploring gender regimes. Her chapter meticulously highlights the ideological and methodological contrasts between social and cultural analyses of technology and the implications for gender analysis. Finally, it discusses the treatment of technology in two leading theoretical fields in the cultural anthropology of modernity and globalisation: the anthropology of technoscience and material culture studies. Bray concludes with a question about the forms of engagement.

Francesca is, similarly, but strikingly surprised on the conspicuous absence of cultural anthropology from the Feminist Technology Studies (FTS), state-of-the-field essays and anthologies. In her view the praxis-oriented interdisciplinary field of FTS has done most among the social sciences to build a vibrant and coherent school of gender and technology studies. She acknowledges, by outlining FTS and its concepts and methods, the shared commitment and overlap between the mainstream cultural anthropology and FTS. Bray strongly recommends a turn to the anthropology of technology, which can offer useful conceptual frameworks and methods for exploring gender regimes. She further highlights the ideological and methodological contrasts between social and cultural analyses of technology and the implications for gender analysis. Her chapter finally discusses the treatment of technology in two leading theoretical fields in the cultural anthropology of modernity and globalisation: the anthropology of technoscience and material culture studies. Bray finally concludes by asking which forms of engagement might be envisaged between the fields.

Women face problems in all stages and phases of the academic life course, are promoted more slowly than men, and are under-represented in high academic rank in most parts of the world. Women's academic career, in other words, remains markedly characterised by strong vertical segregation.[8] Scarcity, invisibility, segregation of women, especially in science, is observed everywhere. In European countries, for example, the percentage of female professors in engineering and technology was 7.2, while their percentage in the humanities and social sciences was 27.0 and 18.6, respectively (she Figures, 2009). Do gender and science issues go beyond race, class, social and geographical distinctions? What is the relation between culture, gender, and science? My own essay (Chapter 2) reviews the extant literature and raises the question whether the nature of gender and science relations are variable cross-culturally. It also suggests that gender and science are culturally constructed, therefore, calls into question the claims of universality that science often makes.

[8] Table A.1 *Percentage of Female by Academic Rank in Selected Countries*

Country	Professor	Assoc. Professor	Asst. Professor
Turkey	28%	34%	46%
France	19%	39%	34%
US	28%	41%	47%
Spain	18%	36%	48%
Sweden	18%	47%	42%
Italy	19%	34%	45%
U.K	17%	37%	47%
Germany	12%	18%	33%

Source: *She Figures*, 2009, European Commission (for the European countries); *Digest of the Educational Statistics*, 2009, National Center for Education Statistics (for US).

Vertical as well as horizontal are the two major types of gender segregation that characterise science all over the world. What are the explanations of gender differences or similarities? Evolutionary psychology states that gender differences are the result of evolution and the product of genetic differences between men and women. In contrast, cognitive social learning theory states that a number of social processes contribute to the development of gender typed behaviour. The learning takes place by observing other's behaviour, analysing those behaviours and modelling them. Role models and socialising agents, as well as perception of gender appropriate behaviour have an important influence on an individual's academic choices. Social structural theory proposes that psychological gender differences are a result of the division of labour by gender and the gender disparity in power. The gender stratification hypothesis maintains that the gender differences are closely related to cultural variations in the opportunity structures for the girls and women. Recent gender similarities hypotheses suggest that men and women are similar for most, but not all, psychological variable. Evidence from meta-analyses supports this hypothesis by indicating that gender differences are small or close to zero in areas such as cognitive abilities, psychological well-being and self-esteem (Hyde, 2005). In the light of the under-representation of women in careers in science, technology, mathematics, and engineering, increasing research attention is being devoted to understanding gender differences in mathematics achievement, attitudes, and affect. Yet, despite of the mounting evidence of gender similarities in the mathematics achievement, often the stereotypes that girls and women lack mathematical ability persist (Hedges and Nowell, 1995; Hyde, Fennema, and Lamon, 1990; Hyde, Lindberg, Linn, Ellis, and Williams, 2008). The ubiquitous presence of gender stratification in science is universally recognised, but certain fundamental questions about its development remain unanswered.

The question of universality of gender and science, the effects of culture and the social structure of science, ultimately seems to have empirical connotations as well. Considerable cross-national variability in the gender gap has been explained by important national characteristics reflecting the status and the welfare of women (Else-Quest et al., 2010). Can cross-national data be taken as a method for establishing the generality of findings and the validity of universal nature of gender inequalities? As gender is always inextricably bound with other societal systems of difference and inequality, one may expect better understanding of gender and science as a social phenomenon on the basis of cross-national comparison. While data from a few countries is little optimistic, in contrast, for many countries it is still a matter of concern.

J. Scott Long in his chapter (Chapter 10), for example, documents some of the truly remarkable changes that have occurred in the representation of women in science and engineering in US. Women are in increasing presence, both in absolute number and as a proportion of all scientists and engineers in all aspects of the career, from the receipt of the PhD, to the entry into the labour force and attaining the rank of full professor. Long Scott finds these changes positive and encouraging, but he shows concern for the substantial differences, which still remain. There are consistent differences that leave women with less achievement, salary and position. Women as a group remain less well-represented and less successful than men in every dimension of the career. While women have clearly made enormous gains in their participation in science and engineering, Long makes it clear that these advances represent neither unconditional success in overcoming gender inequalities nor assurance of continuing progress in the future. He concludes that women have moved from *Scarcity to Visibility.*

But, even in some of the long-industrialised European countries, the story has not been one of automatic growth and progress. For example, the Netherlands happens to be worse than many other European countries. Bosch in her chapter (Chapter 6) raises the question whether there is a 'Dutch case' concerning women's standing in science. She argues that the cause is not to be found in a special brand of Dutch Protestantism, with its strong emphasis on motherhood and the family, and impact on labour patterns and social organisation. Rather, she takes another look at religion, and especially at the specific Dutch segmentation of society along religious and political lines, called *verzuiling*, literally meaning 'pillarisation'. In her view from about 1880 until far into the 1950s, the personal and social life of the Dutch (from schools to sports and ladies' organisations) was organised into four recognised pillars (a Protestant, a Catholic, a socialist, and a liberal pillar), which at the top were represented in political parties. She focuses on the often overlooked fact that between 1880 and 1945 state institutions, such as universities, were thoroughly pillarised, which strongly influenced recruitment and selection for those institutions. That is to say, no woman was appointed to the rank of full professor at any state university in the Netherlands until after 1945. Her essay also reveals that the Dutch case might also be explained by the many reorganisations and down-sizings of universities of more recent years that occurred simultaneously with the expansion of academic feminism. In her view, a newly configured 'pillarisation' has driven deep divides between gender studies scholars, equal opportunities officers, and women scientists.

In contrast to the Netherlands, France ranks among the top European countries in terms of women in science. Claudine Hermann, Françoise Cyrot-Lackmann Jeanne Peiffer and Hélène Rouch (Chapter 11) discuss the French situation. According to these authors, the 'relatively favourable' situation of French women scientists is related to the social structure of the French society, in particular its child-care system, and to the stable permanent positions in academia. French women scientists experience less difficulty than other European colleagues in managing both a private and a professional life.

Women's failure to achieve equality in science has deep cultural roots. It has been repeatedly observed that gender is inextricably bound with the other societal systems of difference and inequality. The importance of cross-cultural studies has long been recognised by the feminist scholars of science, yet cultural singularity remains the reality in science studies with a few exceptions.[9] Societies differ in terms of cultural, political, and economic advantages and disadvantages granted to women and men. Most work on gender and science are based on the personalities and scientific subcultures of Western Europe and North America. It is implicitly or explicitly assumed that the observations and explanations made for those regions will hold true for the rest of the world. The gender stratification hypothesis maintains that such gender differences are closely related to cultural variations in opportunity structures for girls and women. Cross-cultural comparison, beyond Europe and America, are thus desirable. This book, which brings together data and findings from Japan, South Africa, Saudi Arabia, and India besides Europe and USA, will be an effort in this direction.

Mariko Ogawa (Chapter 7) discusses the status of Japanese women scientists. She first traces the historical contexts of Japanese women scientists, then examines their present situation, and finally recommends new strategies for women in S&T in the twenty-first century Japan. Should one have to keep the fingers crossed for the Asian and other developing countries? Luci and Jaamiah (Chapter 9) try to analyse the gender dimensions of the public science system in relation to the emerging knowledge economy context in South Africa. They discuss the role of females in the desired talent required to achieve the objectives of specific innovation strategies. Their chapter is based on: (a) the results of a quantitative study on women's participation in public

[9] In a study comparing Australia, Germany, Hong Kong, Israel, Mexico, Sweden, UK, and USA, the academic profession was found to be gender segregated with women largely absent in many fields of science and engineering (Poole et al., 1997).

sector science, engineering and technology (SET) in South Africa; and (b) an analysis of public policy and strategy aimed at promoting innovation in SET over the period 2005–14. The chapter in the concluding section discusses the need for research to inform institutional and public policy, and the design of measures to effect growth in the Gross Expenditure on Research and Development and other factors related to the performance of the science system.

The data from certain countries are rather encouraging. Saudi Arabia is an important example of educational progress in general and women's education in particular within a short history of education. The modern history of education in Saudi Arabia began in 1926, and the advent of modern university education with the creation of Riyadh University now known as King Saud University in 1957, with an enrolment of 21 students and a staff of nine. In 1960, the Government of the Kingdom of Saudi Arabia undertook the introduction of a national education programme for girls. By the mid-1970s, about half of Saudi Arabian girls were attending school. Five years later, education was available to all Saudi girls. By 1980, there were six universities for women. The government of Saudi Arabia has always recognised the importance of providing educational opportunities to girls as well as boys. Samira I. Islam (Chapter 8) gives surprising statistics about the enrolment of women in natural sciences and medical, and health-related fields in Saudi Arabia. Fifty per cent or more of the Saudi Arabian students enroled in natural science subjects are women. This is remarkable as the formal schooling for girls started in 1960 and Saudi women were admitted to university education in 1973. Saudi women should be praised for having attained such achievements within less than 50 years.

Arpita (Chapter 12) maps the issues and perspectives on gender and science in relation to the Indian context. She derives her conclusions using various Government of India data such as Census, Educational Statistics from the Department of Education, University Grants Commission, Research and Development Statistics of the Department of Science and Technology. In her view the status of women in professions, particularly in science, has still not drawn the attention it deserves in social studies relating to science and technology.

This book is divided into three sections. The first deals with theory, approaches, and methods, and includes chapters that contextualise the work in the second section of the book. The second section provides exemplary case studies. It deals with the gendering of science, scientific institutions, and practices in a few selected countries. The concluding chapter highlights the commonalities, divergences, insights, and challenges which have been presented. This volume brings together the writings of prominent academics and researchers from the various parts of the world, as they discuss gender and science using: (a) empirical evidence in their society; (b) explore the relevance of gender perspectives and theories; and (c) blend theoretical insights with empirical data and findings. The chapters should be able to illustrate cross-national differences in order to move beyond some of the generalisations on women in science that are based on data from relatively few countries. However, though the countries included represent considerable diversity, these may not be representative of all the countries or cultures. The chapters, however, do reveal and discuss some of the universal problems related to gender and science. The value of this volume lies in bringing the key theoretical and empirical issues on gender and science from the diverse cultural perspectives, by the contributors, world leaders in their respective nations, into one place.

References

Bleier, Ruth (1984). *Science and Gender: A Critique of Biology and its Theories on Women*. New York: Pergamon.

Bonner, Thomas Neville (2000). *Becoming a Physician: Medical Education in Britain, France, Germany, and the United States, 1750–1945*. Baltimore: The John Hopkins University Press.

Bynum, W.F. (1994). Science and the Practice of Medicine in the Nineteenth Century. Cambridge: Cambridge University Press.

Catttell, J.M. (1906). *American Men of Science*. New York: Science Press.

Cole, Jonathan R. (1987). *Fair Science: Women in the Scientific Community*. New York: Columbia University Press.

Connell, R.W. (1987). *Gender and Power: Society, the Person and Sexual Politics*. Cambridge: Polity Press.

Delamont, Sara (1987). 'Three Blind Spots? A Comment on the Sociology of Science by a Puzzled Outsider', *Social Studies of Science*, 17 (1): 163–70.

Digest of Education Statistics (2009). National Centre for Education Statistics, US Department of Education.

Else-Quest Nicole M., Janet Shibley Hyde, and Marcia C. Linn (2010). 'Cross-National Patterns of Gender Differences in Mathematics: A Meta-Analysis', *Psychological Bulletin*, 136(1): 103–27.

Etzkowitz, H. and C. Kemelgor (2001). 'Gender Inequality in Science: A Universal Condition?' *Minerva*, Vol.39, No. 2: 153–74.

European Commission (2000). *Science Policies in the European Union: Promoting Excellent through Mainstreaming Gender Equality.* A report from the ETAN Expert Working Group on Women and Science, Research Directorate-General, European Commission.

Fara, Patricia (2002). 'Elizabeth Tollet: A New Newtonian Women', *History of Science,* 40(128): 169–87.

Fausto-Sterling, Anne (1985). *Myths of Gender: Biological Theories about Men and Women.* New York: Basic Books.

—— (2005). 'The Bare Bones of Sex: Part 1 – Sex and Gender', *Signs: Journal of Women in Culture and Society,* 30(2): 1491–1527.

Fox, Mary Frank. 'Women and Scientific Careers.' *In* S. Jasanoff, J. Markle, J. Petersen, and T. Pinch (eds.) (1995) *Handbook of Science and Technology Studies* (pp. 205–223). Newbury Park, California: Sage.

Galton, Francis (1874). *English Men of Science: Their Nature and Nurture.* London: Macmillon & Co.

Haraway, Donna (1991). *Simians, Cyborgs, and Women: The Reinvention of Nature.* New York: Routledge.

Harding, S. (1986). *The Science Question in Feminism.* Ithaca: Cornell University Press.

Harmon, Joseph E. and Gross, Alan G (eds.) *The Scientific Literature: A Guided Tour.* University of Chicago Press, 2007, Chicago.

Hyde, J. S. (2005). 'The Gender Similarities Hypothesis', *American Psychologist,* Vol.60, No. 6, 581–592.

Hyde, J. S., Fennema, E. and Lamon, S. J. (1990). 'Gender Differences in Mathematics Performance: A Meta-analysis'. *Psychological Bulletin,* 107(2), 139–155.

Hyde, J.S., Sara M. Lindberg, Marcia C. Linn, Amy B. Ellis, and Caroline C. Williams (2008). 'Diversity: Gender Similarities Characterize Math Performance', *Science,* Vol.321, No. 5888, 494.

Jordanova Ludmilla (1993). 'Gender and Historiography of Science', *The British Journal for the History of Science,* 26(4) 469–83.

Keller, Evelyn Fox (1978). 'Gender and Science', *Psychoanalysis and Contemporary Thought,* 1(3): 409–33.

—— (1985). *Reflections on Gender and Science.* New Haven, CT: Yale University Press.

McClellan, James III (2003). 'Scientific Institutions and the Organization of Science', in Roy Porter (ed.), *Eighteenth Century Science,* series in Cambridge history of Science, pp. 87–106. Cambridge, UK: Cambridge University Press.

National Science Foundation (1982). *Women and Minorities in Science and Engineering.* Washington, DC: National Science Foundation.

Nye, Robert A. (1997). 'Medicine and Science as Masculine "Fields of Honour"', *Osiris,* 12, 60–79.

Oldenziel, Ruth (1997). 'Boys and Their Toys: The Fisher Body Craftsman's Guild, 1930–1968, and the Making of a Male Technical Domain', *Technology and Culture,* Vol. 38, No. 1, pp. 60–96.

Oldenziel, Ruth (1999). *Making Technology Masculine: Men, Women, and Modern Machines in America, 1870–1945.* Amsterdam: Amsterdam University Press.

Poole, Millicent, Laurel Bornholt, and Fiona Summers (1997). 'An International Study of the Gendered Nature of Academic Work: Some Cross-Cultural Explorations', *Higher Education*, 34(3): 373–96.

Rossiter, M.W. (1982). *Women Scientists in America: Struggles and Strategies to 1940.* Baltimore, MD: Johns Hopkins University Press.

—— (1993). 'The Matthew Matilda Effect in Science', *Social Studies of Science*, 23 (2): 325–41.

Rossiter, Margaret W. (1995). *Women Scientists in America: Before Affirmative Action, 1940–1972.* Baltimore: Johns Hopkins University Press.

Schiebinger, Londa (1989). *The Mind Has No Sex? Women in the Origin of Modern Science.* Cambridge, MA: Harvard University Press.

—— (1999). *Has Feminism Changed Science?* Cambridge, MA: Harvard University Press.

—— (2003). 'Philosopher's Beard: Women and Gender in Science', in Roy Porter (ed.), *Science in the Eighteenth Century*, Vol. 4, the Cambridge History of Science, pp. 184–210. Cambridge: Cambridge University Press.

Scott Long, J. (1992). 'Measures of Sex Differences in Scientific Productivity', *Social Forces*, 70(1): 159–78.

Scott, Joan W. (1986). 'Gender—A Useful Category of Historical Analysis', *American Historical Review*, 91 (5) (December): 1053–75.

She Figures (2009). *Statistics and Indicaters on Gender Equality in Science.* Luxembeurg: Publications Office of European Union.

Snow, C.P. (1993). *The Two Cultures.* Cambridge: Cambridge University Press.

Wennerås, C. and A. Wold (1997). 'Nepotism and Sexism in Peer-review', *Nature*, 387(6631), 341–43.

Whaley, Leigh Ann (2003). Women's History as Scientists: A Guide to the Debates, Santa Barbara, CA.

White, A. (1985). 'Women as Authors and Editors of Psychological Journals: A 10-Year Perspective', *American Psychologist*, 40(5): 527–30.

Section I

Approaches and Perspectives

CHAPTER ONE

Getting More Women into Science

Knowledge Issues*

Londa Schiebinger

INNOVATIONS surrounding women and gender have rocked the world of science and technology in the past three decades. Who, for example, could have predicted that the chief scientist at the National Aeronautics and Space Administration (NASA) would be a woman (France Córdova, who is now chancellor at the University of California, Riverside)? Or who would have thought that geneticists would dethrone the 'master-gene' model – that conceptualised mammalian sex as determined by a single master gene on the Y chromosome – and put in its place an account that emphasises interactions between the testes and ovaries? (Richardson, 2008) Or who would have imagined that an artificial knee would be designed with 19 unique aspects to meet the unique skeletal and load-bearing needs of females? (Feder, 2006).

In my lifetime, the situation in the United States has improved dramatically for intellectual women. We can measure some of these changes iconographically. Anyone growing up in the American consumer culture understands the power of images. Images project messages about hopes and dreams, mien and demeanour, about who should be a scientist and what science is all about. We see some interesting changes in who is imagined to be a scientist in our society. Historically, when prompted to 'draw a scientist', 92 per cent of the

* An earlier version of this paper was published in *Harvard Journal of Law and Gender*, 30, 365–478 (2007).

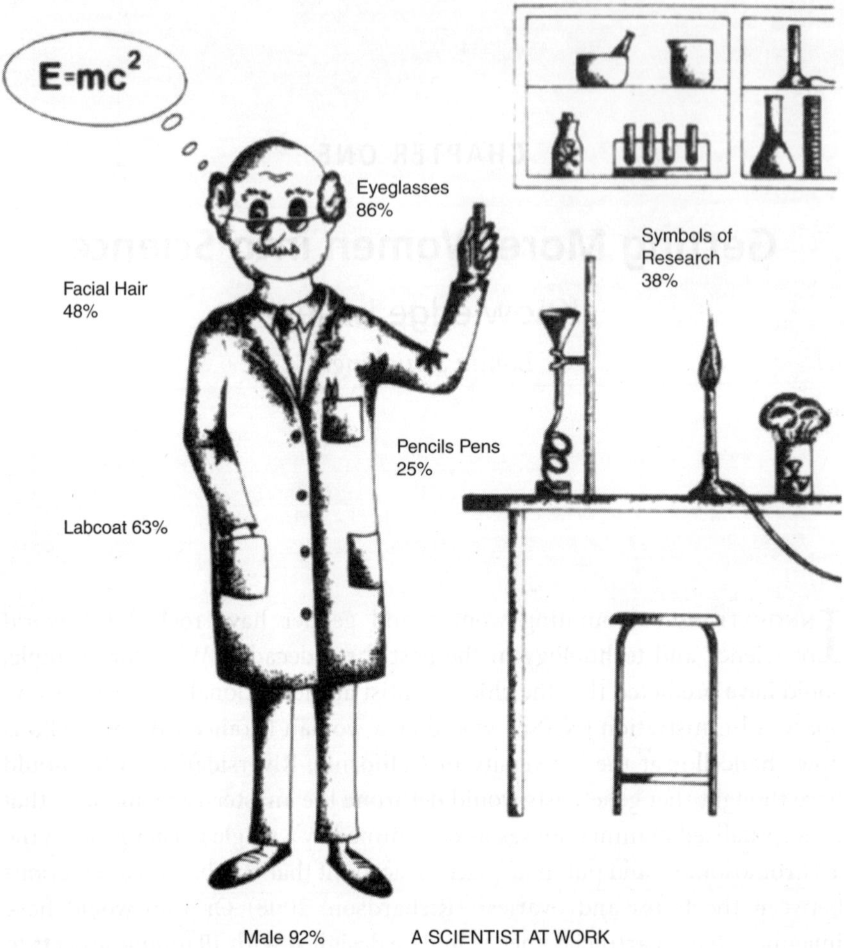

Figure 1.1 Images of Science: The Physicist and the Cowboy
Source: Jane Kahle, 1987

students drew males (Figure 1.1). By the late 1990s, that had reduced to 70 per cent with some 16 per cent of the scientists drawn being clearly female and another 14 per cent ambiguous with respect to sex (Figure 1.2). A remarkable 96 per cent of the scientists continued to be depicted as Caucasian despite the prominence of Asians in science (Rahm Chambonneau, 1997).

We can also see gendered innovations in the content of science, in this case, in understandings of human evolution. Most of us grew up with an image of human evolution as the 'evolution of man'. Evolutionary theory presented

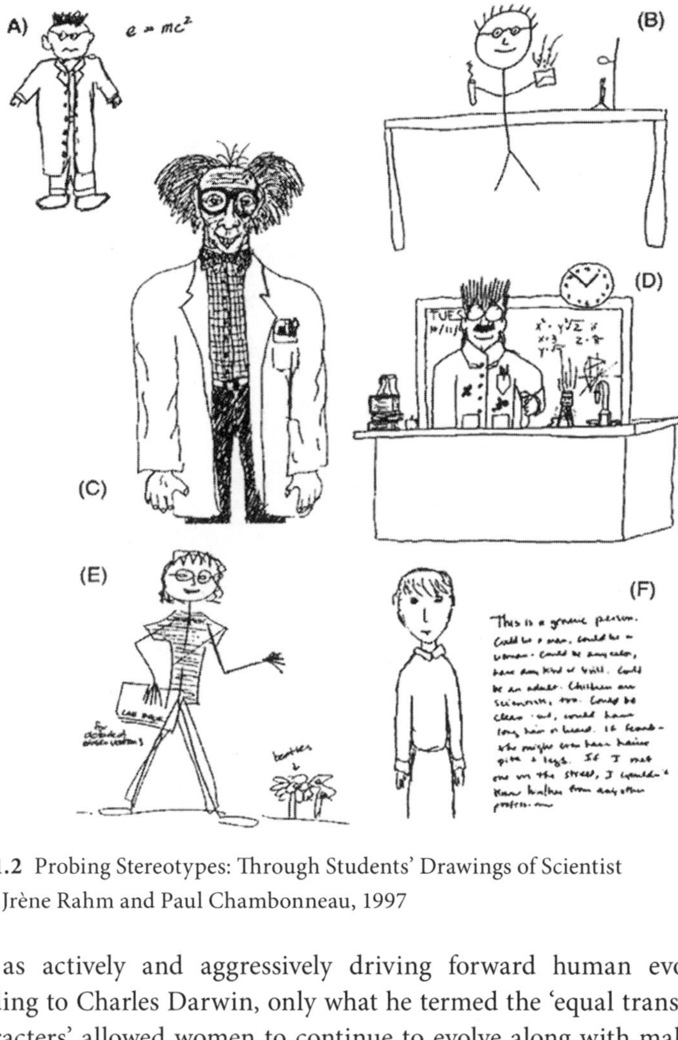

Figure 1.2 Probing Stereotypes: Through Students' Drawings of Scientist
Source: Jrène Rahm and Paul Chambonneau, 1997

males as actively and aggressively driving forward human evolution.[1] According to Charles Darwin, only what he termed the 'equal transmission of characters' allowed women to continue to evolve along with males, who were the ones who embodied the traits of courage and intelligence that kept humans evolving (Fedigan, 1986).[2]

In 1993, a much-heralded new image was produced to correct this picture. In that year, the American Museum of Natural History in New York opened its

[1] See generally Richard and DeVore (1968).

[2] See also Hrdy (1999) (describing how social Darwinism has been used to explain and justify social inequalities between men and women). For early critiques of Darwin, see generally Blackwell (1875); Gamble (1894); Hedwig (1896).

Figure 1.3 Human Biology and Evolution
Source: American Museum of Natural History

new 'Human Biology and Evolution' exhibition featuring this reconstruction of early humans from the 3.5 million-year-old footprints preserved in volcanic ash near Laetoli (Figure 1.3). This diorama clearly gives woman a place in the human evolution, and although the assumptions captured in this image have changed dramatically since the 1960s, the process is still incomplete. The humans embodying the footprints are portrayed as a robust male towering over his smaller female consort, his arm positioned to protect and reassure her. We simply do not know, however, the sex or relationship of the two individuals who made these impressions – footprints cannot be sexed. These early humans might have been a large male and his much smaller mate, but they might also have been a parent comforting his or her adolescent offspring, or just two friends fleeing the volcano.

The purpose of this chapter is to analyse the U.S. theory and practice of creating equality for women in science. As a framework, I set out three distinct levels for analysis: (a) the participation of *women* in science;

(b) gender in the *cultures* of science; and (c) gender in the *results* of science.[3] I will touch on each of these levels but focus on the third, and explore, in particular, how gender analysis, when turned to the sciences, can profoundly affect human knowledge. We will investigate how sharpening the understanding of how gender functions in science and society can open new questions and fields for future research and spark creativity in particular fields of science.

While I think its helpful to distinguish these three analytical levels to the problem of getting more women into science as they are closely tied together as well. I firmly believe, however, that women will *not* become equal participants in science until we have fully investigated and resolved issues at the third level, the level of knowledge.

When considering how bringing women into science might require and/or result in changes in the theories and practices of science, we must remember that modern, academic disciplines are arbitrary ways of cutting up knowledge. They are historical, not natural. Disciplines have developed over the past 200 years when women and under-represented minorities were stringently excluded from the academy.[4] We need to be open to the possibility that human knowledge – what we know, what we value, what we consider important – may change dramatically when women become equal partners in knowledge production. Science is about critical thinking, exploration, and travel into unknown worlds.

We need to seize the moment right now as universities across the country move towards inter-disciplinarity in an unprecedented fashion. Knowledge and the keepers of knowledge – academic institutions – are being remade. Now is the time to make gender an important category of analysis in the new disciplinary configurations.

The Participation of Women in Science

The first and most straightforward level of analysis focuses on increasing the participation of women in science. This level addresses the history and sociology of women's engagement in scientific institutions. Who are the great

[3] See generally Schiebinger (1999).

[4] See generally Schiebinger (1989).

women scientists? What are their achievements? How are the experiences of women in university, industrial, and governmental laboratories?[5]

Since the Sputnik years, the United States and Western European countries have attempted to increase the participation of their populations in science – women as well as men. In the US, this led to foundational legislation, including the Equal Pay Act of 1963,[6] Equal Employment Opportunity Act,[7] and Title IX of 1972,[8] designed to foster equality for women. Beginning in the 1980s, the National Science Foundation (NSF) has attempted to improve the numbers of women in science and engineering by jump-starting their career with extra research money and career support (Rosser, 2008).

The National Academies (the National Academy of Science, National Academy of Engineering, and the Institute of Medicine) have undertaken similar programmes. Founded in 1991, the Academies' Committee on Women in Science and Engineering (CWSE) has worked with the Congress and universities in setting policy aimed at assisting women's careers (Córdova, 2008). These type of programmes have generally aimed to increase the numbers of women in science by 'fixing the women' – that is, making them more competitive – by increasing funding to women's research, teaching them how to negotiate for salary, or, more generally, how to succeed in a man's world.

An interesting new development at this level is the attempt to apply Title IX to science. In spring 2005, the Government Accountability Office issued a report prepared for Senators Ron Wyden and Barbara Boxer on how Title IX of the U.S. Education Amendments of 1972 can be harnessed to increase the number of women and minorities in science (US Government Accountabilty Office, 2004). Work on this issue is still in progress.[9]

Gender in the Cultures of Science

A culture is more than institutions, legal regulations, or a series of degrees or certifications. It consists of unspoken assumptions and values of its members.

[5] See generally Rossiter (1995) (looking at the positions of women in scientific laboratories in the mid-twentieth century); Harriet Zuckerman et al. (1991) (discussing modern research into women in science); Rossiter (1982) (discussing the early history of women in scientific laboratories).

[6] Equal Pay Act of 1963 § 1, 29 U.S.C. § 201 (2000).

[7] Equal Employment Opportunity Act of 1972 § 1, 42 U.S.C. § 2000e (2000).

[8] Title IX of the Education Amendments of 1972 § 1, 20 U.S.C. § 1681 (2000).

[9] See, for example, Michelle R. Clayman Institute for Gender Research (2006).

Despite claims to objectivity and value-neutrality, the sciences have identifiable cultures whose customs and folkways have developed over time. Many of these customs developed historically in the absence of women and, as I have argued elsewhere, also in opposition to their participation.[10] This second level asks how the cultures of science, where success requires at least some mastery of the rituals of day-to-day conformity, codes governing language, styles of interactions, modes of dress, hierarchies of values and practices, have been formed by their predominantly male practitioners. Programmes at this level work to 'fix the institutions'. The NSF's current ADVANCE (Increasing the Participation and Advancement of Women in Academic Science and Engineering Careers) grants, for example, attempt to transform university cultures (NSF, 2005).[11] These efforts range from removing subtle gender biases from hiring practices to restructuring the academic work/life balance by offering parental leave or stopping the tenure clock.[12]

Much remains to be done at this level. To this end, Stanford's Clayman Institute for Gender Research launched a major study of dual-career academic couples in 2006.[13] The growing phenomenon of dual-career couples represents an asymmetry in professional culture that affects women's careers more than men's since women more often than men are partnered with professionals. Among heterosexual couples in the US (and we do not yet have equivalent studies of same-sex couples), women tend to practice 'hypergamy', that is to say, they tend to marry men of higher (or at least not lower) status than their own. This is partly due to the fact that a women's social status was determined historically through marriage, while a man was more often determined by inheritance or success in a profession. Consequently, professional women today are more disproportionately partnered with professionals than are men. To complicate matters further, academics tend to couple within the same discipline. While only 6 per cent of physicists are women, for example, an astonishing 43 per cent of them are married to other physicists (Blondin, 1990). An additional 25 per cent are married to some other type of scientist (Blondin, 1990). A remarkable 70 per cent of women mathematicians and

[10] See generally Schiebinger (1989).

[11] The synopsis of the programme available on the website states 'Institutional Transformation Awards support academic institutional transformation to promote the increased participation and advancement of women scientists and engineers in academe. These awards support innovative and comprehensive programs for institution-wide change.' (NSF, 2005).

[12] See generally Committee on Maximizing the Potential of Women in Academic Science and Engineering (2006); Schiebinger (1999); Valian (1998).

[13] See Michelle R. Clayman Institute For Gender Research (2007).

46 per cent of women chemists are partnered with men in their own fields (Michelle R. Clayman Institute for Gender Research, 2006–07).

Although universities have begun to reform hiring practices to accommodate some, usually outstanding, dual-career couples, being a partner in such a couple makes it difficult to follow the logic of a career and seize opportunities as they arise.[14]

The Stanford study seeks to understand hiring and retention practices involving couples at leading US research universities. This study will culminate in policy recommendations aimed at helping universities recruit and retain top faculty, including women, to leading faculty and administrative positions. Restructuring university practices will help transform the way universities do business and grow academic cultures where women, too, can flourish.

Gender in the Results of Science

Many people are willing to concede that women have not been given a fair chance and the social attitudes and scientific institutions need to be reformed. They are also increasingly willing to concede that women are excluded in subtle and often invisible ways. They stop short, however, from analysing how gendered practices and ideologies have structured knowledge. Is the question of gender in science merely one of institutions and opportunities for women, or does it impact upon the content of these disciplines as well?

Since the enlightenment, science has stirred hearts and minds with its promise of a 'neutral' and privileged vantage point, above and beyond the rough and tumble of political life. Men and women alike have responded to the lure of science: 'the promise of touching the world at its innermost being, a touching made possible by the power of pure thought' (Keller, 2002). The power of Western science – its methods, techniques, and epistemologies – is celebrated for producing objective and universal knowledge, transcending cultural restraints. With respect to gender, race, and much else, however, science is not value neutral. Scholars have begun to document how gender inequalities, built into the institutions of science, have influenced the knowledge issuing from those institutions.[15]

[14] See generally Mcneil and Sher (1998); Wolf-Wendel et al. (2003).

[15] See for example, Gero and Conkey (1991); Hager (1997); Schiebinger (1999, 2004); Spanier (1995).

Here I will discuss several examples of how removing gender bias can open science to new perspectives, new questions, and new missions. Before we turn to these examples, let me say a word first about difference feminism in this regard. Difference feminism – the notion that women do science differently – can be especially unhelpful when applied to knowledge. In the 1980s, much of difference feminism promoted the notion that women had a lot to contribute to science because, it was said, women hold different values and think differently (Belenkey et al., 1986; Goldberger et al., 1996). It is important to understand, however, that gender characteristics often attributed to women – cooperation, caring, cultivating a feeling for the organism, or whatever it may be – date back to the eighteenth century and were produced in efforts to keep women out of science and the public sphere.[16] In romanticising traditional femininity, difference feminism does little to overturn conventional stereotypes of men and women. Women's historically wrought gender differences cannot serve as an epistemological base for new theories and practices in science. There is no 'female style' or 'women's ways of knowing' ready to be plugged in at the laboratory bench or clinical bedside. Women as females of the species do not do science differently; science should not necessarily be 'for women, by women, about women'. Moreover, difference feminism or standpoint theory, as it is sometimes called, can tend to exclude men from understanding how gender operates.[17]

This is not to say that gender bias has not had a huge impact on science. Ignoring these biases is to ignore possible sources of error in past and also future science. What we need is to develop tools of gender analysis. This should include – collecting empirical examples of how gender analysis has changed theory or practice in specific subfields of science; developing frameworks for gender analysis; developing new ways of addressing these issues for sciences such as physics and chemistry, where gender appears not to play a role; and requiring training in gender analysis at the undergraduate and graduate levels. It must be emphasised that gender analysis requires rigorous training; there is no recipe that can be simply plugged into the design of a research project. It must also be emphasised that the tools for gender analysis are as diverse as the variants of feminism and of science. As with any set of tools, new ones will be fashioned and others discarded as circumstances change. Some transfer easily from science to science, others do not. The brilliance

[16] See Schiebinger (1989: 214–64).

[17] See Schiebinger (1999: 4–8).

of their implementation depends, as with other research methods, on the creativity of the research team. Training in gender analysis is something that must become part of undergraduate and graduate education also in the sciences – for everyone. Gender analysis acts as yet another experimental control to heighten critical rigour.

Perhaps the best way to understand how gender analysis works is to study examples where this type of analysis has opened new perspectives or insights in particular areas. The best example of how gender analysis has changed science comes from the biomedical sciences, where we have witnessed a revolution in the women's health research. Since the 1960s, the US has had an active women's health movement – women of all classes and backgrounds demanded better healthcare. By the 1980s, scholars had documented that many medical studies funded by the federal government included few female subjects.[18] Critics found, for example, that a number of large and influential medical studies had omitted women completely – most notably the 1982 'Physicians Health Study of Aspirin and Cardiovascular Disease' performed on 22,071 male physicians and on zero women (Marshall, 2005; Rosser, 1994).[19] Scholars documented that research conducted primarily among men had for many years been generalised to women, even though outcomes for women in terms of disease, diagnosis, prevention, and treatment were not known.

The net effect of gender bias in medical research and education is that adverse drug reactions occur more frequently in women than in men (Mastroianni et al. 1994). For example, over-the-counter antihistamines, initially tested in men, can lead to potentially fatal heart arrhythmias in women (Kaiser, 2005). It is important to point out that no self-correcting mechanism of scientific research kicked in to correct for gender bias in medical research. It seems fairly evident that studying drugs in non-representative populations is just bad science. Correction of the situation in this case required conscious intervention and training. Importantly, in 1993 the Congress passed a federal law requiring that women must be included in clinical drug trials, and that cost could not be used as a justification for excluding them (Primmer, 1997).

In the 1990s, this situation began to improve. The National Institutes of Health (NIH) founded the Office of Research on Women's Health. This

[18] See Ruzek et al. (1997).

[19] Recent studies have shown that while aspirin significantly lowered the risk of heart attack in men, it did not in women.

office has two missions – to increase the number of women in the medical profession and to reconceptualise medical research to include women. In 1991, the federal government announced the Women's Health Initiative, a 14-year, $625 million study of diseases unique to or prevalent in women. This is the largest single study ever undertaken by the NIH (Primmer, 1997: 314). As Bernadine Healy, former head of NIH who oversaw these innovations, put it:

> Let's face it, the way to get scientists to move into a certain area is to fund that area. (Gura, 1995)

Removing gender bias from medicine has helped to improve women's health and well-being. Uncovering gender bias, however, can also lead to improvements in human knowledge more generally. Some tools of gender analysis examine hypothesis-generating metaphors used in science. Take, for example, the wide-spread practice in zoology of calling herds – of horses, antelope, elephant seals, and so on – 'harems'. The assumption is that one mighty male, acting as the *sultan*, safeguards his females and reserves their sexual services for him alone. Gender analysis has called such assumptions into question. It has long been thought that horses, for example, run in harems. Questioning the governing metaphor has led to new scientific understandings of equine breeding and behaviours. Recent Deoxyribonucleic acid (DNA) studies of mustangs, for example, show that a stallion typically sires less than a third of the foals in a band. If he were a *sultan*, he did not guard his females well; the females seemed to have minds of their own and certainly mated outside the band (Schiebinger, 1999: 149).

Second to medicine, biology is a field that has been much transformed by gender analysis. These transformations have not been driven by policies of granting agencies, but by a growing awareness that removing gender bias can improve science. Because biology deals with sex and gender, and because it has been open to women (currently 45 per cent of PhDs are women), biologists have moved swiftly to remove glaring cultural bias. In addition, textbooks have been revised to include the contributions of women scientists and to remove outmoded and sexist metaphors (of the heroic sperm capturing demure and passive eggs, for example).[20]

Philosopher of science Sarah Richardson tells the powerful story of how gender analysis contributed to an overhaul of theories of sex determinism

[20] See Martin (1991).

that guide research in reproductive biology.[21] Tending to privilege things male in the 1980s, geneticists championed a 'master-gene' model of sex determination (the notion that a single gene controls the development of an entire organ system).[22] Further, they saw the Y-chromosome as a trigger gene that, in tandem with sex hormones, drove sexual dimorphism. In this model, a gene on the Y-chromosome initiates testis formation, and testis formation was thought to determine sex.[23] In other words, the Y-chromosome triggered development in males. Females, lacking this trigger, were thought to develop along a default pathway.

Richardson documents how the development of gender analysis in the 1990s, along with an active women's movement in both society and science, dethroned the master gene and put in its place a model of sex determination that takes into account the interactions of testes and ovaries in the co-production of sexual dimorphism.[24] Today biologists see male and female pathways both as highly interactive.[25] Richardson goes on to make the important point that although geneticists gradually became sensitive to gender issues, they did not credit feminism for the many insights it provided.[26] Often when gender becomes one among many tools of analysis in a research programme, its feminist roots are ignored. We need to better understand why it is that when feminism enters the mainstream, it becomes 'business as usual' and is no longer seen as feminist.

A final example comes from the work of paleoanthropologist Lori Hager. Hager has documented how unexamined assumptions about gender have contributed to sexism in sexing fossils – the prized 'finds' for human origins research (Hager, 2008). When viewed uncritically, fossils tend to tell us more about the assumptions of modern-day researchers than about our human ancestors. Working with partial and fragmented skeletons, paleoanthropologists have in the past tended to sex large and robust specimens male, and diminutive specimens female (Hager, 2008). This bias is itself so robust that a whopping 90 per cent of Australian aboriginal skeletal remains have been sexed male – well over the expected sex ratio of about 50 per cent (Hager, 2008). It is scientifically improbable. Hager has discussed how gender analysis helped researchers better recognise that small and large fossils can represent either female and male

[21] See Richardson (2008).
[22] Ibid
[23] Ibid
[24] Ibid
[25] Ibid
[26] Ibid

members of one highly sexually dimorphic species, or individuals belonging to two different species, one robust and the other small (Hager, 2008).

Moving Forward

It is abundantly clear that sexual divisions are seen in physical and intellectual labour structure institutions, knowledge, and everyday objects in our own society, such as artificial hearts and shopping carts. It is also clear that gender analysis has not yielded results uniformly across the various fields of science. While examples of how gender has brought new insights abound in biomedicine, the life sciences, archaeology, primatology, and elsewhere, similar examples are not available for physics or chemistry, for example. The physical sciences have by and large resisted gender analysis. There are several possible reasons why this is so. First, we observe that in disciplines such as biomedicine and biology, with a good number of women practitioners, more progress has been made in knowledge issues. But which came first – the openness of the discipline to new intellectual insights or the greater numbers of women in those disciplines? We do not know. One thing that is true is that the number of people trained in either physics or chemistry and gender studies is extremely small and is something that should be remedied. Second, the lack of interest in gender analysis in the physical sciences may also be due to the fact that objects and processes of the physical sciences are less obviously gendered, if at all. That no gender dimensions exist in physics or chemistry, however, is currently merely a well-formulated hypothesis. We need to run the research.

Once we have made some headway developing gender analysis useful to the natural sciences, how do we mainstream this type of analysis in the day-to-day work of science? There are two steps that need to be followed. First, we need to train students, undergraduate and graduate, along with faculty in how to integrate gender analysis into their research. While most people agree that a student needs to learn molecular biology or particle physics in order to excel in those fields, many believe that one can just 'pick up' an understanding of gender along the way. Understanding gender, however, requires research, development, and training, as in any other field of intellectual endeavour. The NIH programmes I described earlier worked, for example, because a solid body of gender research on these issues was available from a number of leading institutions across the US. New possibilities are also on the horizon. If we are training faculty and graduate students about gender biases in scientific cultures through programmes such as the University of Michigan's highly successful Strategies and Tactics for Recruiting to Improve Diversity and Excellence (STRIDE) programme, adding a knowledge dimension is relatively easy (Stewart et al., 2004).

Second, and this is where policy kicks in, we need to move the NSF along towards requiring that federally funded science and engineering integrate gender analysis into research design, where appropriate. The NSF is lagging behind other federal and international agencies in this regard. At the NIH, the Office of Research on Women's Health (1999) requires proper consideration of sexual differences in medical research (Office for Research on Women's Health, 1999). Also, at the European Commission, the directorate general for research requires that project design address 'systematically whether, and in what sense, sex and gender are relevant in the objectives and methodology of projects'. This is in addition to the requirements that research must address women's needs, as much as men's, and women's participation in research must be encouraged both as scientists/technologists and within the evaluation, consultation, and implementation process (European Commission, 2006).[27] Policy makers need to move the NSF to this position. A number of countries in Europe, such as the Netherlands and Sweden, have been increasing the number of women along with integrating gender analysis into the research design part of their national science policy. Even where this is the case, however, more training is needed in how to incorporate gender analysis into science research. On a recent visit to Sweden, I learned that although such policies are in place, few researchers know what exactly to do.

Let me conclude by suggesting that much work remains to be done. One of the many tasks at hand is to continue to collect empirical examples of how gender analysis has changed theory or practice in specific subfields of science.[28] We need also to continue to develop frameworks of gender analysis that address these issues for sciences, such as physics and chemistry, where gender appears not to play a role in knowledge. Let me emphasise again that this work is crucial to our efforts to recruit and retain women – we will not solve that problem until we solve the knowledge problem. It is intriguing that sciences, such as biomedicine, primatology, archaeology, and biology, where

[27] The requirement that gender analysis be mainstreamed in research design has been removed in the EU 7th framework. Question and Answer Session, German Federal Ministry of Education and Research/European Union Research Commission Conference (Apr. 18, 2007).

[28] See Schiebinger (1999).

gender analysis have flourished, have relatively high numbers of women. In these fields and in many fields in the humanities, employing gender analysis has added spark and creativity by asking new questions and opening new areas to research. Can we afford to ignore such opportunities?

References

Belenky, Mary, Blythe Clinchy, Nancy Goldberger, and Jill Tarule (1986). *Women's Ways of Knowing: The Development of Self, Voice, and Mind.* New York: Basic Books.

Blackwell, Antoinette Brown (1875). *The Sexes throughout Nature.*

Blondin, Pamela Hawkins (1990). *1990 Survey of the Membership of the American Physical Society*, 19: 23–24. American Physical Society.

Córdova, France A. (2008). 'Projects of the National Academies on Women in Science and Engineering', in Londa Schiebinger (ed.), *Gendered Innovations in Science and Engineering*, Stanford University Press.

Committee on Maximizing the Potential of Women in Academic Science and Engineering (2006). *Beyond Bias and Barriers: Fulfilling the Potential of Women in Academic Science and Engineering.* Committee on Maximizing the Potential of Women in Academic Science and Engineering, National Academies.

Dohm, Hedwig (1896). *Women's Nature and Privilege.* Constance Campbell trans., Westport, Ct.: Hyperion (1976).

European Commission (2006). 'Annexure 4: Integrating The Gender Dimension In FP6 Projects', in *Guide for Proposers, Research and Innovation: Structuring the European Research.*

Feder, Barnaby J. (2006). 'Women Get Knees to Call Their Own', *N.Y. TIMES*, May 11.

Fedigan, Linda Marie (1986). 'The Changing Role of Women in Models of Human Evolution', *Annual Review of Anthropology*, 15: 25, 29–33.

Fox Keller, Evelyn (2002). 'Secrets of Life, Secrets of Death: Essays on Language', *Gender and Science*, 78.

Gamble, Eliza Burt (1894). *Evolution of Woman: An Inquiry into the Dogma of Her Inferiority to Men.*

Gero, Joan and Margaret W. Conkey (eds.) (1991). *Engendering Archaeology: Women and Prehistory.* Cambridge, Mass.: B. Blackwell.

Goldberger, Nancy R., Jill Tarule, Blythe Clinchy, and Mary Belenky (eds) (1996). *Knowledge, Difference, and Power: Essays Inspired by Women's Ways of Knowing.* New York: Basic Books.

Gura, Trisha (1995). 'Estrogen: Key Player in Heart Disease among Women', *Science*, 269: 771, 773.

Hager, Lori (ed.) (1997). *Women in Human Evolution.* London and New York: Routledge.

—— (2008). 'Sex Matters: Letting Skeletons Tell the Story', in Londa Schiebinger (ed.), *Gendered Innovations in Science and Engineering.* Stanford University Press.

Hrdy, Sarah Blaffer (1999). The *Woman that Never Evolved*. Cambridge: Harvard University Press.

Kahle, Jane (1987). 'Images of Science: The Physicist and the Cowboy', in Barry Fraser and Geoff Giddings (eds.), *Gender Issues in Science Education* 1, 2 Figure 1.

Kaiser, Jocelyn (2005). 'Gender in the Pharmacy: Does it Matter?' *Science*, 308: 1572.

Lee, Richard B. and Irven DeVore (eds) (1968). *Man the Hunter.* Chicago: Aldine Publications.

Marshall, Eliot (2005). 'From Dearth to Deluge', *Science*, 308: 1570.

Martin, Emily (1991). 'The Egg and the Sperm: How Science has Constructed a Romance Based on Stereotypical Male-Female Roles', *Signs*: *Journal of Women Culture and Society*, 16: 485.

Mastroianni Anna, R. Faden and D. Federman (1994). *Women and Health Research: Ethical and Legal Issues of Including Women in Clinical Studies*, Volume II 37, Institute of Medicine.

Mcneil, Laurie and Marc Sher (1998). 'Dual-Science-Career-Couples: Survey Results'. Available online at http://www.physics.wm.edu/~sher/survey.pdf.

Michelle R. Clayman Institute for Gender Research (2006). *Making Use of Title Ix: Women in Science, Engineering, Technology, and Mathematics.* Michelle R. Clayman Institute For Gender Research, Stanford University. Available online at http://www.Stanford.Edu/Group/Gender/Researchprograms/Title IX.html.

—— (2006–07). *Dual-career Academic Couple Study*. Michelle R. Clayman Institute for Gender Research, Stanford University (unpublished results of a survey of thirteen leading U.S. research universities, on file with author).

—— (2007). 'Dual Career Academic Couples'. Michelle R. Clayman Institute For Gender Research, Stanford University. Description available online at http://www.stanford.edu/group/gender.

NSF (National Science Foundation) (2005). 'Advance: Increasing The Participation And Advancement of Women in Academic Science And Engineering Careers'. Available online at http://www.nsf.gov/pubs/2005/nsf05584/nsf05584.htm.

Office for Research on Women's Health (1999). 2 Agenda for Research on Women's Health for the 21st Century 205–11. Available online at http://orwh.od.nih.gov/research/Agenda_Book_2.pdf.

Primmer, Lesley (1997). 'Women's Health Research: Congressional Action and Legislative Gains: 1990–1994', in Florence Haseltine and Beverly Jacobson (eds), *Women's Health Research: A Medical and Policy Primer*, 301, 309–10.

Rahm, Jrène and Paul Chambonneau (1997). 'Probing Stereotypes through Students' Drawings of Scientists', *American Journal of Physics*, 65(8): 774, 776, Figure 1.

Richardson, Sarah S. (2008). 'When Gender Criticism Becomes Standard Scientific Practice: The Case of Sex Determination Genetics', in Londa Schiebinger (ed.), *Gendered Innovations in Science and Engineering*, pp. 22–42. Stanford University Press.

Rosser, Sue V. (1994). *Women's Health: Missing from U.S. Medicine.* Bloomington: Indiana University Press.

—— (2008). 'Building Two-way Streets to Implement Policies that Work for Gender and Science', in Londa Schiebinger (ed.), *Gendered Innovations in Science and Engineering.* Stanford University Press.

Rossiter, Margaret W. (1982). *Women Scientists in America: Struggles and Strategies to 1940.*

—— (1995). *Women Scientists in America: Before Affirmative Action, 1940–1972.* London: John Hopkins University Press.

Ruzek, Sheryl, Virginia Olesen, and Adele Clarke (eds.) (1997). *Women's Health: Complexities and Differences.* New Brunswick, New Jersey: Rutgers University Press.

Schiebinger, Londa (1989). *The Mind Has No Sex? Women in the Origins of Modern Science.* Cambridge: Harvard University Press.

—— (1999). *Has Feminism Changed Science?* Cambridge: Harvard University Press.

—— (2004). *Nature's Body: Gender in the Making of Modern Science.* New Brunswick: Rutgers University Press.

Spanier, Bonnie B. (1995). *Im/partial Science: Gender Ideology in Molecular Biology.* Bloomington and Indianapolis: Indiana University Press.

Stewart, Abigail J., Danielle LaVaque-Manty, and Janet E. Malley (2004). 'Recruiting Women Faculty in Science and Engineering: Preliminary Evaluation of One Intervention Model', *Journal of Women & Minorities Science and Engineering*, 10: 361.

U.S. Goverment Accountability Office (2004). *Gender Issues: Women's Participation in the Sciences has Increased, but Agencies Need to Do More to Ensure Compliance with Title IX*, GAO (Goverment Accountability Office) Report No. GAO-04–639. Available online at http://www.gao.gov/ new.items/d04639.pdf.

Valian, Virginia (1998). *Why So Slow? The Advancement of Women.* Cambridge, MA: M.I.T. Press.

Wolf-Wendel, Lisa, Susan B. Twombly, and Suzanne Rice (2003). *The Two-body Problem: Dual Career-couple Hiring Policies in Higher Education.* Baltimore: The Johns Hopkins University Press.

Zuckerman, Harriet, Jonathan R. Cole, and John T. Bruer (eds.) (1991). *The Outer Circle: Women in the Scientific Community.*

CHAPTER TWO

Gender Imbalance in Science

Cultural Similarities and Differences

Neelam Kumar

THIS chapter proposes to discuss the relationship between culture, gender, and science. It brings together empirical findings, associated debates, and research to offer an account of the relationship giving examples from various parts of the world. With the help of a review of extant literature on women's access and participation in science, the chapter tries to identify certain similarities and differences across cultures. Culture is an elusive concept. Culture as a way of life gives meaning to things we do, including science, and pervades all social institutions and systems of meanings. Inextricably bound with these societal systems is gender. Out of the social structure and culture, gender and its boundaries are fabricated, which always involves difference and inequality. The recent integrative approach treats gender as a socially constructed stratification system and does not stress the role of biological compulsions like the earlier ones. The feminist perspectives are of special help in changing how we understand the role of social interactions with biological sex differences in shaping feminine and masculine characteristics. Science too is one of the social and cultural realities. Science is a social construction and cultures differ in terms of social processes – thus social processes of science are assumed to differ across cultures. According to the simple syllogistic rule, the consideration of two premises – science is a social construction and cultures differ in terms of social processes – leads to the conclusion that social processes of science differ across cultures. But in the case of gender and science, this rule seems to be defied and many issues related to women in science appear ironically to be marked by a universal pattern.[1] Evidently, there is an increasing need for the discussion about the ways in which gender and science relationship is articulated, embodied, and enacted within and across different cultural, social, political, and economic locations. Gender is a fundamental category to express forms of social difference. Is it a determining factor to the system and processes of science as well? What is, thus, the relation between culture, gender, and science? Does culture influence gender and science relations? Is gender a determining factor to the ways in which science is defined, conducted across cultures? Is science free from gender and cultural biases? There have been debates and discussions on these issues. Yet many caveats remain.

This chapter tries to exemplify the relationship between gender, culture, and science and raises the questions: (a) whether the nature of gender and science relations are variable cross-culturally; (b) as to suggest that they are culturally constructed; (c) therefore, calling into question the claims of universality that science often makes; and (d) why and how science and technology in various societies are underlain by unresolved tensions between gender binaries like masculine and feminine. The social construction of science as well as gender provides a theoretical backdrop. The aim is not to discuss the philosophical or epistemological issues rather to discuss empirical findings and their implications.

Historical Trends

Gender–science relation cannot be seen in isolation from larger historiographical issues. Historically, science has developed within an ideological framework emphasising masculinity and patterned a 'male scientific ethos'. The initial history of women in science was that of exclusion.[2] The institutionalisation

1 Cross-cultural comparisons of Western science and indigenous knowledge systems shed light on the overlaps and disjuncture between them, calling into question the claims of universality that science often makes.

2 From the earliest times women contributed to the development of scientific knowledge, yet we think of the history of science as history of men (Alic, 1986). From the sixteenth century until well into the nineteenth century, the pursuit of expertise in scientific knowledge was considered a heretical alternative to the pursuit of classical knowledge. During this period proponents of anticlassical education actively encouraged women to pursue science and many did. By the end of the nineteenth century as research science began to replace classics, women's representation, particularly their involvement in the central activities of science, began to decline. Once scientific knowledge became culturally dominant after the turn of the twentieth century, women came to be concentrated on its margins (Eisenhart and Finkel, 1998).

of science resulted in the women's marginalisation in science. Till the late nineteenth century, the doors of universities, scientific societies, and research laboratories were not open to women, with a few exceptions. The evidence of women being banned and discouraged from entering into the 'world' of science has been all over the world. Women achieved access to institutions of higher education in US in 1833; Germany in 1908; and Japan in 1913. In India the first graduate degrees were granted to women in 1883. The new scientific societies like Britain's Royal Society which was established in 1662 did not admit women until 1945. Marie Curie was rejected by the French Academy of Sciences. The progress and development of science over the centuries had its own effects on the females' access to it. Segregated education remained the hallmark. Medicine was the first scientific profession to attract females, but only a few top schools opened their doors to women medical students. The leading British and American universities with a few exceptions barred women until the World War I. Engineering and technology remained male dominated for a longer span. Engineering schools were remarkably resistant to women in the early decades of the twentieth century and even after the doors were opened, the trend did not change easily. The very architecture of science, its invisible colleges and practices continue to locate most women in the profession's lower ranks. Do the deeprooted cultural forces within science survive to retain it as 'gendered'?

Contemporary Realities: Points of Comparison

The nature and character of science has changed during the contemporary times. Is there any change in the gender–science relationship in terms of women doing science? Certainly overt discriminatory policies and practices have declined during this period. The access of females to science, their enrollment into it has improved remarkably, yet gender inequality in various forms remains pronounced. Women's lower position in science transcends international boundaries and is a universal, cross-cultural feature of academic careers. Truthfully, one can easily see some implications and continuities of the historical exclusion to the present status and loss of women to science. Gender discrimination in science is still observed in many countries across the world. Scarcity, segregation, and invisibility of women in science is observed in every part of the world. Developed as well as developing, Eastern as well as Western, Southern versus Northern or any other division of the world is unable to reveal a distinct or even a different pattern as far as women in science is concerned. For example, the number of women in professional positions remains higher in many developing countries than in developed countries. Turkey, for example, has more female full professors (23 per cent) than any other European Union (EU) countries. In Germany, the figure stands at 4 per cent, in Denmark 6 per cent, in Sweden 7 per cent, and in the UK 10 per cent. Japan, for example, is one of the richest countries, but it's also one where women have little chance of succeeding in science (Normile, 2006). No other social phenomenon exhibits such similarities and a universal pattern!

Not only the economic development, even a country's historical trend is unable to differentiate it from others as far as gender and science is concerned. Italy has a remarkable history in terms of women in science. Right in the twelfth century it allowed women to study as well as teach science, that too physics and mathematics! Unfortunately if we examine the current scenario, Italy hardly differs from other countries. In the seventeenth century Germany, 14 per cent of all astronomers were women. Today the percentage of women astronomers in Germany is around 8 per cent (Schiebinger, 2008). Variations in law and politics too do not bring a distinct or isolating change to any country in terms of gender, women, and science. Sweden, for example, is known for its egalitarian values but there too nepotism and gender bias characterises science (Weneras and Wold, 1997). Countries with stronger orientations and feminist discourses on science, such as USA, and countries where feminist movements are not strong such as India, appear similar in many ways. Do gender and science issues go beyond race, class, colonisation – social and geographical – distinctions? How are gender and science configured in the different parts of the world as a function of changes in the contexts and cultures? There is a need to examine and explain the gender similarities and differences in science across cultures.[3] The integration of gender, culture, and science within a social constructionist approach may lead to some answer. Cross-national commonalities and variability may help to explore the structural and cultural factors that may underlie the gender–science relationship.

[3] Historians of science have traditionally concentrated on the achievements of scientists in Western Europe and North America while leaving many areas of world less-studied (Koblitz, 2005).

Empirical Illustrations: Scarcity, Segregation, Invisibility

Segregation into certain disciplines has been the most common feature worldwide. Women tend to be concentrated in certain fields and industries such as biology, health, agriculture, and pharmaceuticals, with low representation in physics and mathematics. Other obvious cross-national similarity is uniform under-representation of women in engineering and concentration in the humanities and social sciences. Everywhere gender gap is the most dramatic in engineering and technology. The current enrollment statistics reveal somewhat comparable trends. Women earned half or more of first university degrees in Science and Engineering (S&E) in many countries around the world in 2004, including the United States, Canada, Greece, Portugal, Panama, and several countries in Asia, the Middle East, and Eastern Europe. A number of countries in Europe were little behind, with more than 40 per cent of first university S&E degrees earned by women. In many Asian and African countries, women generally earn about one-third or less of the first university degrees awarded in S&E fields (National Science Foundation [NSF], see statistics at http://www.nsf.gov). Studies point to persistently high levels of segregation in colleges and universities, with women strongly under-represented in science, engineering, and technical programmes even in countries with high overall female enrollment rates (England and Li, 2006; England et al., 2007; Jacobs, 2003; Xie and Shauman, 2003). For example, in Portugal the participation of women in science is exceptionally high, but segregation by scientific field is similar to that in other European countries. Recent comparative analyses suggest, moreover, that some forms of segregation are more, not less, pronounced in the most socially or culturally modern societies (Bradley, 2000; Charles and Bradley 2002, 2006; Charles and Grusky, 2004; Van Langen and Dekkers, 2005).

Under-representation in higher academic rank and slower promotion compared to men constitute another point of similarity (Bayer and Astin, 1975; Cole and Cole, 1973; Fox, 1995; Hurlbert and Rosenfeld, 1992; Kumar, 2001). The vertical distribution in academic careers is skewed everywhere. Women tend to be concentrated in lower-level positions in academics and their percentage in the rank of professor is quite low in most of the countries[4] (Table 2.1).

In general, it has been observed that the female workforce follows a pyramidic structure, with a large base constituting of low-end career profiles

[4] The pattern is worse in science and engineering disciplines.

Table 2.1 *Percentage of Professors who are Female in some Selected Countries*

Country	Per cent of Professors (Total)
Turkey	28
Croatia	26
Bulgaria	24
Finland	23
Switzerland	22
Poland	20
France	19
Italy	19
Sweden	18
United Kingdom	17
Austria	14
Israel	13
Germany	12
The Netherlands	11
Belgium	11

Note: The reference years are as follows
Source: *She Figures*, 2009, European Union: Crotia – 2008; United Kingdom – 2006; France and Israel – 2006; All other countries – 2007.

and significantly fewer numbers as one progresses up the career hierarchy. In all the European countries there are few women in top university and research positions. The prevalence of gendered patterns is more evident in scientific fields and career positions. The possible causes discussed include the lesser number of women entering the field, unequal opportunity, and discrimination, and differences in research performance. Gender difference in scientific productivity has been an important point of discussion, especially for the academic rank. Various studies have attempted to probe the reasons for this differential (Leahey, 2006; Xie and Shauman, 2003). In particular, psycho-cognitive studies exploring verbal (Hyde and Linn, 1988), spatial (Linn and Petersen, 1985; Voyer et al., 1995) and mathematical abilities (Hyde et al., 1990) reveal that, for these abilities, sex differences are found only in few dimensions and, where they do occur, are very limited. Gender gap in publishing productivity has also been examined in terms of the influence of certain demographic variables (Long and Fox, 1995: 56; Xie and Shauman, 2003). But, in general, differences in background measures do not appear to explain gender and scientific publication relationships. Even the interactions of gender with family characteristics such as marriage and motherhood are found to be weak, inconclusive, or varied (Fox, 2005: 146; Xie and Shauman,

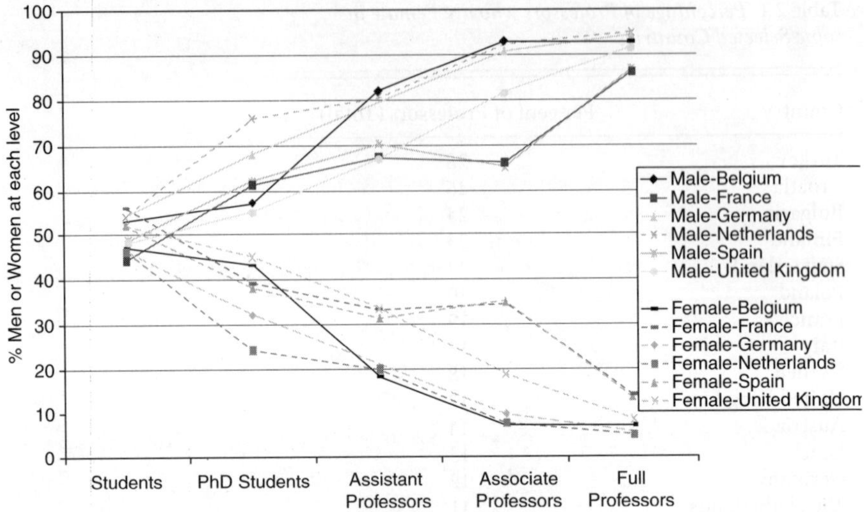

Figure 2.1 'Scissors Diagram': A Typical Academic Career Graph
Source: ETAN, 2000.

2003). Cultural, social, and organisational factors play an important role. This has raised what is called 'productivity puzzle', which several science sociologists have attempted to resolve (Etzkowitz et al., 2000; Fox and Mohapatra, 2007; Palomba and Menniti, 2001).

The low number of professorships is not due to a low representation of female students. An interesting pattern on this was revealed by the famous European Technology Assessment Network (ETAN) report and became popularly known as 'scissors diagram' (Figure 2.1). Women made up only 29 per cent of those employed as scientists and engineers across the European Union in the year 2004. The gender imbalance at the senior grade is greatest in the engineering and technology, where the proportion of women is just 5.8 per cent. The number of female students in engineering and the sciences grew considerably in the 1980s, but this growth has now tapered off at higher level where women are still a minority.

The other interesting figure is the composition of boards that are responsible for making decisions for scientific research directions such as scientific commissions, Research and Development (R&D) commissions, councils, committees, and foundations, academy assemblies, and so on. Of the 15 countries for which data was available, there were only three where the proportion of female members in the scientific boards is over 40 per cent;

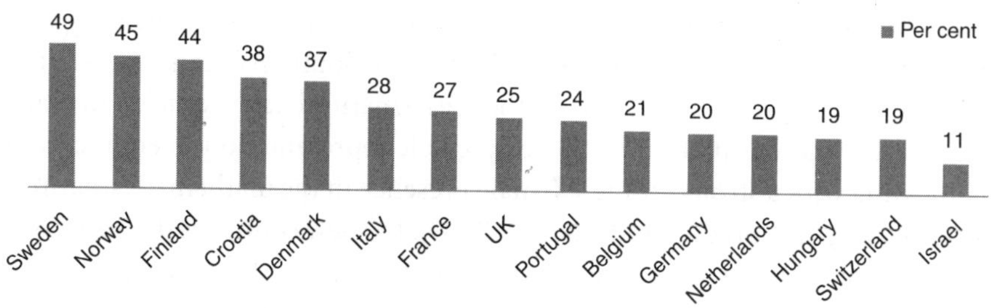

Figure 2.2 Percentage of Female on Boards in the European Union, 2007, Percentage of Females in Scientific Boards

Source: *She Figures*, 2009, European Union.

three in the range 30–39 per cent; and seven in the range 20–29 per cent (*She figures*, 2009). In the French Academy in 2004, out of the 190 members, only fourteen were females; in the Netherlands, one woman among 237 members of the Royal Netherlands Academy of Arts and Sciences; and, in the United Kingdom, 43 women among 1,185 members of the Royal Society of London. Figure 2.2 shows the percentage of women in these boards by country. From left (highest percentage) to right (lowest) we have Sweden, Norway, Finland, Croatia, Denmark, Italy, and France.

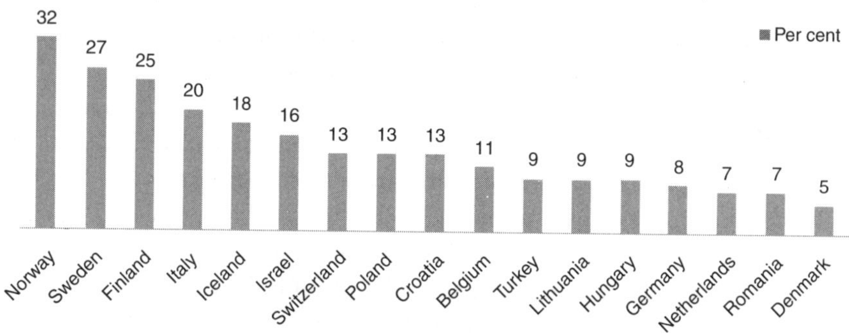

Figure 2.3 Females as Heads of Institution in the Higher Education Sector (HES) in the European Union, 2007, Percentage of Females as Head of Institutions

Source: *She Figures*, 2009, European Union.

Not only similarities, certain differences in terms of gender–science relation also prevail among nations and some of these differences are quite interesting. Countries differ not only in overall levels but also in patterns of sex segregation. For instance, among the three countries with the highest overall index scores,

Finland shows strong female under-representation in both engineering and mathematics/natural science, while Swiss and South African women are slightly over-represented in the mathematics/natural science field. In Italy, we find yet another pattern, with weak female representation in engineering coinciding with a strong absolute female presence in the mathematics/natural science category (Bradley and Charles, 2009). In a few countries (for example, El Salvador, South Korea, and several countries in Eastern Europe), more than 40 per cent of S&E first university degrees earned by women are in engineering, compared to 6 per cent in the United States (NSF, 2008). The level of higher education also reveals differences. The percentage of S&E doctoral degrees earned by women, for example, varies widely by country/economy, from less than 20 per cent in South Korea, Taiwan, Japan, Iran, and Ghana, to more than 50 per cent in Kyrgyzstan, the Philippines, Uganda, Portugal, Latvia, and Lithuania (NSF, 2008).

Moreover, historical and cross-national co-variation among common indicators of female tertiary status (that is, overall enrollments, relative representation at elite levels, and in historically male-dominated fields of study) is quite weak. Dramatic gender differences in distributions across fields of study have been found even in countries where women are well represented in the most elite sectors of higher education. The segregations across fields of study are higher in the United States than in Turkey (Charles and Bradley, 2002). What could be the explanation? Class has been found to be an important factor along with gender in various countries to explain females' access to science. In fact the history of science is always the history of a select group of individuals. The history of women in science is, unfortunately, even more selective. With a few important exceptions these women came from the upper classes and had the resources that enabled them to pursue their studies. Only privileged women could attain education and pursue scientific interests despite their exclusion from educational facilities and the formal and informal fraternities of male scientists. Social structural barriers including economic and social stratification are relevant factors associated with the gender gap in most cultural contexts. In the Indian situation the rural–urban divide also matters (Kumar, 2001).

Indian Example

I will take India as an example for a micro-level study. Science and technology in India has been reported to witness expansion in the post-Independence era. The enrollment pattern in Indian universities for the year 1941–42 shows the total number of girls enrolled for undergraduate course in science is 903 in comparison to 11,217 boys. Only 83 girls were enrolled for a postgraduate course in science in contrast to 1,321 boys. While in medicine their number was 778 against 6,093 boys, in engineering only one was enrolled along with 2,718 boys (Sundaram, 1946). Over the years, women's enrollment has shown significant increase. Figure 2.4 reveals the decadal growth of percentage of women's enrollment in various disciplines.

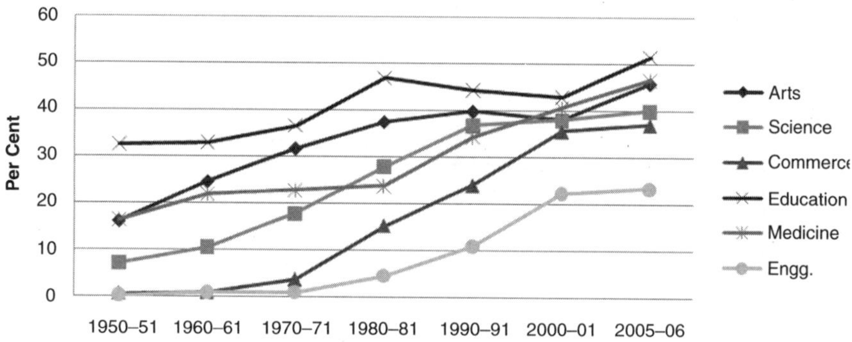

Figure 2.4 Enrollment of Girls as Percentage of Total enrolment in University Education by Discipline
Source: UGC, Annual Report (various years).

Professional areas of engineering and technology still witness a severe imbalance and women's participation which has been limited and confined to junior positions as far as science careers is concerned. Only a few women could make it to senior decision-making positions and get recognition (Figure 2.5).[5] Scientific institutions in India carry essentially masculine ethos and exhibit vertical as well as hierarchical segregation in terms of gender.[6] Traditionally too, women in India have been members of a stratified society, characterised by the ideology and practice of inequality. A patrifocal family structure, which gives precedence to men over women, characterises the Indian society and has been the determining factor

[5] For example, one can see the patterns of fellows elected in Figure 2.5.

[6] Segregation is a fundamental process in social inequality. The characteristics on which groups are sorted symbolise dominant or subordinate status and become the basis of differential treatment. Indeed, segregation facilitates unequal treatment by subjecting groups to differential reward systems (Reskin, 1993: 241). Rank segregation is the differential distribution of the sexes across vertical levels (hierarchy) within organisations. For example, the higher the rank, the lower the proportion of women.

behind women's lives including their educational access and preferences. Incongruities and oppositions can, however, be seen in case of women scientists in India both historically and currently. Women were, for example, denied entry to Raman's laboratory (Sur, 2009).

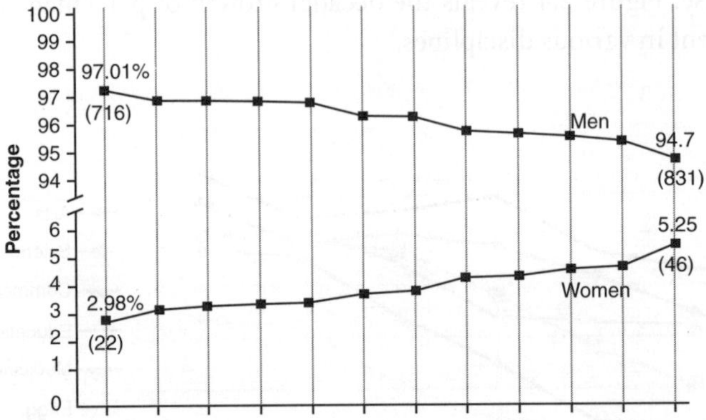

Figure 2.5 Percentage-wise Composition of Fellows for 1995–2006.
Source: Indian Academy of Sciences

Contemporary studies too reveal gender differentiation in Indian scientific institutions. Organisational hurdles outweigh the so-called family constraints (Krishnaraj, 1991). Unequal treatment and subtle discrimination against women scientist and engineers in the behavioural and interpersonal relations also prevail (Jaiswal, 1993); the cultural and social context in which science is learned and practiced contributes to the gendering of science (Mukhopadhyay and Seymour, 1994); and discrimination component is much higher in scientific and technical fields in India than among social sciences and other fields (Duraisamy and Duraisamy, 1998). The prevailing socio-cultural systems in India result in a 'triple burden' for women in academic and scientific careers (Gupta and Sharma, 2002). Women in all professions perform a double role of managing job and domestic responsibilities, which has been commonly referred to as a 'dual burden'. In science, the dual burden is combined with various problems that are specific to the scientific profession.

The policies and programmes related to women's access to higher education (particularly science and technology) and their career growth often face critique and are also held responsible apart from the constraints discussed above. The shortfalls are often attributed to the lack of clarity on the purpose of women's education in educational planning. For instance, the National

Committee on Women's Education (1956) set up to scrutinise the special problems of women's education, on the one hand emphasised the need to bridge the gap between the education of men and women, and on the other hand, reiterated the traditional gender roles in society. Similarly, all the other important commissions such as the University Education Commission (1948–49); the Secondary Education Commission (1952–53); the National Commission on Women's Education (1958); the National Council for Women's Education (1959); the Baktavatsalam Committee (1963); and the National Committee on Women's Education (1970) were hesitant in defining the aims of women's education, and seem to have been caught in contradictory value systems while defining the purpose of female education (Poonacha, 2004). *The Draft National Policy on Education* admitted that although education had expanded in all sectors yet imbalances and inequalities continued to exist.

By the 1970s, however, there was a growing awareness that gender is an important social category, which needs to be taken into account in development planning. The Women's Decade in India began in 1975 with an official report of the Committee on the Status of Women in India. The *Report of the Committee on the Status of Women* (1974), better known as the *Towards Equality Report*, set clear guidelines on the aims of female education. In recent years there are various efforts by government bodies to enhance women's access to science careers. The department of science and technology (DST), for example runs a programme entitled 'Women Scientists Scheme (WOS)' for providing opportunities to women scientists and technologists to return to mainstream science and work as bench-level scientists. The University Grants Commission has been endeavouring to weave gender equality and justice into its policies and university curricula. It is also trying to reduce gender imbalance in the sphere of Engineering and Technology. Further, part-time research fellowship exclusively for women scholars are also going on. To examine the socio-economic and institutional factors that limit women's participation in science, the Indian National Science Academy commissioned a study and has also published a report on *Science Career for Indian Women* (INSA, 2004).

Reflections

Interestingly studies on 'women in science' sometimes explain the disparities in terms of individual characteristics and identities rather than in terms of social systems. Psychological researches especially have been prone to study personality (for example, Eiduson and Beckman, 1973), motivation (Sri Chandra, 1970); and biographical background of scientists (Chambers, 1964; Roe, 1952). Early functionalist psychologists simply assumed that women had less innate talent than men. Research and theories in psychology of science have discussed in great detail the role of cognitive and individual psychological processes in the conduct of science. Social psychology although focuses on the relationships between individuals and their social environments, it too has failed to fully incorporate their social structural positions, such as their location in the social hierarchies of race, gender, and so on (Hollander and Howard, 2000). Despite a growing interest in gender differences in scientific careers, few studies have focused on the impact of research organisation on researchers (Mahlck, 2001).

To conclude we may say:

(a) While there have been efforts to explain horizontal segregation, many a times using gender-essentialist stereotypes, explanations based on economic and social dimensions, the vertical segregation remain largely unexplained.[7] Women's under-representation in top academic career positions persists throughout the European Union. Even in countries where the participation of women in science is exceptionally high, such as Portugal, segregation by rank, is similar to that in the other European countries (ETAN, 2000). Certainly the very processes related to science or the 'culture of science' is crucial. Many feminist scholars have described how core science characteristics, for example, rationality and objectivity, are culturally aligned with masculine rather than feminine qualities.

(b) Patterns of cross-national variability in segregation are at most weakly related to measures of women's status in other social spheres. The phenomenon cannot be understood as a simple function of the 'status of women' or the level of modernity in any given national or historical context or gender-essentialist stereotypes.

Does science as an institutionalised system of social practices stick to its own ethos? Or is it gendered? Why are there still so few women scientists, especially at the upper levels of the scientific professions? Gender–science relationships are invariant across cultures. The state of women in science even in contrasting

[7] Horizontal segregation refers to distributional inequalities that are not explicitly hierarchical, while vertical segregation refers to inequalities in rank or prestige within the scientific community.

social, political, and economic systems, such as Europe, North America, South America, and Asia appears to be similar. A common harsh reality persists in science – higher the hierarchy, fewer the women. The persistence of women's under-representation and unequal position in science and engineering over the years and across cultures, poses a challenge to the explanations offered. However it is clear and obvious that the phenomenon cannot be adequately explained by the theories based on biological characteristics and differences, gender-role stereotypes or socialisation patterns. Gender inequality remains an ongoing debate. Does this chapter dispute the view that systems of gender identity are so extraordinarily variable cross-culturally as to suggest that they are totally culturally constructed, and, therefore, not open to universal explanation?

References

Alic, Margaret (1986). *Hypatia's Heritage*. London: The Women's Press.

Bayer, Alan E. and Astin, Helen S. (1975). Sex Differentials in the Academic Reward System. *Science*, Vol. 188. No. 4190, pp. 796–802.

Bradley, Karen (2000). The Incorporation of Women into Higher Education: Paradoxical Outcomes?, *Sociology of Education*, 73, No 1: 1–18.

Bradley, Karen and Maria Charles (2009). 'Indulging Our Gendered Selves? Sex-Segregation by Field of Study in 44 Countries', *American Journal of Sociology*, 114(4): 924–76.

Chambers, Jack A. (1964). 'Creative Scientists of Today', *Science*, 145(3637) (11 September): 1203–05.

Charles, Maria and Karen Bradley (2002). Equal but Separate? Across-National Study of Sex Segregation in Higher Education. *American Sociological Review*, Vol. 67, No. 4: 573–579.

—— (2006). A Matter of Degrees: Female Underrepresentation in Computer Science Programs Cross-Nationally. In Joanne McGrath Cohoon and Bill Aspray (eds.) *Women and Information Technology: Research on the Reasons for Underrepresentation*, MIT Press, Cambridge, MA, pp. 183–203.

Charles, Maria and David Grusky (2004). *Occupational Ghettos: the Worldwide Segregation of Women and Men*, Stanford, CA: Stanford University Press.

Cole, J. R. and Cole S. (1973). *Social Stratification in Science*. The University of Chicago Press, Chicago.

Duraisamy Malathi and P. Duraisamy (1998). 'Sex Segregation and Discrimination among Scientific Personnel' in Maithreyi Krishnaraj, Ratna M Sudarshan and Abusaleh Shariff (eds), *Gender, Population and Development*, Oxford University Press, Delhi, 159–84.

Eiduson, Bernice T. and Linda Beckman (eds.) (1973). *Science as a Career Choice; Theoretical and Empirical Studies*. New York: Russell Sage Foundation.

Eisenhart, Margaret A. and Elizabeth Finkel (1998). *Women's Science*. Chicago: The University of Chicago Press.

England, Paula and Su Li (2006). "Desegregation Stalled: The Changing Gender Composition of College Majors, 1971–2002." *Gender & Society*, 20: 657–677.

England, Paula, Paul Allison, Su Li, Noah Mark, Jennifer Thompson, Michelle Budig, Han Sun (2007). "Why Are Some Academic Fields Tipping Toward Female?" The Sex Composition of U.S. Fields of Doctoral Degree Receipt, 1971–2002. *Sociology of Education*, 80: 23–42.

ETAN (2000). *Science Policies in the European Union: Promoting Excellence through Mainstreaming Gender Equality – A Report from the ETAN (European Technology Assessment Network)*, Expert Working Group on Women and Science, European Communities, Brussels.

Etzkowitz, Henry, Carol Kemelgor, and Brian Uzzi (2000). *Athena Unbound. The Advancement of Women in Science and Technology*. Cambridge: Cambridge University Press.

Fox, M. F. (2005). 'Gender, Family Characteristics, and Publication Productivity among Scientists', *Social Studies of Science*, 35(1): 131–50.

Fox, M. F. and S. Mohapatra (2007). 'Social-organizational Characteristics of Work and Publication Productivity among Academic Scientists in Doctoral-granting Departments', *Journal of Higher Education*, 78(5): 543–71.

Fox, Mary Frank (1995). "Women and Scientific Careers." In S. Jasanoff, J. Markle, J. Petersen, and T. Pinch, eds., *Handbook of Science and Technology Studies* (pp. 205–223). Sage: Newbury Park, California.

Government of India (1975). *Towards Equality: Report of the Committee on the Status Stam of Women*. New Delhi: Ministry of Human Resource Development.

—— (1979). *Draft on National Policy on Education*. New Delhi: Government of India, Ministry of Education and Social Welfare.

Gupta, N. and A.K. Sharma (2002). 'Women Academic Scientists in India', *Social Studies of Science*, 32(5–6): 901–15.

Harding, Sandra (1986). *The Science Question in Feminism*. London: Open University Press.

Hollander, Jocelyn A. and Judith A. Howard (2000). 'Social Psychological Theories on Social Inequality', *Social Psychology Quarterly*, Vol. 63, No. 4: 338–51.

Hyde, J.S., E. Fennema, and S.J. Lamon (1990). 'Gender Differences in Mathematics Performance: A Metaanalysis', *Psychological Bulletin*, Vol. 107, No. 2: 139–155.

Hyde, J.S., and M.C. Linn (1988). 'Gender Differences in Verbal Ability: A Meta-analysis', *Psychological Bulletin*, Vol. 104, No. 1: 53–69.

Hurlbert, Jeanne, and Rachel A. Rosenfeld (1992). 'Getting a good job: Rank and institutional prestige in academic psychologists' careers', *Sociology of Education*, Vol. 65, No. 3: 188–207.

INSA (Indian National Science Academy) (2004). *Science Career for Indian Women: An Examination of Indian Women's Access to and Retention in Scientific Careers*. Indian National Science Academy (INSA), Delhi.

Jacobs, J. A. (2003). 'Detours on the Road to Equality: Women, Work and Higher Education', *Contexts*, Vol. 2, No. 1: 32–41.

Jaiswal, Rajendra Prasad (1993). *Professional Status of Women: A Comparative Study of Women in Science and Technology*. Jaipur: Rawat Publications.

Koblitz, Ann Hibner (2005). 'Gender and Science: Where Science Is on the Margins', *Bulletin of Science, Technology & Society*, 25(2): 107–14.

Krishnaraj, Maithreyi (1991). *Women and Science: Selected Essays*. Bombay: Himalaya Publishing House.

Kumar, Neelam (2001). 'Gender and Stratification in Science: An Empirical Study in Indian Setting', *Indian Journal of Gender Studies*, 8(1): 51–67.

Leahey, E. (2006). 'Gender Differences in Productivity: Research Specialization as a Missing Link', *Gender and Society*, 20(6): 754–80.

Linn, M.C. and A.C. Petersen (1985). 'Emergence and Characterisation of Sex Differences in Spatial Ability: A Meta-analysis', *Child Development*, Vol. 56, No. 6: 1479–98.

Long, J. S. (1992). 'Measure of Sex Differences in Scientific Productivity', *Social Forces*, 71(1): 159–78.

Long, J. S. and Fox, M. F. (1995). Scientific Careers: Universalism and Particularism. *Annual Review of Sociology*, 21: 45–71.

Mahlck, P. (2001). 'Mapping Gender Differences in Science Careers in Social and Bibliographic Space', *Science, Technology and Human Values*, 26(2): 167–90.

Mukhopadhyay, C. and S. Seymour (1994). 'Introduction and Overview', in C. Mukhopadhyay and S. Seymour (eds), *Women, Education and Family Structure in India*, pp. 1–33. San Francisco: Westview Press.

Normile, Dennis (2006). 'Women in Science: Getting Women Scientists Back on the Career Track in Japan', *Science*, 311(5765): 1235–36.

NSF (National Science Foundation) (2008). 'Science and Engineering Education'. Available at http://www.nsf.gov/statistics/seind08/start.htm. 28th March, 2010.

Palomba, R. and A. Menniti (eds) (2001). *Minerva's Daughters*. Rome, Italy: Institute for Research on Population and Social Policies.

Poole Millicent, Laurel Bornholt and Fiona Summers (1997). 'An International Study of the Gendered Nature of Academic Work: Some cross-cultural Explorations', *Higher Education*, Vol. 34, No. 3: 373–96.

Poonacha, Veena (2004). 'Uncovering the Gender Politics of Science Policies, Education and Institutional Practices', paper presented at Nistads International conference on *Women in Science: Is the Glass Ceiling Disappearing?* New Delhi, 8–10 March.

Reskin, Barbara, 'Sex Segregation in the Workplace.' *Annual Review of Sociology* 19: 241–70.

Roe, Anne (1952). *The Making of a Scientist*. New York: Dodd, Mead and Co.

Schiebinger, Londa (ed.) (2008). *Gendered Innovations in Science and Engineering*. Stanford: Stanford University Press.

She Figures (2009). *Women and Science Statistics and Indicators* (Luxembourg: Office for Official Publications of the European Communities, European Commission).

Sri Chandra (1970). *Scientists: A Social Psychological Study.* Delhi: Oxford and IBH.

Sundaram, M.S. (1946). 'Education in British India', *The Journal of Negro Education*, 15(3): 513–25.

Sur, Abha (2009). 'Dispersed Radiance: Women Scientists in C.V. Raman's Laboratory', in Neelam Kumar (ed.), *Women and Science in India*. New Delhi: Oxford University Press, pp. 98–134.

Traweek, Sharon (2006). 'Women in Science: Under-represented and Under-measured' *UIS Bulletin on Science and Technology Statistics*, Issue No. 3, November.

University Grants Commission (UGC) (1988–89): Annual Report, UGC, New Delhi.

University Grants Commission (UGC) (1994–95): Annual Report, UGC, New Delhi.

University Grants Commission (UGC) (2005–06): Annual Report, UGC, New Delhi.

van Langen, Annemarie and Hetty Dekkers (2005). Cross-National Differences in Participating in Tertiary Science, Technology, Engineering and Mathematics Education, *Comparative Education*, Vol. 41, No. 3, pp. 329–350.

Voyer, D., S. Voyer, and M.P. Bryden (1995). Magnitude of Sex Differences in Spatial Ability: A Meta-analysis and Consideration of Critical Variables, *Psychological Bulletin*, 117: 250–70.

Wenneras, C. and A. Wold (1997). 'Nepotism and Sexism in Peer Review', *Nature*, Vol.387. No 6631: 341–43.

Xie, Y. and K. A. Shauman (2003). *Women in Science: Career Processes and Outcomes.* Massachussets: Harvard University Press.

CHAPTER THREE

Gender and Technology*

Francesca Bray

One fundamental way in which gender is expressed in any society, is through technology. Technical skills and domains of expertise are divided between and within the sexes, shaping masculinities and femininities – maybe the iconic womanly skill is basket making, whereas men should excel at hunting (MacKenzie, 1991); or boys must learn to clean their fathers' tools to get a feel for grease before they are taught to use them (Mellström, 2004); or poor women raise silkworms and sell the cocoons to rich households where the mistress organises the tasks of reeling, spinning, and weaving among her servants (Bray, 1997); or boys huddle around the computer screen, practicing hacking skills, while girls develop new communication codes using emoticons (Lægran, 2003b; Miller, 2004). In the contemporary world, or at any rate in the Western nations which pioneered industrialisation and have thus been able to dominate worldwide production of material and intellectual goods, services, and desires, for so long, technology is firmly coded male. Men are viewed as having a natural affinity with technology, whereas women supposedly fear or dislike it. Men actively engage with machines – making, using, tinkering with, and loving them. Women may have to use machines, in the workplace or in the home, but they neither love nor seek to understand

* 'Reprinted, with permission, from the *Annual Review of Anthropology*. Volume 36 © 2007 by Annual Reviews'. www.annualreviews.org.

them – they are considered passive beneficiaries of the inventive flame. The modernist association of technology with masculinity translates into everyday experiences of gender, historical narratives, employment practices, education, the design of new technologies, and the distribution of power across a global society in which technology is seen as the driving force of progress.

'Since technology and gender are both socially constructed and socially pervasive, we can never fully understand one without also understanding the other' (Lohan and Faulkner, 2004: 319). A dense web of debate within the field of gender and technology studies, or feminist technology studies (FTS), catalyses continual advances in studying what FTS terms the coproduction of gender and technology. Explorations of 'constructive' tensions in FTS (Lohan, 2000) aim to develop innovative analyses of the material worlds we are creating through technology, and of technology's role in shaping local and global configurations of power, forms of identity, and ways of living. Although expressed in different terms, this debate shadows current anthropological concerns with the transformative role and destabilising potential of technology in emergent configurations of *oikos* (what are the forms of human community?) and *anthropos* (what is a human being?) (Collier and Ong, 2005). Yet curiously the two debates are not in dialogue but remain largely unconnected.

Theoretical debates around the gender-and-technology pair principally engage feminist sociologists and historians working in critical technology studies. Nordic social anthropologists and one or two representatives of the Anglophone and French school of the anthropology of technology also contribute to the debates. These scholars argue with each other, collaborate, and contribute to the same collections. The FTS scholars draw on feminist philosophers of science and technology such as Harding (1986) and Haraway (1991), and gender theorists such as Butler (1993), who are also regular sources of theoretical inspiration to anthropologists. Yet the absence of overlap between FTS and cultural anthropology is striking. The latter is conspicuous by its absence from FTS state-of-the-field essays (Lerman et al., 1997; Lohan and Faulkner, 2004; Wajcman, 2002) and important FTS anthologies (Lerman et al., 2003; Mackenzie and Wajcman, 1999). Conversely, most cultural anthropologists grappling with flows and subjectivities in the contemporary world, even when they put 'technology' at the heart of their research, ignore FTS scholarship and define, delineate, and articulate their key questions and objects of inquiry in subtly different terms.

This chapter begins by looking at FTS, its origins and goals, and the concepts and methods it has developed for relating gender and technology. It then turns to the anthropology of technology, which does not highlight gender to the same degree but nevertheless offers useful conceptual frameworks and methods for exploring gender regimes. Gender–technology relations also feature in the anthropology of work, labour, and development, but unfortunately space precludes discussing them here (see Freeman, 2001; Mills, 2003; Ortiz 2002). Rather, to highlight the ideological and methodological contrasts between social and cultural analyses of technology and the implications for gender analysis, we discuss the treatment of technology in two leading theoretical fields in the cultural anthropology of modernity and globalisation – the anthropology of technoscience and material culture studies. It concludes by asking what forms of engagement might be envisaged between the fields.

Feminist Technology Studies: The Coproduction of Technology and Gender

Feminist technology studies have developed in dialogue with the history and sociology of technology, disciplines in which feminist critiques have played a central part in overturning grand narratives and developing new analytical models (Faulkner, 2001; Lerman et al., 1997; Wajcman, 2004). Feminist sociologists and historians based in the Netherlands, the United Kingdom, and Australia, and a network of Norwegian scholars that includes social anthropologists, have played a prominent role in developing the field.

Arguing that in the modern world an effective engagement with technology is essential to feminist praxis, FTS strives to develop the theoretical and methodological tools to analyse technology and gender simultaneously in equal depth (Faulkner, 2001; Lohan, 2000). Unlike other feminist research on technology, which tends to treat technological artefacts as readymades, FTS looks to the production of technology as a point of political leverage.

One influential narrative of modernity, a 'standard view' (Pfaffenberger, 1992) still in common currency today, designates science as the purest and most powerful form of knowledge, the driving force of modernity; technology is essentially the application of science to practical problems. Technology studies long ago rejected this model, insisting that technology must be studied in its own right as a distinctive practice; in the 1980s science studies also came to acknowledge the critical role of technology and its epistemologies in shaping the production of scientific knowledge. Despite exploring the political, cultural, and even cosmological dimensions of technical projects, technology studies long remained gender-blind, focusing on modern industrial and military technologies and reflecting the social realities of the engineering and business worlds in foregrounding Man the Machine-Maker (Staudenmaier, 1985).

In the 1970s, radical feminists and eco-feminists initiated a critique of the inherently patriarchal nature of technology and of technoscience more generally. Here the perils of essentialisation surfaced. Some feminists condemned all technology as intrinsically oppressive of women; others perpetuated stereotypes of women as inherently nurturing. Socialist feminists generally tried to be more contextual in their work, pushing Marxist analysis beyond class to ask why and how modern Western technology had become a male domain; to address the gendering impact of modern divisions of labour and of the assignation of women to the domestic sphere; to expand the spectrum of significant technologies to include refrigerators as well as space probes and suspension bridges; and to explore the reproductive and ethical as well as the productive effects of labour organisation or of technological design (Cockburn, 1983; Corea et al., 1985; Kramerae et al., 1988; Oakley, 1974; Wajcman, 1991). Cowan's landmark study of household technologies (1983) undermined the common belief that technology makes our lives easier, showing how mechanisation served to raise cultural standards of cleanliness rather than freeing women from domestic drudgery. Through interrogating concepts such as technological efficiency and significance (Stanley, 1993), FTS has broadened the scope of technology studies to include such assemblages as the brassiere, the closet, and the white collar (McGaw, 1996). Feminist studies of the engineering profession charted the institutional, social, and cultural barriers against women (Arnold and Faulkner, 1985; Bucciarelli, 1994; Cockburn, 1985). The FTS agenda was both intellectual and political: while undermining gender stereotypes and masculinist accounts of modernity, the ultimate goal of feminist technology studies was, and remains, the translation of scholarship into feminist praxis (Faulkner, 2001, Wajcman, 2004). FTS follows the technology studies agenda in studying technology as a distinctive domain, but like feminist science studies (Harding, 1986) it interrogates its gendering at every level (Cockburn and Ormrod, 1993).

In the late 1980s, constructivist approaches emerged in technology studies that shifted theoretical and empirical attention from engineers' decisions to the complex social negotiations and contestations, the heterogeneity of expertise, of interest groups, and of material or institutional networks involved in technological innovation and in the stabilisation or redesigning of artefacts (Bijker et al., 1987). The concept of 'sociotechnical systems' reflected the principle that the social and the technological are inseparable, a 'seamless web' (Hughes, 1986). Marxist scholars unmasked the politics embodied or encoded in the design of technological artefacts (Feenberg, 1999; Winner, 1986). Actor network theorists proposed treating artefacts as having agency: these non-human actors may resist enrollment into our technological projects; furthermore we may delegate to non-human actors moral as well as material roles, inscripted into their design (Akrich, 1992; Latour, 1992).

A core interest of constructivist studies of technology is how artefacts (mass-produced bicycles, electrical supply systems) come to be as they are (Hughes, 1983; Pinch and Bijker, 1987). This approach initially tended to keep the focus of analysis upstream, looking at the processes of conceptualisation and the marshalling of resources that go into design, production, and marketing. As feminist critics noted, in modern industrial societies, an upstream focus may exclude women. However, the artifact itself, or its representation through instruction manuals, advertisements, marketing, or the media, can often be shown to incorporate 'configurations of the user', including 'gender scripts', for instance, shaver models that inscript male desires to tinker versus female preferences for simplicity (van Oost, 2003) or cars marketed to men as powerful, to women as reliable (Hubak, 1996).

The FTS scholar Cowan first brought attention to the importance of the consumer in determining the success or failure of technologies. She defined the 'consumption junction' as 'the place and time at which the consumer makes choices between competing technologies' (1987: 263). Once consumers (or rather users), like producers, were treated as rational actors embedded in complex socio-technical and cultural systems, it became easier to explain their decisions to adopt or to refuse a technology, as well as the degrees of 'interpretive flexibility' to which they might subject it (Lægran, 2003a; Parr, 1999).

This shift of attention downstream, to consumers, mirrored a broader trend in social and cultural analysis towards studying consumption as the principle site for the production of meaning and the reproduction of power relations in modern societies. In technology studies, however, the role of consumer is more complex, interesting, and powerful than is usually the case in cultural studies. In technology studies consumers are users (or refusers), engaging actively sometimes positively, sometimes negatively with the physical as well as the symbolic dimensions of the artifact (Oudshoorn and Pinch, 2003).

New technologies are often threatening and unfamiliar. To be incorporated into our lives they must be successfully 'domesticated' (Lie and Sørensen,

1996; Silverstone and Hirsch, 1992; Sørensen and Berg, 1991). At one level, we learn to adapt to the technologies, acquiring and communicating technical skills and developing uses and meanings including gendered subjectivities within 'communities of practice' (Mellström, 2004; Paechter, 2006; Wenger, 1998). Equally important is the feedback upstream of intended and unintended uses. So-called 'user-centred design' is now routine in many industries (Oudshoorn et al., 2004), and the choices and subjectivities of non-users are becoming just as important to industry (and to social scientists) as those of users (Kline, 2003; Wyatt, 2003).

In the introduction to the second edition of their influential collection on the social shaping of technology, Mackenzie and Wajcman urge researchers to continue to examine 'the specific ways in which this shaping takes place... [for] if the idea of the social shaping of technology has intellectual or political merit, this lies in the details' (1999: xvi). But how might case studies best be connected to cast light on broader political configurations? The FTS does not share the current obsession of Anglophone anthropology with theorising globalisation. Rather, it proposes the concept of integration as an approach to processes of interpenetration and patterns of homogenisation or heterogeneity within a community, nation, region, or global network. On one level, technological integration hinges on the effective interconnection of technical hardware and expertise; on another level, it is a political, social, and cultural process (Arnold, 2005; Misa and Schot, 2005). Although 'users' remain a key focus in FTS, one recent integrative approach, the 'mediation junction' (Oldenziel et al., 2005), locates stakeholder interactions, coalitions, and contestations within overarching contexts of regulation or policy, and of state, market, and civil society (see also Oudshoorn and Pinch, 2003: pp. 101–90). Oldenziel et al. (2005) highlight the importance of consumer organisations in the post-war incorporation of American-style kitchens into European homes, consumption styles, and social values and also into safety regulations, systems of energy supply, and brand rankings. Other studies compare patient activism around cancer testing in the United States and the United Kingdom (Parthasarathy, 2003) or the impact on regulatory policy of global coalitions supporting or contesting genetically modified crops (Bray, 2003).

Another prominent concern in current FTS is the exploration of femininities and masculinities, their performance through technology, and issues of practice, skill, and embodiment, including emotions, pleasure, sexuality, and eroticism (Law, 1998; Law and Singleton, 2000). Together with

Butler's (1993) analysis of gender as performance, Connell's (1995) concept of 'hegemonic masculinity', 'the configuration of gender practice which embodies the currently accepted answer to the problem of the legitimacy of patriarchy' (p. 77), serves FTS scholars as a tool to explore how particular gendered identities are attributed, achieved, and performed and their place within broader configurations of power.

Wajcman has noted a distinction between two expressive and constitutive forms of masculinity, both connected to the mastery of technology. One is based on toughness and practical skills (for example, the mechanic), the other on intellectual acuity (for example, the software designer) (Wajcman, 1991). Horowitz's collection *Boys and their Toys?* (2001) examines 'manhood in the workplace', 'learning to be men', and 'manhood at play'. Faulkner and her colleagues explore different ways in which men and women talk about their technical aptitude, setting these self-representations against actual practice (Faulkner, 2000; Kleif and Faulkner, 2003). Mellström (2003) has studied the relation between technologically configured masculinities and state ideologies of modernity in Malaysia; how the embodied 'learned dispositions' of mechanics are fostered and transmuted from father to son (2002); and the uses of leisure artifacts such as motor-bikes in male bonding in Sweden and Malaysia (2004). Although the equation between masculinity and technology in Western societies is durable, there are often huge mismatches between image and practice so that fractured and contradictory constructions of masculinity often coexist (Faulkner, 2000). Meanwhile research on non-Western societies challenges these associations. Lagesen's research in Malaysia, for example, shows that young women enter the profession of software engineering in roughly equal numbers to men and believe that their different practices of problem-solving are equally conducive to excellence (Lagesen, 2005).

The FTS scholars use the term coproduction to designate the dialectical shaping of gender and technology. The concept is intended to highlight the performative, processual character of both gender and technology and to avoid the analytical and political pitfalls of essentialising either (Berg, 1997; Faulkner, 2001; Grint and Gill, 1995). In modern societies, gender is constitutive of what is recognised as technology, determining whether skills are categorised as important or trivial (Bowker and Star, 1999). An electric iron is not technology when a woman is ironing clothes, but it becomes technology when her husband mends it. A woman engineer who tests microwave ovens is told by her male colleagues that her job is really just cooking (Cockburn and Ormrod, 1993). In the 1970s computers were thought of as 'information technologies' and coded male; it was widely assumed that women would have problems with them. By the 1990s computers had also become 'communication technologies'; now it was presumed that women would engage with them enthusiastically. 'New technologies spur processes of boundary work and renegotiations of what is to be considered masculine and feminine' (Lie, 2003a: 21; Lohan, 2001).

In terms of praxis, the overarching goal of FTS is to analyse how technology is implicated in gender inequalities to work towards more democratic forms of technology. Noting the relatively limited potential of consumer intervention for democratising technologies from the outside, some FTS scholars suggest that rather than continuing to focus predominantly on consumption, identity, and representation, FTS should return to production and work, or to the gendering of design processes and the gender subjectivities of designers, as research sites (Oudshoorn et al., 2004; Wajcman, 2004). An important paper by Suchman (1999) based on an anthropological consultancy for technology design in a large industrial enterprise, draws on Haraway and on labour theory to propose new modes of feminist objectivity, rooted in densely structured and dynamic landscapes of working relations that destabilise the boundaries between producer and user. Documenting the masculinist ideologies of the engineering world and exposing prevalent stereotypes about women and technology may both contribute to democratising technology from the inside out. Eventually they might inflect prevailing ideologies of technology. More modestly, given that gender systems are more difficult to change than are material technologies, they suggest ways to encourage more women to become engineers or to reshape state or industry policies of training and employment (Gansmo, 2003; Kvande, 1999).

Anthropology of Technology, Anthropology of Techniques

Within the American tradition of cultural anthropology, technology has generally been viewed 'as a context for, rather than a central part of, culture' (Wilson and Peterson, 2002: 450). Pfaffenberger (1992) lays out a melancholy history of neglect, dating back to Malinowski's declaration that the study of technology alone was scientifically sterile (1935: 460) and to Kroeber and Kluckhohn (1952: 65), who rejected the term material culture on the grounds that the culture was the idea behind the artifact. Technology continued to be studied by archaeologists, cultural ecologists (including Geertz at an early point of his career [1963]), and development anthropologists; feminist archaeologists have been particularly productive in rethinking gender – technology relations (Gero and Conkey, 1991; Wright, 1996). Yet within mainstream cultural anthropology in the United States, technology was not an object of analysis in its own right, and no recognised field of anthropology of technology emerged (Pfaffenberger, 1992; Suchman, 2001). This antimaterialist aversion was less marked in British social anthropology, but despite some distinguished studies and original theoretical claims (Gell, 1992; Goody, 1971, 1986; Ingold, 2000; Sillitoe, 1988), there too anthropological interest in technology as a theorisable category has remained muted.

In 1992, Pfaffenberger published an impassioned call to anthropologists to take technology seriously. Anthropology was uniquely qualified, he argued, to answer important questions about technology as a universal human activity. He proposed translating the concept of 'sociotechnical systems', borrowed from technology studies, into a template for anthropological study, laying a basis for comparative analysis of the place of technologies in the generation of meaning in pre-capitalist as well as capitalist societies. In 2001, Pfaffenberger once again lamented 'the enormous cost of Anglo-American anthropologists' penchant to ignore technological activities' (p. 84). His paper appears in a wide-ranging collection of perceptive and original essays on technology by archaeologists and anthropologists. But theoretically and methodologically they sprawl: a noble attempt by the editor to extract a coherent agenda for an anthropology of technology reads like a list, not a programme, and gender is not mentioned (Schiffer, 2001b).

Among the few American anthropologists to take technology seriously as technology are Suchman and Downey. Both work among engineers, focusing on the design and production of technologies, the business contexts in which they are developed, and the material incorporation of values and worldviews into artifacts such as bridges or computer-aided design and computer-aided manufacturing (CAD/CAM) technology (Downey, 1992, 1998; Suchman, 2001). In an essay advocating 'cyborg anthropology', Downey et al. (1995) propose close anthropological attention not only to representations or consumption of technology, but also to the cultures of the technical communities that produce technologies and to the specific material effects of technology on perception, communication, and identity. The authors propose cyborg anthropology as an action-oriented agenda, aligned with FTS, that would engage the general public and unmask the material as well as cultural dimensions of domination by race, class, and gender.

From her uncharacteristic perspective as an anthropologist working with industry, Suchman (2001) distinguishes three aspects of research on contemporary technology: (a) ethnographic studies of sites of technology production; (b) studies of technologies-in-use; and (c) ethnographically based design interventions. Although aspect (c), rooted firmly in aspects (a) and (b), would be the goal of feminist technology studies, anthropological studies of technologies are usually limited to aspect (b). In the absence of sustained debate around technology as a distinctive category of material activity, rather than just another source of metaphors, it is not surprising that most anthropologists prefer just to look at the dimensions that are most obviously cultural productions. As Axel (2006) notes, anthropologists writing on emergent technologies, for example, information and communication technologies (Escobar, 1994; Hakken, 1993; Wilson and Peterson, 2002), invariably claim that anthropology is a discipline that is particularly well suited to charting their emergence. Yet these are accounts not of technology per se but of specific technologies, and it is not clear that they offer anything distinctive from analyses produced in other branches of cultural studies.

Over decades of intensive debates in the pages of *Techniques et culture* (Techniques and Culture) and other francophone journals, the French school of anthropology of techniques, which also includes archaeologists, economists, engineers, historians, and sociologists, has developed specific theoretical and methodological repertories for the comparative study of technologies. The convention of defining technique to include bodily practices (*techniques du corps*) as well as the use of tools dates back to Mauss, who saw *techniques du corps* as distinctive cultural practices, and to Leroi-Gourhan, who treated tool and anatomy as inseparable in his analysis of the logic of technical action. The French approach begins with detailed attention to 'operational sequences' or *chaines operatoires*, 'the series of operations involved in any transformation of matter (including our own body) by human beings' (Lemonnier, 1992: 25). From systematic observation of the operational sequences of production or use, analysis proceeds to what Lemonnier calls the 'social representation of technologies': this denotes not only the kinds of meaning that usually attract the attention of cultural anthropologists, but also the ideas governing the construction and use of tools and artefacts, an ethnoscience of material nature and action.

Skills (*savoir-faire*), documented through operational sequences, are a key focus in which material, mental, social, and cultural resources converge (d'Onofrio and Joulian, 2006). The analysis of technological choices or styles goes beyond, but must account for, the relevant material affordances or constraints and systems of technical skill and understanding (Lemonnier, 1993). The core observational and analytical methods may be deployed within a variety of overarching frameworks, including actor-network theory (Latour, 1993), modes of production (Guille-Escuret, 2003), or anthropology of ritual (Lemonnier, 2004). The approach spans high-tech, low-tech, and no-tech, from the design of high-speed urban transportation systems (Latour, 1996), through the rocky negotiations of technology transfer (Akrich, 1993), to gender differences in Indian pottery-making (Mahias, 1993) or the place of posture in Chinese femininities (Flitsch, 2004).

Similar to the American anthropologists of technology, the French school views technology as a universal human activity and emphasises the need to build strong analytical and empirical bridges between upstream and downstream, artefact production and use. Its conceptual frameworks and methods are designed to apply equally to old or new technologies. Scholars such as Mahias (2002) have deployed them brilliantly to illuminate the interpenetration of 'traditional' and industrial, local and 'global' technologies and technological cultures. Although gender-technology relations are not as prominent or sustained a theme as in FTS, the methods lend themselves to finely textured studies of gendered identity, some focused on individual technologies or bodily practices (Darbon, 2002; Desrosiers, 1997; Pardo, 2004), others on gendered repertories of technical skills (Mahias 2002). Although Latour's study of Aramis (1996) has been criticised for gender blindness (Wajcman, 2004), it offers rich materials for the study of masculinities. In a study of imperial China, Bray (1997) documents the historical dynamics of a 'gynotechnics', mutually shaping technologies of dwelling, production, and reproduction central to hegemonic and pragmatic gender identities. Refining the concept of *techniques du corps*, Ingold (2000) proposes treating the skills of craft and of art under the same heading and highlights their ontogenetic nature. Far from being added onto a preformed body, skills grow with the body: '[T]hey are fully part and parcel of the human organism, of its neurology, musculature, even anatomy, and so are as much biological as cultural' (p. 360). This approach suggests bridges to recent FTS researches, inspired by Butler (1993), on the 'achievement' of gender (Lie, 2003a).

Anthropology and Technology

Classic anthropological monographs, including Malinowski's, are rich in materials on technical activities and their meanings (Malinowski, 1935; Pfaffenberger, 2001). Examining the articulations of work, production, and skills with exchange, ritual, kinship dynamics, and social differentiation, they address, as does FTS but implicitly, sociotechnical systems, 'seamless webs' of material, social, and symbolic practices and relations. Although not expressed in these terms, classic anthropology contributed some fine precursors to the study of technology and gender, for instance in studies of sexual divisions of labour (for example, Hugh-Jones, 1979; Richards, 1939).

Once the concept of gender became a specific analytical focus, feminist scholars focused on technical practices, old and new, to retheorise core anthropological concepts radically, including kinship (Strathern, 1992), exchange (Weiner, 1992), or space (Moore, 1986). As the anthropology of gender fused with the anthropology of modernity and of globalisation, attention turned to the role of technoscience in reshaping gender regimes. And with the broader cultural turn emphasising the importance of consumption as the constitutive site of subjectivities and power, the new field of material culture studies contrived a radical new antiessentialist perspective on technologies.

Anthropology of Technoscience

Technology and such derived concepts as 'technoscapes' or 'technonature' figure prominently in recent anthropological theories of the place of technoscience in modernity and/or globalisation. Key concerns of anthropological studies of technoscience, as of FTS, are the formation of the modern subject and the distribution of power through emerging global networks. However, Escobar (1994) explicitly distinguishes the agenda of the anthropology of technoscience from that of the sociology of technology: 'For anthropologists, inquiry into the nature of modernity as the background for current understanding and practice of technology is of paramount importance. In this anthropology is closer to the philosophy than to the new sociology of technology' (p. 213). The culturalist approach to technoscience, like the 'standard view', is interested first and foremost in science, powerful knowledge instrumentalised through technology. Technologies are of anthropological interest as phenomena emerging from particular cultural contexts, contributing to new cultural worlds such as 'cyberculture' or 'techno-nature' (Escobar, 1994, 1999).

In destabilising boundaries between the human and the natural or between human and machine, promoting new, troubling relations of intimacy, or facilitating new forms of governmentality, emergent technologies such as in-vitro fertilisation, transnational organ transplants, stem-cell research, or data-banks raise new questions of 'how to live' (Collier and Lakoff, 2005). New technologies may be conceptualised as prostheses, elements of cyborg fusions between human and machine that extend our capacities and permit enhanced modes of being and relating; new forms of interpenetration of zones of space and time; and new possibilities for action at a distance, for connection, coalition, or control (Axel, 2006; Rafael, 2003; Wright, 2001). They may figure as tools for both research and accumulation, concentrating capital or biocapital in certain sites while providing the material procedures and equipment for the domestication of new life forms such as stem cells (Franklin, 2005). The term global assemblages has been proposed to address the spatial and political dynamics of these restless flows and concentrations of material and symbolic resources (Ong and Collier, 2005).

Most work within the anthropology of technoscience that explicitly attends to gender–technology relations addresses biopower and its new subjects: the new masculinities or femininities achieved through remakings and resexings of the body; or through cross-class, transnational, or interethnic reconfigurations of kinship and reproduction (Kaufman and Morgan, 2005). Analysis focuses on the potentialities and interpellations inherent in the new science and its representations; on users as 'ethical pioneers'; on interactions between experts and technicians and the 'lay' users (or refusers) of biomedical services; and on 'lay' appropriations or contestations of new disciplinary regimes (Greenhalgh, 2005; Rapp, 1998). However the technological apparatus itself is usually left as a black box. Despite Downey's cyborg manifesto, there are few anthropological studies of the material production or design of the technologies of biopower, cybercultures, or techno-natures. Rabinow's illuminating biographies of technology, studying the coproduction of technological apparatus, technocracy, research agendas, and scientific imaginaries, are rare anthropological analyses of the power inherent in the nuts and bolts of technology (Rabinow, 1996; Rabinow and Dan-Cohen, 2005). Traweek's classic upstream study of the mechanical foundations of high-energy physics (1988), which explicitly explores the gendering of technocratic production and practice, is another exemplary rarity.

Material Culture Studies

The anthropology of technoscience engages with heroic technologies, such as DNA sequencing or organ transplantation, that promise to transform what it means to be human. Material Culture Studies (MCS) currently takes up the challenge of decoding the mundane technologies of everyday life such as kitchen equipment or cars, analysing the role of material artefacts in producing subjectivities and social relations. As a counterbalance to classical Marxist analyses that treated work and production as the loci where identity and meaning were produced, the cultural Marxism of MCS prioritises meaning and identity production through the social processes of consumption (Miller, 1995). One theoretical concern of MCS is to critique the reification of globalisation by demonstrating that the 'global' is always manifested and experienced as a 'local' phenomenon. Widely viewed as global in nature, yet intrinsically cultural in their use, the new communications technologies offer irresistible test cases.

MCS studies of the Internet in Trinidad (Miller and Slater, 2000) or of cell-phones in Jamaica (Horst and Miller, 2005) generate richly textured analyses of how technology use intertwines with sociality, including the expression and affirmation of gendered identities and forms of intimacy and relatedness. They also document the gratifying extension of Jamaican or Trinnie styles of communication across transnational spaces, transforming the experiences of migration or diaspora. The point is convincingly made that Caribbean Internet users are not reacting to globalisation but creating it. By insisting that the new technologies facilitate but do not determine these cultural extensions, these studies reflect the MCS position on 'materiality'.

MCS proposes the concept of materiality to transcend the object-subject divide, viewed as an enduring weakness of Western thought. One might have thought this would open up very interesting possibilities for theorising technology, skills, and subjectivity. However, in repudiating reification of the object, MCS specifically dismisses technology as an analytical category. Although Miller develops methods for charting the extension of technology use that correspond to the specific ways in which the Internet or cell-phones work, he insists that the primary interest is how they are brought into being as cultural artefacts. It is correct, as Miller asserts, that the Internet is in constant flux, its features continually reworked by its users. Yet even the Internet involves a framework of technical design, costing, and regulation (local or transnational) that channels and constrains the forms of communication and sociality it allows (Wilk, 2005; Wilson and Peterson, 2002). Miller's studies of communications technologies are actually rich in detail on the political-economic context within which they were launched and adopted, and on user skills, technical as well as social. Generally speaking, however, MCS is open to criticism for excessive culturalism: 'while the demolition of the essentialised object was an urgent necessity, the declaration of objects' and images' emptiness has become a proof for an anthropology committed to the victory of the cultural over the material, and of the discursive over the figural' (Pinney, 2002: 259).

Fruitful Exchanges?

The interdisciplinary field of feminist studies of technology has done more than any other social science to build a vibrant and coherent school of gender and technology studies. FTS has drawn heavily on ideas and methods developed within anthropology: the integrity of social action and culture; the 'micro-macro' linkage of everyday skills and techniques and political-economic activities; and detailed empirical observation and broad ranging comparative analysis. Could we now envisage more explicit and sustained forms of engagement among different branches of anthropology and FTS, to strengthen our understanding of gender–technology relations in a rapidly changing world?

Philosophically, FTS and the anthropology of technology share a strong materialism in their approach to culture-technology dialectics. Exchange between the fields therefore presents few epistemological problems. FTS lacks research on gendered dimensions of technical skills (Faulkner, 2001), and here methods developed by the French school for documenting operating sequences and *savoir-faire* might prove helpful. In considering the full spectrum of gender subjectivities achieved or imposed through technology in different contexts, another obvious lack in FTS at present is studies of non-Western societies, past as well as present. The anthropology of technology, by theorising technology as a universal human activity, offers not only a rich spectrum of non-Western and pre-modern case studies, but also analytical frameworks for reinterpreting historical and ethnographic documents from FTS perspectives.

In its attention to the materialities of everyday life, the French school of anthropology of technology shares common ground with MCS, but fundamental disagreement about whether technology constitutes an analytical category is a serious barrier to dialogue. It is not totally insurmountable, however. Dant (2005) argues for the value of incorporating more attention to technical skills and practices into MCS analysis; some contributors to *Material Culture Studies* focus on technological goods as technologies (Shove and Southerton, 2000); and French practitioners of MCS have successfully borrowed from the anthropology of techniques, integrating analysis of production and skills into their studies of consumer culture (Faure-Rouesnel, 2001; Warnier, 1999). Were Anglophone MCS to tread a similar path it might have to abandon some ambitious idealist claims about materiality. Yet valuable new insights into the coproduction of technology and gender might result if the strengths of MCS in charting the coproduction of global and local culture were extended to acknowledge technology. This would also provide a neat way for MCS to incorporate global flows of financial, corporate, and regulatory power more fully into their analyses.

The anthropology of technoscience attends closely to these global flows of power, and despite significant philosophical differences with FTS, there is a strong case to be made for closer dialogue between the fields. Concepts such as sociotechnical systems, stabilisation, and integration allow FTS to explore how technologies and the associated politics of gender travel across space and time and how they consolidate into systems that resist change. These approaches, along with FTS methods for studying the design and production of technologies, could enhance technoscience studies of biopower and of global assemblages. Attention to the gendering of technical design would be particularly valuable in advancing understanding of biopower. Conversely, in focusing so closely on the gender-technology nexus itself FTS sometimes neglects deeper lying ideogical dimensions within which any regime of truth concerning gender and technology must ultimately be understood, and which the anthropology of technoscience takes as its object, namely emergent configurations of *oikos* and *anthropos*.

References

Akrich, M. (1992). 'The Description of Technical Objects', in W. Bijker and J. Law (eds.), *Shaping Technology/Building Society*, pp. 205–24. Cambridge, MA: MIT Press.

—— (1993). 'A Gazogene in Costa Rica: An Experiment in Techno-sociology', in P. Lemonnier (ed.), *Technological Choices: Transformations in Material Cultures since the Neolithic*, pp. 289–337. London/New York: Routledge.

Arnold, D. (2005). 'Europe, Technology, and Colonialism in the Twentieth Century', *History of Technology*, 21: 85–106.

Arnold, E. and W. Faulkner (1985). 'Smothered by Invention: The Masculinity of Technology', in W. Faulkner and E. Arnold (eds.), *Smothered by Invention: Technology in Women's Lives*, pp. 18–50. London: Pluto.

Axel, B.K. (2006). 'Anthropology and the New Technologies of Communication', *Cultural Anthropology*, 21: 354–84.

Berg, A.J. (1997). *Digital Feminism*. Trondheim: Centre for Technology and Society, Norwegian University.

Bijker, W., T.P. Hughes, and T. Pinch (eds.) (1987). *The Social Construction of Technological Systems*. Cambridge, MA: MIT Press.

Bijker, W. and J. Law (eds.) (1992). *Shaping Technology/Building Society*. Cambridge, MA: MIT Press.

Bowker, G.C. and S.L. Star (1999). *Sorting Things Out: Classification and its Consequences*. Cambridge, MA: MIT Press.

Bray, F. (1997). *Technology and Gender: Fabrics of Power in Late Imperial China*. Berkeley: University of California Press.

—— (2003). 'Genetically Modified Foods: Shared Risk and Global Action', in B.H. Harthorn and L. Oaks (eds.), *Revising Risk: Health Inequalities and Shifting Perceptions of Danger and Blame*, pp. 185–207.Westport, CT: Praeger.

Bucciarelli, L.L. (1994). *Designing Engineers*. Cambridge, MA: MIT Press.

Butler, J. (1993). *Bodies that Matter: On the Discursive Limits of Sex*. New York: Routledge.

Cockburn, C. (1983). *Brothers: Male Dominance and Technological Change*. London: Pluto.

—— (1985). *Machinery of Dominance: Women, Men and Technical Know-How*. London: Pluto.

Cockburn, C. and S. Ormrod (1993). *Gender and Technology in the Making*. London/Thousand Oaks: Sage Publications.

Collier, S.J. and A. Lakoff (2005). 'On Regimes of Living', in A. Ong and S.J. Collier (eds.), *Global Assemblages: Technology, Politics, and Ethics as Anthropological Problems*, pp. 22–39. New York/Oxford: Blackwell.

Collier, S.J. and A. Ong (2005). 'Global Assemblages, Anthropological Problems', in A. Ong and S.J. Collier (eds.), *Global Assemblages: Technology, Politics, and Ethics as Anthropological Problems*, pp. 3–31. New York/Oxford: Blackwell.

Connell, R. W. (1995). *Masculinities*. Berkeley: University Calif. Press.

Corea, G., D. Klein, J. Hanmer, H.B. Holmes, B. Hoskins, et al. (1985). *Manmade Women: How New Reproductive Technologies Affect Women*. London: Hutch inson.

Cowan, R.S. (1983). *More Work for Mother: The Ironies of Household Technology from the Open Hearth to the Microwave*. New York: Basic Books.

—— (1987). 'The Consumption Junction: A Proposal for Research Strategies in the Sociology of Technology', in W. Bijker, T.P. Hughes, and T. Pinch (eds.), *The Social Construction of Technological Systems*, pp. 261–80. Cambridge, MA: MIT Press.

Dant, T. (2005). *Materiality and Sociality.* Maidenhead, UK: Open University Press.

Darbon, S. (ed.) (2002). 'Pour une anthropologie des pratiques sportives. Propriétés formelles et rapport au corps dans le rugby a XV', *Techniques et Culture*, 39: 1–27.

Desrosiers, S. (1997). 'Textes techniques, savoir-faire et messages codés dans les textiles des Andes', *Techniques et Culture*, 29: 155–73.

d'Onofrio, S. and F. Joulian (2006). 'Dire le savoir-faire: gestes, techniques et objets', *Cahiers d'anthropology', Sociale*, 1: 9–12.

Downey, G.L. (1992). 'CAD/CAM Saves the Nation: Toward an Anthropology of Technology', *Knowledge Society: Anthropology Science and Technology*, 9: 143–68.

—— (1998). *The Machine in Me: An Anthropologist Sits among Computer Engineers.* New York: Routledge.

Downey, G.L., J. Dumit, and S. Williams (1995). 'Cyborg Anthropology', *Cultural Anthropology*, 10: 264–69.

Escobar, A. (1994). 'Welcome to Cyberia: Notes on the Anthropology of Cyber-culture', *Current Anthropology*, 35: 211–31.

—— (1999). 'After Nature: Steps to an Antiessentialist Political Ecology', *Current Anthropology*, 40: 1–30.

Faulkner, W. (2000). 'The Power *and* the Pleasure? A Research Agenda for 'making Gender Stick' to Engineers', *Science, Technology, & Human Values*, 25: 87–119.

—— (2001). 'The Technology Question in Feminism: A View from Feminist Technology Studies', *Women's Studies International Forum*, 24: 79–95.

Faure-Rouesnel, L. (2001). 'French Anthropology and Material Culture', *Journal of Material Culture*, 6: 237–47.

Feenberg, A. (1999). *Questioning Technology.* London: Routledge.

Flitsch, M. (2004). *Der Kang. Eine Studie zur materiellen Alltagskultur bäuerlicher Gehöfte in der Manjurei. Wiesbaden.* Germany: Harrassowitz.

Franklin, S. (2005). 'Stem Cells R Us: Emergent Life Forms and the Global Biological', in A. Ong and S.J. Collier (eds.), *Global Assemblages: Technology, Politics, and Ethics as Anthropological Problems*, pp. 59–78. New York/Oxford: Blackwell.

Freeman, C. (2001). 'Is Local: Global as Feminine: Masculine? Rethinking the Gender of Globalization', *Signs*, 26: 1007–37.

Gansmo, H.J. (2003). 'Limits of State Feminism: Chaotic Translations of the 'girls and computing' Problem', in M. Lie (ed.), *He, She and IT Revisited: New Perspectives on Gender in the Information Society*, pp. 135–72. Oslo: Gylendal.

Geertz, C. (1963). *Agricultural Involution: The Processes of Ecological Change in Indonesia.* Berkeley: University of California Press.

Gell, A. (1992). 'The Technology of Enchantment and the Enchantment of Technology', in J. Coote and A. Shelton (eds.), *Anthropology, Art and Aesthetics*, pp. 40–63. Oxford: Clarendon.

Gero, J.M. and M.W. Conkey (eds.) (1991). *Engendering Archaeology: Women and Prehistory.* Oxford: Blackwell.

Goody, J. (1971). *Technology, Tradition and the State in Africa.* Oxford: Oxford University Press.

—— (1986). *The Logic of Writing and the Organization of Society.* Cambridge, UK: Cambridge University Press.

Greenhalgh, S. (2005). 'Globalization and Population Governance in China', in A. Ong and S.J. Collier (eds.), *Global Assemblages: Technology, Politics, and Ethics as Anthropological Problems*, pp. 354–72. New York/Oxford: Blackwell.

Grint, K. and R. Gill (eds.) (1995). *The Gender-Technology Relation.* London: Taylor & Francis.

Guille-Escuret, G. (2003). 'Retour aux modes de production, sans contrôle philosophique' *Techniques et Culture,* 40: 81–105.

Hakken, D. (1993). 'Computing and Social Change: New Technology and Workplace Transformation, 1980–1990', *Annual Review of Anthropology,* 22:107–32.

Haraway, D. (1991). 'Situated Knowledges: The Science Question in Feminism and the Privilege of Partial Perspective', in D. Haraway (ed.), *Simians, Cyborgs, and Women*, pp. 183–201. New York: Routledge.

Harding, S. (1986). *The Science Question in Feminism.* Ithaca, NY: Cornell University Press.

Horowitz, R. (ed.) (2001). *Boys and their Toys? Masculinity, Class, and Technology in America.* New York/London: Routledge.

Horst, H. and D. Miller (2005). 'From Kinship to Link-up: Cell Phones and Social Networking in Jamaica', *Curent. Anthropology,* 46: 755–88.

Hubak, M. (1996). 'The Car as a Cultural Statement: Car Advertising as Gendered Socio-technical Scripts', in M. Lie and K.H. Sørensen (eds.), *Making Technology Our Own? Domesticating Technology into Everyday Life*, pp. 171–200. Oslo: Scandaniavian University Press.

Hughes, T. P. (1983). *Networks of Power: Electrification in Western Society 1880–1930.* Baltimore: Johns Hopkins University Press.

—— (1986). 'The Seamless Web: Technology, Science, Etcetera, Etcetera', *Social Studies of Science,* 16: 281–92.

Hugh-Jones, C. (1979). *From the Milk River: Spatial and Temporal Processes in Northwest Amazonia.* Cambridge, UK: Cambridge University Press.

Ingold, T. (2000). 'Of String Bags and Birds' Nests: Skill and the Construction of Artefacts', in T. Ingold (ed.), *The Perception of the Environment: Essays on Livelihood, Dwelling and Skill*, pp. 349–61. London/New York: Routledge.

Kaufman, S.R. and L.M. Morgan (2005). 'The Anthropology of the Beginnings and Ends of Life', *Annual Review of Anthropology,* 34: 317–41.

Kleif, T. and W. Faulkner (2003). '"I'm No Athlete but I Can Make This Thing Dance"—Men's Pleasures in Technology', *Science, Technology, & Human Values,* 28: 296–325.

Kline, R. (2003). 'Resisting Consumer Technology in Rural America: The Telephone and Electrification', in N. Oudshoorn and T. Pinch (eds.), *How Users Matter: The Co-construction of Users and Technology,* pp. 51–66. Cambridge, MA: MIT Press.

Kramerae, C. (ed.) (1988). *Technology and Women's Voices: Keeping in Touch.* London: Routledge & Kegan Paul.

Kroeber, A. and C. Kluckhohn (1952). *Culture: A Critical Review of Concepts and Definitions.* Cambridge, MA: Harvard University Press.

Kvande, E. (1999). '"In the belly of the beast": Constructing Femininities in Engineering Organizations', *European Journal of Women's Studies*, 6: 305–28.

Lægran, A.S. (2003a). 'Escape Vehicles? The Internet and the Automobile in a Local-global Intersection', in N. Oudshoorn and T. Pinch (eds.), *How Users Matter: The Co-Construction of Users and Technology*, pp. 81–100. Cambridge, MA: MIT Press.

—— (2003b). 'Just Another Boys' Room? Internet Cafes as Gendered Technosocial Spaces', in M. Lie (ed.), *He, She and IT Revisited: New Perspectives on Gender in the Information Society*, pp. 198–227. Oslo: Gylendal.

Lagesen, V.A. (2005). 'A Cyber-feminist Utopia? Perceptions of Gender and Computer Science among Malaysian Women Computer Science Students', in *Extreme make-over: the Making of Gender and Computer Science*, pp. 155–94. PhD thesis, Centre for Technology and Society, Nor. University Science and Technololgy.

Latour, B. (1992). 'Where Are the Missing Masses? The Sociology of a Few Mundane Artifacts', in W. Bijker and J. Law (eds.), *Shaping Technology/Building Society*, pp. 225–57. Cambridge, MA: MIT Press.

—— (1993). 'Ethnography of a "high-tech" Case', in P. Lemonnier 2003, pp. 372–98.

—— (1996). *Aramis, or the Love of Technology.* Cambridge, MA: Harvard University Press.

Law, J. (1998). 'Machinic Pleasures and Interpellations', in B. Brenna, J. Law, and I Moser (eds.), *Machines, Agency and Desire*, pp. 23–49. Oslo: Centre for Technology and Culture, University of Oslo.

Law, J. and V. Singleton (2000). 'Performing Technology's Stories: On Social Constructivism, Performance, and Performativity', *Technology and Culture*, 41: 765–75.

Lemonnier, P. (1992). *Elements for an Anthropology of Technology.* Ann Arbor: Museum Anthropology, University Michigan.

—— (ed.) (1993). *Technological Choices: Transformations in Material Cultures since the Neolithic.* London/New York: Routledge.

—— (2004). 'Mythiques chaînes opératoires', *Techniques et Culture*, 43–44: 25–43.

Lerman, N.E., A.P. Mohun, and R. Oldenziel (1997). 'The Shoulders We Stand On and the View from Here: Historiography and Directions for Research', *Technology and Culture*, 38: 9–30.

Lerman N.E., R. Oldenziel, and A.P. Mohun (eds.) (2003). *Gender and Technology: A Reader.* Baltimore: Johns Hopkins University Press.

Lie, M. (2003a). 'Gender and ICT—New Connections', in M. Lie (ed.), *He, She and IT Revisited: New Perspectives on Gender in the Information Society*, pp. 9–33. Oslo: Gylendal.

—— (ed.) (2003b). *He, She and IT Revisited: New Perspectives on Gender in the Information Society.* Oslo: Gylendal.

Lie, M and K. H. Sørensen (eds.) (1996). *Making Technology Our Own? Domesticating Technology into Everyday Life.* Oslo: Scandaniavian University Press.

Lohan, M. (2000). 'Constructive Tensions in Feminist Technology Studies', *Social Studies of Science*, 30: 895–916.

—— (2001). 'Men, Masculinities and "Mundane" Technologies: The Domestic Telephone', in E. Green and A. Adam (eds.), *Virtual Gender: Technology, Consumption and Identity*, pp. 189–205. London/New York: Routledge.

Lohan, M. and W. Faulkner (2004). 'Masculinities and Technologies: Some Introductory Remarks', *Men and Masculinity*, 6: 319–29.

Mackenzie, D. and J. Wajcman (eds.) (1999). *The Social Shaping of Technology*. Maidenhead, UK: Open University Press.

MacKenzie, M. (1991). *Androgynous Objects: String Bags and Gender in Central New Guinea*. Chur, Switzerland: Harwood Academy.

Mahias, M.C. (1993). 'Pottery Techniques in India: Technical Variants and Social Choice', See P. Lemonnier 2003, pp. 157–80.

—— (2002). *Le Barattage du Monde. Essais d'Anthropologie des Techniques en Inde*. Paris: Editions MSH.

Malinowski, B. (1935). *Coral Gardens and their Magic: A Study of Tilling the Soil and of Agricultural Rites in the Trobrian Islands*. London: Routledge & Kegan Paul.

McGaw, J.A. (1996). 'Reconceiving Technology: Why Feminine Technologies Matter', in R.P. Wright (ed.), *Gender and Archaeology*, pp. 52–79. Philadelphia: University Penn. Press.

Mellström, U. (2002). 'Patriarchal Machines and Masculine Embodiment', *Science, Technology, & Human Values*, 27: 460–78.

—— (2003). *Masculinity, Power and Technology: a Malaysian Ethnography*. Aldershot, UK: Ashgate.

—— (2004). 'Machines and Masculine Subjectivity: Technology as an Integral Part of Men's Life Experience', *Men and Masculinity*, 6: 362–82.

Miller, D. (ed.) (1995). *Acknowledging Consumption: A Review of New Studies*. London: Routledge.

Miller, D. and D. Slater (2000). *The Internet: An Ethnographic Approach*. Oxford: Berg.

Miller, L. (2004). 'Those Naughty Teenage girls: Japanese Kogals, Slang, and Media Assessments', *Journal of Linguistic Anthropology*, 14: 225–47.

Mills, M.B. (2003). 'Gender and Inequality in the Global Labor Force', *Annual Review Anthropology*, 32: 41–62.

Misa, T.J. and J. Schot (2005). 'Inventing Europe: Technology and the Hidden Integration of Europe', *History of Technology*, 21: 1–19.

Moore, H. (1986). *Space, Text and Gender: An Anthropological Study of the Marakwet of Kenya*. Cambridge, UK: Cambridge University Press.

Oakley, A. (1974). *The Sociology of Housework*. London: Martin Robertson.

Oldenziel, R., A.A. de la Brohèze, and O. deWit (2005). 'Europe's Mediation Junction: Technology and Consumer Society in the Twentieth Century', *History of Technology*, 21: 107–39.

Ong, A. and S.J. Collier (eds.) (2005). *Global Assemblages: Technology, Politics, and Ethics as Anthropological Problems*. New York/Oxford: Blackwell.

Ortiz, S. (2002). 'Laboring in the Factories and in the Fields', *Annual Review of Anthropology*, 31: 395–417.

Oudshoorn, N., E. Rommes, and M. Stienstra (2004). 'Configuring the User as Everybody: Gender and Design Cultures in Information and Communication Technologies', *Science, Technology, & Human Values*, 29: 30–63.

Oudshoorn, N. and T. Pinch (eds.) (2003). *How Users Matter: The Co-construction of Users and Technology*. Cambridge, MA: MIT Press.

Paechter, C. (2006). 'Power, Knowledge, and Embodiment in Communities of Sex/gender Practice', *Women's Studies International Forum*, 29: 13–26.

Pardo, V. (2004). 'Le récit des deux tisseuses', *Techniques et Culture*, 43–44: 277–87.

Parr, J. (1999). 'What Makes a Washday Less Blue? Gender, Nation and Technology Choice in Postwar Canada', *Technology and Culture*, 38: 153–86.

Parthasarathy, J. (2003). 'Knowledge is Power: Genetic Testing for Breast Cancer and Patient Activism in the United States and Britain', in N. Oudshoorn and T. Pinch (eds.), *How Users Matter: The Co-construction of Users and Technology*, pp. 133–50. Cambridge, MA: MIT Press.

Pfaffenberger, B. (1992). 'Social Anthropology of Technology', *Annual Review of Anthropology*, 21: 491–516.

—— (2001). 'Symbols Do Not Create Meanings—Activities Do: or, Why Symbolic Anthropology Needs the Anthropology of Technology', in M.B. Schiffer (ed.), *Anthropological Perspectives on Technology*, pp. 77–86. Albuquerque: University of New Mexico Press.

Pinch, T. and W. Bijker (1987). 'The Social Construction of Facts and Artifacts', in W. Bijker, T.P. Hughes, and T. Pinch (eds.), *The Social Construction of Technological Systems*, pp. 17–51. Cambridge, MA: MIT Press.

Pinney, C. (2002). 'Visual Culture', in V. Buchli (ed.), *The Material Culture Reader*, pp. 81–104. Oxford: Berg.

Rabinow, P. (1996). *Making PCR: A Story of Biotechnology*. Chicago: University Chicago Press.

Rabinow, P. and T. Dan-Cohen (2005). *A Machine to Make a Future: Biotech Chronicles*. Princeton, NJ: Princeton University Press.

Rafael, V. (2003). 'The Cell Phone and the Crowd: Messianic Politics in the Contemporary Philippines', *Public Culture*, 15: 399–425.

Rapp, R. (1998). 'Refusing Prenatal Diagnosis: The Meanings of Bioscience in a Multicultural World', *Science, Technology, & Human Values*, 23: 45–70.

Richards, A.I. (1939). *Land, Labor and Diet in Northern Rhodesia*. Oxford: Oxford University Press.

Schiffer, M.B. (ed.) (2001a). *Anthropological Perspectives on Technology*. Albuquerque: University of New Mexico Press.

—— (2001b). 'Toward an Anthropology of Technology', in M.B. Schiffer (ed.), *Anthropological Perspectives on Technology*, pp. 1–15. Albuquerque: University of New Mexico Press.

Shove, E. and D. Southerton (2000). 'Defrosting the Freezer: From Novelty to Convenience, a Narrative of Normalization', *Journal of Material Culture*, 5: 301–19.

Sillitoe, P. (1988). *Made in Niugini. Technology in the Highlands of Papua New Guinea.* London: British Museum.

Silverstone, R. and E. Hirsch (eds.) (1992). *Consuming Technologies: Media and Information in Domestic Spaces.* London: Routledge.

Sørensen, K.H. and A.J. Berg (eds.) (1991). *Technology and Everyday Life: Trajectories and Transformations.* Oslo: Nordic Research Council of Science and Humanity.

Stanley, A. (1993). *Mothers and Daughters of Invention: Notes for a Revised History of Technology.* New Brunswick, NJ: Rutgers University Press.

Staudenmaier, J.S. (1985). *Technology's Storytellers: Reweaving the Human Fabric.* Cambridge, MA: MIT Press.

Strathern, M. (1992). *Reproducing the Future: Anthropology, Kinship and the New Reproductive Technologies.* London: Routledge.

Suchman, L.A. (1999). 'Working Relations of Technology Production and Use', in D. Mackenzie and J. Wajcman (eds.), *The Social Shaping of Technology*, pp. 258–65. Maidenhead, UK: Open University Press.

—— (2001). 'Building Bridges: Practice-based Ethnographies of Contemporary Technology', in M.B. Schiffer (ed.), *Anthropological Perspectives on Technology*, pp. 163–78. Albuquerque: University N.M. Press.

Traweek, S. (1988). 'Inventing Machines that Discover Nature', in S. Traweek (ed.), *Beam Times, Life Times: The World of High Energy Physics*, pp. 46–73. Cambridge, MA: Harvard University Press.

Van, Oost E. (2003). 'Materialized Gender: Configuring the User during the Design, the Testing, and the Selling of Technologies', in N. Oudshoorn and T. Pinch (eds.), *How Users Matter: The Co-construction of Users and Technology*, pp. 193–208. Cambridge, MA: MIT Press.

Wajcman, J. (1991). *Feminism Confronts Technology.* Cambridge, UK: Polity.

—— (2002). 'Reflections on Gender and Technology Studies: In What State is the Art?' *Social Studies of Science*, 30(3): 447–64.

—— (2004). *Technofeminism.* Cambridge, UK: Polity.

Warnier, J.P. (1999). *Construire la culture matérielle—l'homme qui pensait avec ses doigts.* Paris: PUF.

Weiner, A.N. (1992). *Inalienable Possessions: The Paradox of Keeping-While-Giving.* Berkeley: University of California Press.

Wenger, E. (1998). *Communities of Practice: Learning, Meaning and Identity.* Cambridge, UK: Cambridge University Press.

Wilk, R. (2005). 'Comment on Horst & Miller', *Current Anthropology*, 46: 772.

Wilson, S.M. and L.C. Peterson (2002). 'The Anthropology of Online Communities', *Annual Review Anthropology*, 31: 449–67.

Winner, L. (1986). 'Do Artifacts Have politics?' in L. Winner (ed.), *The Whale and the Reactor: A Search for Limits in an Age of High Technology*, pp. 19–39. Chicago: University Chicago Press.

Wright, M.W. (2001). 'Desire and the Prosthetics of Supervision: A Case of Maquiladora Flexibility', *Culture and Anthropology*, 16: 354–73.

Wright, R.P. (ed.) (1996). *Gender and Archaeology*. Philadelphia: University of Pennsylvania Press.

Wyatt, S. (2003). 'Non-users Also Matter: The Construction of Users and Non-users of the Internet', in N. Oudshoorn and T. Pinch (eds.), *How Users Matter: The Co-construction of Users and Technology*, pp. 67–80. Cambridge, MA: MIT Press.

CHAPTER FOUR

Gender, Science, and the Psychology of Science

Gregory J. Feist

GIVEN its stature as a transformative influence on society, science is and ought to be an object of intense study. Philosophers, historians, and sociologists devote systematic attention to questions such as what distinguishes scientific from non-scientific knowledge; what is the historical context to great scientific discoveries, such as the theory of evolution or quantum mechanics; and what are the sociological and political forces behind becoming a 'have or a have-not' in science. What is conspicuously absent – at least until the mid 1980s – from these studies of science (metasciences) is psychology (Feist, 2006; Feist and Gorman, 1998). This is all the more puzzling, given the fact that philosophers, historians, and even sociologists of science often touch on inherently psychological processes in their writings on science and scientists, such as imagination, creativity, thought processes, social influence, and motivation. As recently as 1985, there was little accumulated knowledge concerning topics in psychology of science. As Mahoney wrote in the late 1970s: 'In terms of behaviour patterns, affect, and even some intellectual matters, we know more about alcoholics, Christians, and criminals than we do about the psychology of the scientist' (Mahoney, 1979: 349). In contrast, other disciplines like philosophy, history, and sociology have spawned clearly identifiable and established sub-disciplines devoted to science studies. Because science deserves more attention from psychologists, one aim of this chapter is to show how the psychology of science can contribute to important questions concerning scientific and mathematical behaviour; in this case how gender affects interest in and ability towards math and science.

Gender, Science, and Psychology

One of the more contentious and polemical questions in the psychology of science concerns the role that gender plays in science in general, and in scientific and mathematical ability and achievement in particular. As Evelyn Fox Keller, among others, has pointed out, the history of science is replete with associations, both implicit and explicit, between science and male; with male scientists frequently trying to 'tame' or 'control' the feminine 'Mother Nature' (Keller, 1985; cf. Nosek et al., 2002). The topics of gender and science and gender differences in scientific achievement could be and have been the focus of a chapter or a book, and I leave the more exhaustive review of this literature to others. There are three questions on gender and science that, I believe, have accumulated enough literature in psychology to warrant attention: interest-attrition, mathematical ability, and productivity.

Interest-Attrition

One of the more entrenched influences on the development of scientific interest appears to be gender. The research over the last few decades did suggest gender differences in science or mathematics, either in the form of explicit attitudes (Eccles, 1987; Hyde, Fennema, Ryan, Frost and Hopp, 1990), or implicit attitudes (Nosek et al., 2002), performance on aptitude tests (Benbow and Stanley, 1983; Benbow et al., 2000; Geary, 1998; Halpern, 2000), or actual graduation and career data (Cole, 1987; Cole and Zuckerman, 1987; Farmer et al., 1999; Jacobwitz, 1983; Long, 2001; National Science Foundation (NSF), 1999; O'Brien et al., 1999; Reis and Park, 2001; Stuessy (1989); Subotnik et al., 1993). The general conclusion from this body of research is that men are more likely than women to view science positively, be more interested in science and mathematics as a career, and less likely to drop out of science (Benbow et al., 2000; Feist, 2006; Long, 2001; Subotnik and Steiner, 1994; Webb et al., 2002). Moreover, although there is no overall gender difference in intelligence, while in a few studies there appeared some systematic differences in the mathematical domain (males being higher) and in the verbal domain (females being higher) (Benbow and Stanley, 1983; Geary, 1998; Halpern, 2000; Kimura, 1999), in other the gender similarities hypothesis, which holds that males and females are similar on most, but not all, psychological variables was supported (Hyde, 2005).

There are, however, at least two important qualifications to any generalisations: gender differences are less apparent in childhood and adolescence than adulthood and they are less apparent in the social sciences than in the physical sciences, with biological sciences being in the middle. For instance, in terms of courses taken, the 'gender gap' in science is not evident at the high school or undergraduate level or in the social sciences. High school male and female students were equally likely to take advanced mathematics courses (trigonometry and calculus) and almost as likely to take advanced science courses (biology, chemistry, and physics). In advanced science courses there were a slightly higher percentage of females taking biology and chemistry, and a slighter higher percentage of males taking physics. As students progress through their academic careers, however, there is an increasing gender disparity in interest in science and math (Long, 2001; NSF, 1999; Rosser, 1988; Subotnik et al., 1993). At the undergraduate level, the percentage of women who earned science or engineering degrees in 1995 was 46 per cent (after being about 38 per cent ten years earlier). At the graduate level, a more obvious gender gap exists, even in the biological and social sciences, with 39 per cent of the master's degrees in science and engineering and 33 per cent of the doctoral degrees in science and engineering being awarded to women. And finally in terms of career, the disparity widens even more, with only 4–6 per cent of the full professors in science and math being women. The most extreme gender difference is seen at the most elite level (Long, 2001). On average, only 2 per cent of the members of the National Academy of Sciences are female.

The most exhaustive and extensive study of PhD scientists over a 22-year-period (1973 to 1995) by the National Research Council has documented progress but not yet equality for women in science. When the appropriate controls (such as rank, field, and institution) are made, the gender disparity is not so extreme, but it still exists. For instance, men hold a 14 percentage-point advantage in holding tenure-track positions, but this difference approaches zero once the career age is held constant. This suggests that the gender disparity in tenure-track science positions should continue to decline as more and more women become eligible. Also, salary differences diminish once rank is controlled for, but they do not disappear completely, suggesting that men do get paid a bit more for the same position. Moreover, marriage and family does affect men's and women's rank and productivity differently, but not necessarily in the manner one might expect. Long (2001), for instance, reports that women who interrupted their careers for marriage and family in 1979 were less likely to obtain a tenure-track position, but there was no effect in 1995. For men, on the other hand, the effect of getting married and having children had a positive effect on productivity and this effect increased between 1979 and 1995. The question of gender differences in mathematics achievement, attitudes, and effect is a continuing concern as scientists seek to address the under-representation of women at the highest levels of science, technology, mathematics, and engineering (Halpern et al., 2007; National Academy of Sciences, 2006).

The other qualification is that not all fields of science are equally gender biased in their distributions. Unequal distributions are most striking in the physical sciences, less striking in the biological sciences, and least striking in the social sciences. Only 17 per cent of the engineering degrees and 35 per cent of both the mathematics and physical-earth science degrees were awarded to women, whereas nearly 50 per cent of the biological and social science degrees, and 73 per cent of the psychology degrees were awarded to women in 1995 (NSF, 1999). Similarly, Long's (2001) analysis of trends in national samples showed that from 1973 to 1995 women went from being 2 per cent to 12 per cent of the engineering PhD graduates and from 21 per cent to 51 per cent of the social-behavioural science PhD graduates. All other scientific fields were in between these two ends of the continuum. In addition, in a sample of mathematically precocious students who immediately after high school said that they intended to major in mathematics or science, five years later men were more likely to have received engineering and physical science degrees and women more likely to have received biological science and medical science degrees (Webb et al., 2002).

One goal of the psychology of science, therefore, is to unpack some of the factors behind why women decide disproportionately to leave science, even those who are demonstrably among the most promising young scientists and mathematicians in the nation. Some previous work suggests a few possible explanations: number of hours worked per week, self-image, stereotype threat, parental behaviour, and opting for and having greater talent for 'people-oriented' rather than 'thing-oriented' professions. For instance, Camilla Benbow and colleagues report that gender differences in math achievement dissipate once number of hours worked per week is controlled for, suggesting that work-load rather than gender is the underlying cause (Benbow et al., 2000). Stereotypes that girls and women lack mathematical ability persist, though there is evidence of gender similarities in math achievement (Hedges and Nowell, 1995; Hyde, Fennema and Lamon, 1990; Hyde, Lindberg, Linn, Ellis and Williams, 2008). A few studies suggest that core cognitive and motivational patterns in women and men are much the same in contrast to the traditional view that girls are more interested in people than in objects (Spelke, 2005).

Another explanation focuses on self-image. One's self-perceived career identity no doubt carries a lot of weight in career choice, with people most likely to pursue those careers that match their identity and self-perceived ability. If one cannot imagine oneself in a career then there is little chance one will even attempt to pursue it. Can one easily envision and imagine oneself as a 'scientist'? For some (males) the label fits well with their projected self-image and for others it does not (females). A recent study based on undergraduate research by Susannah Paletz reported an interaction between gender, self-image, and interest in science. In a study of 211 college students at a selective liberal arts college, men with either high or low science self-images were still slightly more interested in science than women with the same high science self-image (Feist et al., 2004). Similarly, math skills have been seen as part of the male domain, and if one is female there is less congruency between self-image and a career in mathematics or science (Kelly and Nelson, 1999; Nosek et al., 2002; Tobin et al., 1995). Moreover, occupational interest research has demonstrated that congruency between talent, performance, self-perception, and drive is the best predictor of career interest. Research has demonstrated the power of self-efficacy training to increase math and science self-efficacy both for course work and career attainment, at least over a short period of time (Bandura, 1982; Luzzo et al., 1999).

Related to self-image, stereotype threat theory and research also sheds some light on the phenomenon of gender differences in science (Davies et al., 2002; Nosek et al., 2002; Shih et al., 1999; Steele, 1997; Steele and Aronson, 1995). Davies and colleagues conceptualise stereotype threat as 'The risk of being personally reduced to a negative stereotype can elicit a disruptive state among stigmatized individuals that undermines performance and aspirations in any alleged stereotype-relevant domain – a situational predicament termed "stereotype threat"' (Davies et al., 2002: 1616). Stereotype threat suggests that achievement of stereotyped individuals can be adversely affected by being reminded of negative stereotypes. Moreover, stereotype threat can both hinder and help performance through negative and positive identity. Margaret Shih and colleagues (1999), for instance, demonstrated that when Asian women doing a math task have their ethnic identity primed, and consistent with the stereotype that Asians are good at mathematics, they perform better than controls. When, however, they have their gender identity primed, consistent with the stereotype that women are not good at mathematics, they perform worse.

Finally, a more nativist explanation comes from various lines of research suggesting that there is a general gender effect in dispositional interest towards people and things (Achter et al., 1996; Connellan et al., 2000; Lippa, 1998; Lubinski and Humphreys, 1990; Prediger, 1982; Schmidt et al., 1998; Webb et al., 2002). The People-Thing dimension might well be an underlying influence on the kind of science one becomes interested in, especially physical versus social science (Holland, 1992; Lippa, 1998; Prediger, 1982). The 'People' end of the dimension is mapped onto 'Social' careers that involve informing, training, enlightening other people. On the other hand, the 'Thing' end of the dimension involves 'Realistic' careers that involve manipulating things, machines, objects, tools, and animals. Building on Prediger's work (1982), Richard Lippa (1998) reported gender ratios of roughly four or five to one of males in 'Thing' rather than 'People' careers, and ratios of two or three to one of females in People rather than Thing careers. Interestingly, there were no gender differences on the Ideas-Data dimension. The work of Simon Baron-Cohen has demonstrating high proportions of high functioning autism (Aspergers) as well as high proportions of males in the physical sciences and engineering (Baron-Cohen et al., 1999, 2001). Autistic children are also more likely to have engineer for father or grandfather (Baron-Cohen et al., 1997, 1998). Baron-Cohen and colleagues also have reported evidence that this gender difference in people versus thing orientation is present in neonates as young as three days old, suggesting a biological origin (Connellan et al., 2000). The mathematical and science professions are conceptualised as being investigative but on the thing side of Holland's vocational hexagon.

Mathematical Ability

One of the more contentious findings in the gender difference literature involves mathematical ability. Many studies reported that male score higher than females at the low, medium, and high ends of the distribution (Achter et al., 1996; Astin, 1975; Backman, 1972; Benbow, 1988; Benbow and Stanley, 1980, 1983; Benbow et al., 2000; Deaux, 1985; Fischbein, 1990; Fox, 1976; Geary, 1998; Holden, 1987; Keating, 1974; Leahey and Guo, 2002; Maccoby and Jacklin, 1974; Moore and Smith, 1987; Stanley, 1988; Webb et al., 2002). As both Stanley and his student Camilla Benbow have reported, one of the biggest surprises in collecting these data, however, was the large and consistent gender difference among the extreme scores – ultimately reaching as high as a twelve to one ratio in favour of males (Benbow and Stanley, 1980, 1983). Furthermore, Benbow's (1988) target article in *Behavioural and Brain Sciences* was commented on by more than forty experts, and although virtually none of the commentaries took issue with whether a gender difference exists, there was little agreement concerning the potential causes of this gender difference (Bleier, 1988; Eysenck, 1988; Farmer, 1988; Sternberg, 1988; Vandenberg, 1988). However, Eleanor Maccoby and Carol Jacklin's review of the literature sparked a couple of qualifications to this generalisation. First, there is little to no gender difference before adolescence, although the exact timing is somewhat disputed. Second, at least up through early adolescence, girls achieve higher grades than boys in math classes. The question of gender differences in mathematics achievement however becomes important to address due to some evidence of gender similarities in math achievement (Hedges and Nowell, 1995; Hyde, Fennema, and Lamon, 1990; Hyde et al., 2008). A gender gap in mathematics achievement persists in some nations but not in others (Else-Quest et al., 2010). Indeed, the gender difference in math raises the important, rather unsettled, question of what its cause is.

What are some of the possible explanations of this gender difference? Benbow (1988) reviewed the evidence for seven of the more common environmental explanations: attitudes towards math, perceived usefulness of math, confidence and self-efficacy, encouragement from parents and teachers, sex-typing, differential course-taking, and career and achievement motivation. She found that some of these environmental influences do distinguish males and females. For example, females do like mathematics less, find it less useful for their future goals, and have less confidence in their ability than males. Furthermore, mathematics is somewhat sex-typed as a 'masculine' enterprise, parents and teachers are more encouraging of male than female mathematical achievement, differences in math courses do not explain aptitude differences, and finally male career motivation is more independent of parent or teacher support than female.

However, these explanations neither rule out nor are inconsistent with biological explanations. As Hans Eysenck's (1988) commentary pointed out, the situational findings could result from either genetic or environmental origins. Benbow (1988) also more directly addressed the biological explanations, and offered four possibilities: hemispheric laterality, allergies, hormonal influences, and myopia. For instance, based on a high incidence of left-handedness in the mathematically precocious and in particular the precocious males, and the greater bilateral or diffuse cognitive functioning of left-handed individuals, she concluded that bilateral and/or a strong right hemispheric functioning may be implicated in extreme mathematical ability. Furthermore, prenatal exposure to testosterone has been postulated to influence handedness and immune disorders, and therefore could be an indirect influence on mathematical ability (Geschwind and Behan, 1982). To quote Benbow: 'In sum, the above physiological correlates, especially the possibility of prenatal testosterone exposure, lend credence to the view that sex differences in extremely high mathematical reasoning ability may be, in part, physiologically determined…'. Of course, some of the above discussion on physiological correlates is speculative' (Benbow, 1988, p. 182). Suffice it to say that the physiological explanations were the focus of most of the criticism in the commentaries. However, many criticisms did not take issue with the fact that biological explanations may play a role, but rather that their mechanisms are more complex than, and the evidence is not as solid as, Benbow's (1988) presentation. The real answer will almost certainly be found in a more complex explanation arguing that our interests and abilities, whatever they may be, require both some biological givens and some environmental training and encouragement to be fully expressed and actualised. However these precise biological and environmental mechanisms need further investigation.

Productivity

Comparing publication rates of men and women has consistently shown that men produce more works than women, although the difference has declined from the 1960s to the 1990s and afterwards (Cole, 1979, 1987; Cole and Zuckerman, 1987; Guyer and Fidell, 1973; Helmreich et al., 1980; Long, 1992, 2001; Pasewark et al., 1975; Xie and Shauman, 1998; Zuckerman and Cole, 1975). This gender difference appears to hold for the total number of publications as well as yearly average. However, there is some contradictory evidence regarding whether this gender difference increases or decreases across the course of one's career. Some have reported that the gender gap on productivity increases (Cole, 1987), whereas others reported it decreases over the course of one's career (Guyer and Fidell, 1973; Long, 1992, 2001; Xie and Shauman, 1998). For instance, Long's analysis of large national data sets in 1995 showed roughly a 30 per cent greater total publication rate for men compared to women when looking across all ranks. As the data are broken down within rank, however, the male advantage drops to 13 per cent at the tenure-track (Assistant Professor) rank and to 6 per cent at the tenured rank (Associate Professor) and 7 per cent at the full-professor rank. In short, productivity differences clearly do drop when rank is controlled for but they do not disappear entirely.

As with age and productivity, and interest in science and mathematics, the question with gender and productivity that begs to be addressed is that of cause. Once again, and as is somewhat inevitable, explanations are more contentious and less consensual than the description of the phenomenon. Differences in marital status, family obligations, prestige of institution, rank of position, training, and motivation have each been investigated, but with negative or inconsistent results. One consistent difference from the years 1935 to 1965 between male and female scientists was their marital status, with about 50 per cent of female scientists and 90 per cent of male scientists on average being married (Long, 2001). Female scientists seemed to be 'sacrificing' marriage for a career in science. This gender difference, however, is approaching equality, having dropped to slightly more than 70 per cent of the men and slightly less than 70 per cent of the women scientists being married by 1995.

The intuitively appealing answer that women are hindered by multiple roles of scientist, wife and mother, and are relegated to marginal departments has inconsistent empirical support. In fact, some research has shown that married women tend to slightly outproduce single women, and women with one or two children tend to outproduce women with no or more than three children (Cole 1979, 1987; Cole and Zuckerman, 1987). This research, however, was conducted only with women who were employed full-time in academic positions. Other research has examined the more general question of the likelihood of being in the workforce, and here one sees that the percentage of men is above 90 per cent in each of the following four categories: single (94 per cent), married (96 per cent), married with older children (97 per cent) and married with younger children (97 per cent) (Long, 2001). But for women these same four categories show a steady decline in the percentage in the workforce. For single women it is 94 per cent, for married women it is 89 per cent, for women with older children it is 80 per cent, and for women with younger children it drops to 71 per cent. By these national results, it is clear that marriage and family affect male and female scientists differently. This makes sense in light of another finding reported by Long: in 1995, 51 per cent of the women who were employed part time in science said they were working part time because of 'family obligations', whereas only 3 per cent of the part-time employed men gave that reason. Long concludes: 'Overall, marriage and family are the most important factors differentiating the labor force participation of male and female scientists and engineers' (Long, 2001: 91).

Sociologists of science in general have presented evidence that gender differences in productivity cannot fully be explained by differences in type of institution (college versus university) or prestige of department (Cole and Cole, 1973; Long, 2001). For instance, the Cole brothers reported that when both institutional and departmental variables were entered first in a regression equation and thereby held constant, the relationship between gender and productivity still persisted. One explanation, however, that does have some support is hours worked per week (Long, 2001). For example, in research on career outcomes of the precocious mathematics group of Julian Stanley, once hours worked per week is controlled for the gender difference in productivity and achievement disappeared (Benbow et al., 2000). This suggests that the number of hours working rather than gender per se may explain productivity differences. An important fact to keep in mind when comparing men and women's productivity levels is that female PhDs are less likely to be employed full-time.

Conclusions

By making clear the motivational, personality, attitudinal, and emotional forces behind interest in and talent for science, the psychology of science has much to contribute to the understanding of how gender affects scientific and mathematical interest and behaviour. I have reviewed some of the recent research on the psychological forces behind differences and similarities in male and female interest and talent for science as well as differences in attrition from mathematics and science. Among the more intriguing psychological explanations for the increasing gender difference in interest in science and mathematics with age and in the physical sciences in particular are self-image, stereotype threat, and being either 'people' or 'thing' oriented. If we, as a culture, are to achieve more gender equity in all domains of science and if the 'glass-ceiling' is to truly be broken, psychologists and psychologists of science in particular are going to have to become even more active in the enterprise of empirically investigating and theoretically explaining gender and science.

References

Achter, J.A., D. Lubinski, and C.P. Benbow (1996). 'Multipotentiality Among the Intellectually Gifted: "It was Never There and Already it's Vanishing"', *Journal of Counseling Psychology*, 43(1): 65–76.

Astin, H.S. (1975). 'Sex Differences in Mathematical and Scientific Precocity', *Journal of Special Education*, 9(1): 79–91.

Backman, M.E. (1972). 'Patterns of Mental Abilities: Ethnic, Socioeconomic, and Sex Differences', *American Educational Research Journal*, 9(1): 1–12.

Bandura, A. (1982). 'Self-efficacy Mechanism in Human Agency', *American Psychologist*, 37(2): 122–47.

Baron-Cohen, S. Bolton, P., Wheelwright, S., Short, L., Mead, G., Smith, A., and Scahill, V. (1998). 'Autism Occurs More Often in Families of Physicists, Engineers, and Mathematicians', *Autism*, 2(3): 296–301.

Baron-Cohen, S., S. Wheelwright, C. Stott, P. Bolton, and I. Goodyer (1997). 'Is There a Link between Engineering and Autism?' *Autism*, 1(1): 101–9.

Baron-Cohen, S., S. Wheelwright, R. Skinner, J. Martin, and E. Clubley, E. (2001). 'The Autism-Spectrum Quotient (AQ): Evidence from Asperger Syndrome/ High-functioning Autism, Males and Females, Scientists and Mathematicians', *Journal of Autism and Developmental Disorders*, 31(1), 5–17.

Baron-Cohen, S., S. Wheelwright, V.E. Stone, and M. Rutherford (1999). 'A Mathematician, a Physicist, and a Computer Scientist with Asperger yndrome: Performance on Folk Psychology and Folk Physics Tests', *Neurocase*, 5(6): 475–83.

Benbow, C.P. (1988). 'Sex Differences in Mathematical Reasoning Ability in Intellectually Talented Preadolescents: Their Nature, Effects, and Possible Causes', *Behavioural and Brain Sciences*, 11(2): 169–83.

Benbow, C.P. and J.C. Stanley (1980). 'Sex Differences in Mathematical Ability: Fact or Artifact?' *Science*, 210(4627): 1262–64.

—— (1983). 'Sex Differences in Mathematical Reasoning Ability: More Facts', *Science*, 222(4475): 1029–31.

Benbow, C.P., D. Lubinski, D.L. Shea, and H.E. Eftekhari-Sanjani (2000). 'Sex Differences in Mathematical Reasoning Ability at Age 13: Their Status 20 Years Later', *Psychological Science*, 11(6): 474–80.

Bleier, R. (1988). 'The Plasticity of the Human Brain and Human Potential', *Behavioural and Brain Sciences*, 11(2): 184–85.

Cole, J.R. (1979). *Fair Science: Women in the Scientific Community*. New York: Free Press.

—— (1987). 'Women in Science', in D. Jackson and P.J. Rushton (eds.), *Scientific Excellence*, pp. 359–75. Beverly Hills, CA: Sage Publications.

Cole, J.R. and H. Zuckerman (1987). 'Marriage, Motherhood, and Research Performance in Science', *Scientific American*, 256(2): 119–25.

Cole, J.R. and S. Cole (1973). *Social Stratification in Science*. Chicago: The University of Chicago Press.

Connellan, J., S. Baron-Cohen, S. Wheelwright, A. Batki, and J. Ahluwalia (2000). 'Sex Differences in Human Neonatal Social Perception', *Infant Behaviour and Development*, 23(1): 113–18.

Davies, P.G., S.J. Spencer, D.M. Quinn, and R. Gerhardstein (2002). 'Consuming Images: How Television Commercials that Elicit Stereotype Threat Can Restrain Women Academically and Professionally', *Personality and Social Psychology*, 28(12): 1615–1628.

Deaux, K. (1985). 'Sex and Gender', *Annual Review of Psychology*, 36: 49–81.

Eccles, J.S. (1987). 'Gender Roles and Women's Achievement-related Decisions', *Psychology of Women Quarterly*, 11(2): 135–72.

Else-Quest, Nicole M., Janet Shibley Hyde, and Marcia C. Linn (2010). 'Cross-National Patterns of Gender Differences in Mathematics: A Meta-Analysis', *Psychological Bulletin*, 136(1): 103–27.

Eysenck, H.J. (1988). 'O tempora, o mores!' *Behavioural and Brain Sciences*, 11(2): 189–90.

Farmer, H.S. (1988). 'Predicting Who Our Future Scientists and Mathematicians Will Be', *Behavioural and Brain Sciences*, 11(2): 190–91.

Farmer, H.S., J.L. Wardrop, and S.C. Rotella (1999). 'Antecedent Factors Differentiating Women and Men in Science/nonscience Careers', *Psychology of Women Quarterly*, 23(4): 763–80.

Feist, G.J. (2006). *The Scientific Mind: An Introduction to the Psychology of Science*. New Haven, CT: Yale University Press.

—— (2006). 'The Development of Scientific Talent in Westinghouse Finalists and Members of the National Academy of Sciences', *Journal of Adult Development*, Volume 13, Number 1, 23–35.

Feist, G.J. and M.E. Gorman (1998). 'Psychology of Science: Review and Integration of aNascent Discipline', *Review of General Psychology*, 2(1): 3–47.

Feist, G.J., S. Paletz, W. and Weitzer (2004). 'Predicting Scientific Interest in College Students: The Influence of Quantitative Skills, Gender, Self-image, and Personality'. In preparation.

Fischbein, S. (1990). 'Biosocial Influences on Sex Differences for Ability and Achievement Test Results As Well As Marks at School', *Intelligence*, 14(1): 127–39.

Fox, L.H. (1976). 'Sex Differences in Mathematical Precocity: Bridging the Gap', in D.P. Keating (ed.), *Intellectual Talent: Research and Development*, pp. 32–54. Baltimore, MD: Johns Hopkins University Press.

Geary, D.C. (1998). *Male, Female: The Evolution of Human Sex Differences*. Washington, DC: American Psychological Association.

Geschwind, N. and P. Behan (1982). 'Left-handedness: Association with Immune Disease, Migraine, and Developmental Learning Disorder'. Proceedings of the National Academy of Sciences, 79(16): 5097–100.

Guyer, L. and L. Fidell (1973). 'Publications of Men and Women Psychologists: Do Women Publish Less?' *American Psychologist*, 28(2): 157–60.

Halpern, D.F. (2000). *Sex Differences in Cognitive Abilities* (3rd ed.). Mahwah, NJ, US: Lawrence Erlbaum Associates.

Halpern, D.F., C.P. Benbow, D.C. Geray, R. Gur, J.S. Hyde, and M.A. Grensbacher (2007). 'The Science of Sex Difference in Science and Mathematics', *Psychological Science in the Public Interest*, 8(1): 1–51.

Hedges, L.V. and A. Nowell (1995). 'Sex Differences in Mental Test Scores, Variability, and Numbers of High-scoring Individuals', *Science*, 269(5220): 41–45.

Helmreich, R.L., J.T. Spence, W.E. Beane, G.W. Lucker, and K.A. Matthews (1980). 'Making It in Academic Psychology: Demographic and Personality Correlates of Attainment'. *Journal of Personality and Social Psychology*, 39(5): 896–908.

Holden, C. (1987). 'Female Math Anxiety on the Wane', *Science*, 236(4802): 600–01.

Holland, J.L. (1992). *Making Vocational Choices* (2ndedition). Odessa, FL: Psychological Assessment Resources.

Hyde, J.S. (2005). 'The Gender Similarities Hypothesis', *American Psychologist*, 60(6): 581–92.

Hyde, J.S., E. Fennema, M. Ryan, L.A. Frost, and C. Hopp (1990). 'Gender Comparisons of Mathematics Attitudes and Affect: A Meta-analysis', *Psychology of Women Quarterly*, 14(3): 299–324.

Hyde, J.S., E. Fennema, and S.J. Lamon (1990). 'Gender Differences in Mathematics Performance: A Meta-analysis', *Psychological Bulletin*, 107(2): 139–55.

Hyde, J.S., S.M. Lindberg, M.C. Linn, A.B. Ellis, and C.C. Williams. (2008). 'Gender Similarities Characterize Math Performance', *Science*, 321(5888): 494–95.

Jacobwitz, T. (1983). 'Relationship of Sex, Achievement, and Science Self-concept to the Science Career Preferences of Black Students', *Journal of Research in Science Teaching*, 20: 621–28.

Keating, D.P. (1974). 'The Study of Mathematically Precocious Youth', in J.C. Stanley, D.P. Keating, and L.H. Fox (eds.), *Mathematical Talent: Discovery, Description, and Development*, pp. 23–46. Baltimore, MD: Johns Hopkins University Press.

Kelly, K.R. and R.C. Nelson (1999). 'Task-specific Occupational Self-efficacy Scale: A Predictive Validity Study', *Journal of Career Assessment*, 7(4): 381–92.

Kimura, D. (1999). *Sex and Cognition*. Cambridge, MA: MIT Press.

Leahey, E. and G. Guo (2002). 'Gender Differences in Mathematical Trajectories', *Social Forces*, 80(2): 713–32.

Lippa, R. (1998). 'Gender-related Individual Differences and the Structure of Vocational interests: The Importance of the People-things Dimension', *Journal of Personality and Social Psychology*, 74(4): 996–1009.

Long, J.S. (1992). 'Measures of Sex Differences in Scientific Productivity'. *Social Forces*, 71(1), 159–178.

—— (ed.). (2001). *From scarcity to visibility: Gender differences in the careers of doctoral scientists and engineers*. Washington, DC: National Academy Press.

Lubinski, D., and L.G. Humphreys (1990). 'A Broadly Based Analysis of Mathematical Giftedness', *Intelligence*, 1(3)4, 327–355.

Luzzo, D.A., P. Hasper, K.A. Albert, M.A. Bibby, and E.A. Martinelli (1999). 'Effects of Self-efficacy-enhancing Interventions on the Math/science Self-efficacy and Career Interests, Goals, and Actions of Career Undecided College Students', *Journal of Counseling Psychology*, 46(2): 233–43.

Maccoby, E.E., and C.N. Jacklin (1974). *The Psychology of Sex Differences.* Stanford, CA: Stanford University Press.

Mahoney, M.J. (1979). 'Psychology of the Scientist: An Evaluative Review', *Social Studies of Science*, 9(3): 349–75.

Moore, E.G.J. and A.W. Smith (1987). 'Sex and Ethnic Group Differences in Mathematics Achievement: Results from the National Longitudinal Study', *Journal for Research in Mathematics Education*, 18(1): 25–36.

National Academy of Sciences (2006). *Beyond Bias and Barriers: Fulfilling the Potential of Women in Academic Science and Engineering.* Washington DC: National Academies Press.

National Science Foundation (NSF) (1999). *Women, Minorities, and Persons with Disabilities in Science and Engineering: 1998* (NSF 99–87.) Arlington, VA: National Science Foundation.

Nosek, B. A., M.R. Banaji, and A.G. Greenwald (2002). 'Math = Male, Me = Female, Therefore Math ≠ Me', *Journal of Personality and Social Psychology*, 83(1): 44–59.

O'Brien, V., M. Martinez-Pons, and M. Kopala (1999). 'Mathematics Self-efficacy, Ethnic Identity, Gender, and Career Interests Related to Mathematics and Science', *Journal of Educational Research*, 92(4): 231–35.

Pasewark, R. A., B.J. Fitzgerald and R.N. Sawyer (1975). 'Psychology of the Scientist: XXXII. God at the Synapse: Research Activities of Clinical, Experimental, and Physiological Psychologists', *Psychological Reports*, 36(2): 671–74.

Prediger, D.J. (1982). 'Dimensions Underlying Holland's Hexagon: Missing Link between Interests and Occupations?' *Journal of Vocational Behaviour*, 2(3): 259–87.

Reis, S. M. and S. Park (2001). 'Gender Differences in High-achieving Students in Math and Science', *Journal for the Education of the Gifted*, 25(1): 52–73.

Rosser, S. (1988) (ed.). *Feminism within the Science and Healthcare Professions: Overcoming Resistance.* Exeter, England: A. Wheaton and Co.

Schmidt, D. B., D. Lubinski, and C.P. Benbow (1998). 'Validity of Assessing Educational-vocational Preference Dimensions among Intellectually Talented 13-year olds', *Journal of Counseling Psychology*, 45(4): 436–53.

Shih, M., T. L. Pittinsky, and N. Ambady (1999). 'Stereotype Susceptibility: Identity Salience and Shifts in Quantitative Performance', *Psychological Science*, 10(1): 80–83.

Spelke, E. S. (2005). 'Sex Differences in Intrinsic Aptitude for Mathematics and Science? A Critical Review', *American Psychologist*, Vol. 60, No. 9, 950–958.

Stanley, J. C. (1988). 'Some Characteristics of SMPY's "700–800 on SAT-M before age 13 group": Youths Who Reason *extremely* Well Mathematically', *Gifted Child Quarterly*, 32(1): 205–09.

Steele, C. M. (1997). 'A Threat in the Air: How Stereotypes Shape Intellectual Identity and Performance', *American Psychologist*, 52(6): 613–29.

Steele, C. M. and J. Aronson (1995). 'Stereotype Threat and the Intellectual Test Performance of African Americans', *Journal of Personality and Social Psychology*, 69(5): 797–811.

Sternberg, R.J. (1988). 'The Male/female Difference is There: Should We Care?' *Behavioural and Brain Sciences*, 11(2): 210–11.

Stuessy, C.L. (1989). 'Path Analysis: A Model for the Development of Scientific Reasoning Abilities in Adolescents', *Journal of Research in Science Teaching*, 26(1): 41–53.

Subotnik, R.F. and C.L. Steiner (1994). 'Adult Manifestations of Adolescent Talent in Science: A Longitudinal Study of 1983 Westinghouse Science Talent Search Winners', in R. Subotnik and K.D. Arnold (eds.), *Beyond Terman: Contemporary Longitudinal Studies of Giftedness and Talent. Creativity Research*, pp. 52–76. Norwood, NJ: Ablex Publishing Corp.

Subotnik, R.F., R.A. Duschl, and E.H. Selmon (1993). 'Retention and Attrition of Science Talent: A Longitudinal Study of Westinghouse Science Talent Search Winners. *International Journal of Science Education*, 15(1): 61–72.

Tobin, K., D.J. Tippins, and K.S. Hook, (1995). 'Students' Beliefs About Epistemology, Science, and Classroom Learning: A Question of Fit', in S.M. Glynn and D. Reinders (eds.), *Learning Science in the Schools: Research Reforming Practice*, pp. 85–108. Hillsdale, NJ: Erlbaum.

Vandenberg, S.G. (1988). 'Could These Sex Differences be Due to Genes?' *Behavioural and Brain Sciences*, 11(2): 212–14.

Webb, R.M., D. Lubinski, and C.P. Benbow (2002). 'Mathematically Facile Adolescents with Math-science Aspirations: New Perspectives on their Educational and Vocational Development', *Journal of Educational Psychology*, 94(4): 785–94.

Xie, Y. and K.A. Shauman (1998).'Sex Differences in Research Productivity: New Evidence About an Old Puzzle', *American Sociological Review*, 63(6): 847–70.

Zuckerman, H. and J.R. Cole (1975). 'Women in American Science', *Minerva*, 13(1): 82–102.

CHAPTER FIVE

Women and Minorities in Science

Discrimination and the Solution

Henry Etzkowitz

WOMEN and minorities face problems in all stages and phases of their scientific life course. The taken-for-granted norms, the structure of society, especially the relationship between family and work, and the organisation of scientific institutions create a series of interrelated dilemmas for women in science that must be addressed in a comprehensive fashion. The solution surely includes changes in the society at large, for example, in gender relations of family and work that create a 'triple burden' for women in science (Gupta, 2002). But there can be no diversion of attention or shifting the blame from science to society. Any action to include women and minorities must start at home, within science and technology itself, rather than waiting for societal or generational change.

When scientists and technologists take up the issues of women in science, typically in response to the concerns about impending human capital shortages, the focus tends to be on the front end of the so-called 'pipeline' of scientific career progression. Too often, when the issue of lack of women and minorities is raised, the tendency within scientific institutions is to view the problem solely in terms of recruitment. Numerous programmes have been started to recruit students from elementary and high schools, and to encourage women and minorities to take an interest in science and engineering careers. However, many of those recruited do not pursue technical careers or face conditions that do not allow them to perform at their highest level (Lovitts, 1996). Having developed intellectual capital, these scientists face a persisting 'social capital gap', the failure to fully include them in the informal networks that encourage scientific achievement at the highest level.

The negative experiences of women and under-represented minorities in institutes and organisations of science and technology must be addressed. Otherwise, much of the good work done in recruiting them will turn out to be futile, even counter-productive, if the way in which research institutes, and graduate departments and training programmes work to deter women are not examined and restructured. Barriers created by a 'male model' of scientific research and training that do not take the needs of women into account must be removed. The focus of our research has been on these institutional and organisational barriers to women and minorities in science, and that is what will be discussed in this chapter.

The Social Capital Gap

Our finding on the effects of the taken-for-granted workings of science and technology institutions and organisations on women in science may be summarised as follows:

(a) Informal social practices in graduate programmes isolate women, giving them an educational difference significantly different than their male peers, opening up a social capital gap that negatively affects women's scientific careers.

(b) Given a relative lack of social capital, women tend to rely more on their personal intellectual capital in developing their research programmes. This has the effect of channelling them into slower moving fields that are less dependent upon collaborative research.

(c) These negative effects of ingrained cultural practices on women in science are hidden behind a normative structure that proclaims universality.

(d) The contradiction between the ideology and practice of science is exemplified by inflexible rules that while they vary in different societies, uniformly reduce women's chances for scientific achievement.

(e) When the status of a field is low, women will be found in large numbers; as the status increases; the number of women declines. Conversely, when status and pay declines, women are allowed into the field.

(f) Social capital gaps are reduced by measures that bring women and minorities into the informal networks of scientific practice.
(g) Exposure of the social structure of how science works is an important first step to recognising and changing exclusionary practices.

The Dilemmas of Women in Science

The question, 'Why is a man studying women in science?' is often raised. My answer is personal history and serendipity.[1] My mother received an undergraduate degree in Geology from Hunter College in New York, with very high honours, but never found a career in the field. The apparent explanation was that the degree was awarded during the Depression. I took it that the reason she did not have a career in science was because of the economic downturn and lack of jobs. However, at the Geology department of the University of Texas, there are plaques honouring the men who had made contributions to the field of Geology, who had graduated from that university during that era. Despite the Depression it proved possible at least for some men to pursue careers in Geology. So it was necessary to think again. Pnina Abir-Am, the historian of science, raised the question of science policy for women in science, at a European science policy workshop several years ago (Abir-Am, 1989). At this meeting also the question came up as to why there were so few male science policy analysts studying issues of women in science. A female science policy analysts suggested that if a women studies women in science and does policy research analysis, it may reduce her status in the eyes of her male colleagues. There are, of course, notable, for example, Laurel Smith-Doerr, who has studied university–industry relation issues and has also worked on women in science and technology (Smith-Doerr, 2004).

Several years ago, as part of a study on the organisation of academic research groups in the US university system, we explored the hypothesis that these groups operate as 'quasi-firms', with an entrepreneurial dynamic built into their structure. The entrepreneurial dynamic of US academic research groups was a possible explanation for the relative lack of a gap between academic life and a start-up firm. Professors in the sciences at research universities have to go out and to get funds and establish

[1] See Wesseling (2004) for an illuminating discussion of the role of personal background in scientific discourse.

themselves individually at a very early academic age. These research groups are in many ways like business firms – the so-called individual investigator has to organise the funding, manage the personnel, and so on (Etzkowitz, 1992). Carol Kemelgor, then a senior project student at SUNY Purchase, was doing interviews in the genetics department at an area medical school. One day she came back with the observation that the female professors were organising their research groups differently than the men; more collegially and less hierarchically, with limits on time at the workplace. These mostly young women professors were trying to carve out a personal sphere while operating their laboratories on a 9 to 5 schedule, at least with respect to their own on-site participation. These were intriguing findings about significant anomalies. However, we were studying the organisation of research groups as entrepreneurial organisations, not gender issues, so we put these observations aside. However, it happened that Eric Bloch, the director of the National Science Foundation (NSF), at that time was concerned about an imminent shortfall of human resources in science and technology of the states. As in wartime, the idea was raised that the US should try to recruit women into scientific careers. Bloch, the NSF head, asked Phylis Moen, the director of Sociology Programme in NSF 'What do sociologists know about women in science?' Moen called, me, as head of the Science, Knowledge, and Technology (SKAT) section in the American Sociological Association, and asked for the suggestions of researchers. I nominated several people who had done research on the topic and then mentioned Carol Kemelgor's observations of the non-traditional approaches which some female scientists were taking in organising their work. Dr Moen replied, 'You are in the field; you should do a study.' This was the start of the Athena Unbound Project to understand the factors that promoted and inhibited women's participation in academic science.

The Athena Unbound Studies

Our first study was of five disciplines in one department, followed by a second department in the same discipline at another university. Then we undertook a large project that involved 400 in-depth interviews with female and male graduate students and faculties in twelve research universities in the Carnegie One category of leading research universities spread across the US East coast, West coast, the Southeast and the Rocky Mountain region. The study was carried out in five disciplines: biology, physics, chemistry, computer science, and electrical engineering.[2] We identified *unofficial* and *official* graduate training programmes in the same department, with male students included in informal social networks by their mentors while female were largely left out of information and opportunity circles.

Our next finding came from the sampling strategy itself. Our statistician from the American Physical Society designed a sampling frame of department, from the National Research Council survey of doctorates, selecting on three criteria: (a) largest number of PhDs granted to women; (b) smallest number of PhDs granted; and (c) the departments which were the most advanced. However, when Carol Kemelgor went to do the first wave of interviews she could not find any difference in the experience of women in different categories of departments. We soon determined that the doctoral survey did not take account of the national origins of the female PhDs surveyed. Most of the women at the departments, which had a large number of female graduates, were non-US citizens. These women, from the different national groups, were well integrated with the men from their own countries and were getting support. The American women were not similarly integrated with their national peers, so it did not make a difference whether a school had been graduating few female PhDs or a large number. The relatively small number of American women tended to be isolated in US academic science departments.

The Paradox of Critical Mass

There has long been an expectation that even if equality was not reached, if 15 per cent of so-called 'critical mass' could be attained, then discrimination against women in science would fall away. However, as numbers move from very few where 'invisibility' obtains, to a larger number, and there is a more visible presence of women, discrimination increases. The expectation that critical mass would solve the problem is also disconfirmed because as their numbers increase, women typically do not form a unified group. Some continue to follow the male model, others try to carve out a female model of doing science. Unfortunately, these two groups usually are at odds with each other. For example, rather than experiencing identification with some women faculty, a young woman said, 'I'm never going to be like them. If I

[2] See Etzkowitz et al. (2000).

know I cannot be like them, then what am I going to become?' In recent years, some of the senior women 'male modelers' have come to see that they have experienced discrimination in their own careers and revised some of their thinking. Women in a department also separate into different research groups and may not be in touch.

Even more frustrating than the relative lack of positive effects from critical mass is the suppression of negative effects as a consequence of the acceptance of the sociological premise of universalism as an inherent feature of science (Merton, 1942[1973]). Robert K. Merton held that for science to be fertile and productive, scientists must be judged only by their work and win status and membership within the scientific community on these criteria alone. The scientific enterprise could not afford to base the 'acceptance or rejection of [scientific] claims.... on the personal or social attributes of their protagonists: his [*sic*] race, nationality, religion, class and personal qualities are as such irrelevant' (Merton, 1942[1973]: 553). Although sexism was not specified as an area of discrimination at that time, it was taken for granted. This has been exemplified by the unanswered letter the historian of science Helene Metzger wrote to Emile Meyerson, her senior contemporary in 1933, demanding to be accepted as a professional equal, 'In the republic of Minds, we are all equal and you must prove that reason is on your side, not impose it by force or by intimidation' (Chimisso and Fruedenthal, 2003).

Normative Fallacies

Universalism promotes the belief among some male scientists that we could not possibly be discriminating against women because our enterprise is based upon the principle that social characteristics are not taken into account in making judgements. Some of the worst practices have been exposed recently. One example was the Massachusetts Institute of Technology (MIT) self study. Female full professors found to their surprise that their pay was lower and their labs smaller than those of their male colleagues. One change is that 25 years ago the studies of inequality of women at MIT were done by female graduate students. Today, they are carried out by full professors.

The other recent study was done of the Space Telescope Institute, which identified a chilly climate, marginalisation, and men denying the problem. Senior women were leaving the institute for jobs elsewhere because nothing was being done. In this case the supervising authorities of this lab demanded that something be done and the process is still ongoing. It is only the exceptional event of President Vest's promise to address the findings from a study conducted by senior female faculty identifying invisible discrimination at MIT (Wilson, 1999) and the departure of senior women researchers from the Space Telescope Institute, in response to deprecation of their contribution, that breaks through the veil of silence (www.aura-astronomy.org).

Dual Hidden Male and Female PhD Programmes

In many departments, pursuing the same course and meeting similar requirements to obtain PhD degree, senior women realised that they had different experience than men towards the end of the programme. The men were included in informal social relations with the faculty members, invited to pubs, and played basketball games together. It was at these informal occasions that men picked up what it is that they really have to do to get the PhD. Without these clues women have to go by the book and what is written down and so professional acculturation takes longer as a result of being excluded from the social capital transmitted through informal social relations.

The kula ring is the Melanesian Island practice of giving away your most prized possessions and the one who gives away the most gains the most in status. In academic science it is giving away your prized bright students to a colleague at another university. This process shows the esteem with which you hold the receiving colleague; it also exhibits status gained by offering the 'gift'. These exchanges are typically negotiated in small workshops where outstanding graduate students are introduced to each other and each others' professors. The senior women students who felt they were in a different PhD programme from their male peers, eventually became aware that that they had largely been excluded from this process of asexual reproduction in which scientific research groups renew themselves.

Given a relative lack of social capital, women have to work much harder to prove themselves. When women confront a scientific establishment that fails to support their ability to function optimally, they naturally seek alternative paths to pursue their professional goals. They become immersed in perfectionism rather than taking risks and assertively getting preliminary findings out. Moreover, female scientists may not be included in collaborations important to creating networks and building 'social capital'.

Adoption of a niche approach is a resilient response to exclusion from professional networks. This over-reliance on human capital has some negative consequences as well. The myth of scientific individualism perpetuates an impossible standard that does not exist for the majority of men, but is frequently the core experience for women. Ironically, it is the women who are the 'rugged individualists', forced to fend for themselves. Lack of full membership in the scientific community creates uncertainty and self-protectiveness. The cumulative production of stress in women's scientific careers is expressed as a feeling of loneliness and helplessness in a hostile environment (Trinational Conference, 2003).

Surplus Anxiety among Women in Science

Apprentice male scientists are typically welcomed to full membership in the 'club'. They need not 'watch their back' beyond what is common for all of us in competitive work situations. They are able to attend conferences or begin their first academic appointment without undue concern or any level of awareness that they may be disregarded because of their sex. The experience for many women is one in which there is frequently an underlay of anxiety as a response to the threat of not being taken seriously. In hostile environments, along with intellectual demands and the everyday stresses of a scientific career, there is an additional expenditure of emotional resources in response to the anxiety produced by discrimination.

Women in science produce a 'surplus of anxiety', in consequence of the triple burden of gendered discrimination, 'role overload' and social capital depletion that interferes with scientific production and recognition of accomplishment. Surplus anxiety, like surplus repression (Marcuse, 1974), impedes the realisation of institutional and personal goals. The production of surplus anxiety inevitably displaces, or makes unnecessarily difficult, some of the production of 'good science' that might otherwise occur in a less stressful environment. Some stress is inevitable and even productive, the so-called 'creative tension' of the best scientific workplaces. Conversely, too little, and especially too much, stress is counterproductive.

Of course, anyone doing scientific work is subject to stress: to get results, raise funds, and arrange publications. It is often argued that all scientists face increased anxiety due to the funding exigencies. In a number of countries, the conditions for careers in science are deteriorating due to lack of public funding; severe budget cuts at universities and public research institutes. Even in the US, where budgets have been increasing, there is a perceived shortfall. This is in part due to increased competition for funds from universities and regions that were formerly viewed as less competitive or even non-competitive in the scientific research arena, until the connection between research success and economic development became broadly apparent.

Thus, as a vocation or career, science is becoming less and less attractive to men *and* women, especially if they belong to younger cohorts that, from a labour market perspective, are in 'the void'. Deteriorating conditions may affect individual investment decisions and expected rates of investment over the course of careers in science by both the junior and senior scientists. These changes in structural and institutional factors increase stress and anxiety experienced by scientists of both the sexes. Nevertheless, heightened anxiety for men and women due to funding exigencies is, for women, merely a modest overlay on the 'surplus anxiety' induced by gender discrimination.

For women there is an extra burden since they need to deal with: (a) expectations of not being taken seriously; (b) the experience of exclusionary practices; (c) being dependent on colleagues who are hostile and who are trying to undermine you; (d) the emulation of the male model; (e) role overload: too many activities to do – teaching and research, administrative – and playing female role; and (f) being put on many committees, so equality can be said to exist. A junior faculty member describes a role that far exceeds her male colleagues; 'I've been asked to do a lot. To talk to undergraduates, graduate students, women's groups, all sorts of things. But if I'm not here in three years, if I'm not retained, then I'm not going to do anybody any good. So I can only do so much because I'm more worried about my retention at this point'. It subverts the ability of women to pursue their research if they have to be on all of these committees and then one is not expected to raise the issues in science surrounding the women and, as already noted, discrimination hidden by the assumptions of universalism.

The Negative Effects of Inflexible Rules

In different countries, and in different academic cultures, the problems faced by women in science appear to be different. However, the third common principle that we can derive is: 'If an academic cultural rule is inflexible, it works against women; regardless of specific content'.

In the US, the rule is exogamy – to move ahead at each level, an individual is expected to move to a different university, always choosing the best professor, department, and the highest status school that is available. Several moves are expected, from undergraduate college to graduate school, to postdoctoral fellowship and to initial professorial appointment. For women, it is typically less easy than men to make multiple moves, given their family and relationship responsibilities. If one has to stay in the same area, but is expected to change universities – unless it is a very large metropolitan area with many good schools – the chances are that a person who is geographically constrained will be pushed down the academic ladder by the operation of this rule. There is also the related 'two-body problem', of a couple having to find two jobs in a same place. When an academic couple moves, the man tends to have preference in the choice of job and the woman is expected to find whatever opportunity she can.

In Sweden, the opposite rule of endogamy applies. When a leading Swedish professor was told about the US model, he said, '[T]hat's crazy; why would you want to send your best student away. Your best student is the one who is going to take over from you. That's who you keep.' This rule also works against women as an inflexible rule, if a woman decides to move, because her spouse is moving. She loses the opportunity to continue in her original research group which is her best chance for moving ahead in the Swedish context. As a general principle, if the rule is inflexible, it does not work for women.

Here one may refer to Robert Kohler's analysis of Thomas Morgan and his drosophila research team at Columbia University in the early twentieth century (Kohler, 1994). In the early years of the research, when the field of genetics lacked status, there were women working with Morgan on the research projects. As the importance of the investigation was recognised, its status rose; men came into the field and the women disappeared. This is a process that can still be identified in recent years, for example, in the field of computer software, as the importance of that field increased in comparison to hardware, which had high status in the early years of the discipline, men started taking it up seriously. Thus, a first principle of institutional deterrence of women from science can be formulated: 'When the status of a field is low, women will be found in large numbers; as the status increases, the number of women declines'. The corollary to this principle is that it also works the other way round: 'when the status and pay declines, women are allowed into the field.' Either way women lose.

The implication of this relationship is that merely getting an increase in numbers in a field does not mean that the problem is solved. Indeed, more women entering a field may be the result of a change in the field, making it less attractive to men. This has been the experience in academic computer science in Mexico. As salaries stayed low, men left to join various industries and women were allowed into academic jobs. At the Biological Sciences Division at Lund University in Sweden, the Dean said, 'The problem is going away. We just had this huge increase in women going into dentistry.' On the other hand, he also mentioned that the science of dentistry is declining in status in Sweden so the very indicator of a solution is also an indicator of a continuing problem.

The Male Model of Science

The traditional male path is one of full-time devotion to research, emphasis on early achievement, and exclusive identification of oneself with science. In other professions such as medicine and law, it is easier for women to drop out and then re-enter the career line later. In science, women find this difficult and careers off the beaten track are looked upon unfavourably. With limited exceptions for breakthrough individuals,[3] science is characterised by a 'sexual separation of labour' in which male and female scientists often pursue different career paths (Etzkowitz, 1971).[4] The paradox of women in science is that when they are excluded from traditional career paths, they may reappear in non-traditional career paths at the margins of the traditional scientific enterprise but at the crossroads of new links between science and society, such as in intellectual property management and science journalism. We call these non-traditional paths, the 'Rachael Carson' track after the female scientist whose path to the PhD and academia was thwarted but who found a career in government service, where her official duties explicating ecological research for government publications led her to synthesise a broad range of data into a new environmental paradigm (Lear, 1997).

Another part of the taken-for-granted male model of science that has deleterious consequences for women is the strongly held belief that research achievement is primarily a province of the young. The obvious source of negative side effect of an assumption of youthful achievement for women in science is the coincidence of the childbearing years. However, Harriet Zuckerman and Robert K. Merton in their work on Age Structure in Scientific Research show that this is not the case; that older scientists continue to contribute at a high level (Merton and Zuckerman, 1973). It stands to reason. After all, many are at the peak of their careers and have amassed significant financial and social resources and are located in central networks. They can attract

3 See Hargittai and Hargittai (2003).

4 See also Jacobs (2001).

graduate students and post-doctorates who work with them and expand their research productivity. Nevertheless, the belief remains strong that scientific productivity is disproportionately associated with youth. The emphasis on the early part of the career has been impervious to sociological findings.

This taken-for-granted belief has significant implications for women in science, especially since the early part of an academic career coincides with childbearing years. Thus, the issue of the relationship between the seven-year tenure clocks versus the biological clock for childbearing has arisen but remains unresolved in the US. This is the case even when some universities have made provisions to allow a candidate with young children the leeway to stop the clock for a year prior to tenure review. Very few female faculties are willing to take advantage of these offers because they know it would be taken negatively, if they any way distinguish themselves from their male colleagues. They may not even be aware of the possibility. For example, a computer science professor said: 'They told me that I couldn't have a course reduction when I was about to have my baby. A colleague asked me why I didn't use my buy-out. I never knew about it. I went to the head of the department and he said I couldn't use it because I had never asked for it.'

Even attempts to redress specific problems can have negative consequences for women unless similar issues are addressed for men. A requirement that has been introduced in Swedish universities points the way – male as well as female faculty with young children are required to take time off from the workplace.

Invisible Women in Science

I was recently invited to give a talk on *Athena Unbound* at the oceanography department of University of Rhode Island. After the main talk, I was invited to a monthly 'brown bag' lunch meeting of women in science in oceanography. In addition to the professors and graduate students, there was a large group of women who were research associates. In the research associate model, people in these positions worked for someone else. However, these female research associates were basically working for themselves, getting their own grants and raising their own research funds. They were nominally associated with a faculty member but they had authority over their own projects. Given the nature of the academic system, they preferred this position without job security or career mobility because they wanted to balance family with work. They wanted to work two-thirds of the time, no more. It is not possible to be on the faculty tenure track and operate that way.

The advantage of their creative restructuring of the research associate status was that they could balance work and family; the disadvantage is the lack of a career ladder. Given their values and an inflexible academic system, their alternative as women who wish to seriously pursue science was to be invisible, separate, and unequal. Carrying out their research at a relaxed pace in conjunction with family life, these female oceanographers were pursuing 'off the book' scientific careers. These relatively invisible female researchers had creatively transformed the traditional subordinate female associate position into a relatively independent status, albeit one with no career mobility.

Time Bind

There is a common time bind for women in science, whether in US, Mexico or Eastern Europe. If you are going to have a family you could find time to do your research, you could balance these two, but to do politicking and networking to move up and become a high-level administrator, women say that is too much. There is just not enough time to do three things. Two could be managed at best. A female faculty member described conditions in her department, 'Women who are making it in our department are doing so at the expense of their personal and familial ties. Another female advisor is new to this department and untenured, so I'm willing to bet she won't be having any kids.'

Change Strategies

The 'kula ring of scientific success' inducts persons of similar social characteristics in a process of asexual reproduction. Women and minorities often experience 'social capital gaps', exclusion from social networks, conducive to performance and recognition at the highest levels that reduces future career chances. Two strategies have been proposed to close social capital gaps: (a) programmes focused on the excluded group, but participants may be 'ghettoized', stigmatised and isolated in their own world; despite the solidarity attained and (b) identification and reform of exclusionary practices. Change strategy is largely dependent upon the leadership. In a hierarchical academic system, pressures from below can only call attention to a problem but cannot resolve it. We identified a few departments where there had been revolutions from above by reorganising the social practices of the departments. These departments, some of them, though in the second ranking universities, became well known as female friendly among the prospective female graduate students and were able to attract the best female students. At the University of California, Berkley Computer Science Department, the female graduate student leadership organised a programme to bring back those female students who have been out of the educational market for sometime through the re-entry program. This was institutionalised with the support from the department.

Programmes for Minorities in Science

The second study I am going to draw upon is the evaluation of the Leadership Alliance Programme (http://www.theleadershipalliance.org). This is a programme that brings students from under-represented minorities, which in the US context is African-Americans, Hispanics, and Native Americans. This programme brings the students from the smaller colleges – especially the historically African-American universities – to the leading research universities. These students become members of the research groups during summer and carry out small projects or a part of the larger projects. This experience provides them with much needed social capital. Although these programmes reduce the social capital gap, they do not eliminate it.

Graduate admissions at the elite universities work not solely by applications and formal credentials. Many people have good grades and fine recommendations on paper. So how do you distinguish among them? It is the telephone call from a faculty member, whether it is a research university or undergraduate college, known by the professor seeking new students, that often makes the difference. It is through the phone call that the admitting professor, in effect, demands and receives a 'personal guarantee' of the quality of the applicant. If the two professors are not known to each other, the applicant may fall by the wayside. The summer programme helps reduce that 'social capital gap' by making the applicant personally known to the graduate school professor. Nevertheless, without a connection to the undergraduate school professor; the ultimate guarantee is still lacking and much of the social capital gap remains in place. A summer research programme for high school students at Rockfeller University points the way to fill this gap by bringing the teachers as well as the students into the lab for the summer.

There have traditionally been two sets of schools in this programme: (a) sending schools, small colleges and (b) receiving schools, research universities. An in-between level of schools is also emerging. They were formerly teaching institutions but are now becoming research institutions at least in a few fields and some of them by getting large grants of the National Institute of Health (NIH), have been able to establish the PhD programmes. Hunter College in the City University of New York and the University of Maryland, Baltimore County have new PhD programmes, often with women playing the leading role. These are typically women who are highly committed to research but have been discriminated against by major universities. As these previously teaching schools opened up PhD programmes, they became a credible alternative to pursue a scientific career, combining research and mentoring. These new programmes may be more attractive to women and minorities than going to a major school and so these schools have that advantage in recruiting. I predict that as these schools emerge as institutes of PhD programmes, they will likely have more impact on improving the condition of women and minorities in science than the existing leading universities.

Remedying Social Capital Deficits for Women and Minorities in Science

The experiences of women scientists begin and end with the consequences of social exclusion in an activity, which necessitates, perhaps demands, community. All too frequently the consequences of social isolation and loneliness have been attributed to inherent deficits within the women themselves. The argument has been that they lack the right human capital for physically demanding and mathematically intensive scientific work, whether by nature's wisdom, which has divided the gene pool, or by self-selection into softer fields that permit greater attention to the family. However, the experience of separateness and stigma makes the tendencies for self-blame, lack of self-confidence, fear of risk-taking, and role confusion at the highest faculty level more understandable.

Cultural prescriptions and proscriptions regarding gender are not immutable. We may postulate an alternate model of conducting science, which can incorporate cooperation and communal qualities, reduce over-reliance on task orientation and instrumentality. Many a times the younger group of female and male faculty members in science does emphasise collaboration and community in their working environment. Some of the male advisors are excellent mentors to all their students regardless of their gender. These type of emergent groups are more sympathetic and have the understanding of the problems faced by the women graduate students. Their 'relational' approach is a contrast to the singularly focused, aggressive, and 'instrumental' approach of most of their colleagues.

The expectation built into science, that the cognitive content of science is rational, has often been carried over into the assumption that the social institutions of science are, as a matter of course, constructed upon similar premises. However, as rationality does not always carry the day in resolving research issues, it fails to explain fully the operation of the scientific workplace. There is a need for wider organisational and departmental policies. For example, women should be recruited into pivotal positions and placed in important committees so that they develop a network of contacts, access 'inside' information on organisational politics, and display their administrative competence. Organised mentoring programmes will be helpful in changing the academic culture and raising the social capital. Department review (with outside referees) for diversity issues should evaluate the departments' recruitment, hiring, and retention of women. Such reviews could also be used for ranking departments.

A 'supply side' approach, by itself, is insufficient. One of the keys to mitigating the triple burden on women academics lies in university-wide policies on childcare, parental leave, and slowing of the tenure clock, including flexible work schedules and financial help with childcare. The offbeat career track due to an attempt to combine family and career commitments should not be held against women in recruitment and retention. More research is needed to identify the anomalous career paths of female scientists that have both negative and positive consequences. Some women, seemingly lost to science, have turned their scientific interests away from bench research in academia to positions such as marketing director for a biotechnology firm, venture capital firm partner, and independent inventor.

Marginalisation, isolation, and a demand for autonomous, independent functioning in graduate school, within a highly social and socialising activity available to men, places women in a double bind. Discrimination is exacerbated when adaptive attempts for affiliation through women's groups are labelled as indicating women having 'special needs' It is further compounded when similarly isolated women faculty are offered up as a solution to institutional problems as they too grapple with issues of isolation. Underneath all of these confused and confusing experiences is, frequently, a covert understanding, that to survive one must get through in silence.

The women who have reached high levels within the scientific community are clearly individuals who have, heretofore, been able to exercise and develop their capacity to fulfill their potentials and experience a sense of their own self-agency without suffering defeat. Made to feel weak and vulnerable by an unexpected discriminatory assault, women must then grapple with their own hostile feelings while continuing to rely and depending on those who may diminish and rebuff them. It is this un-level playing field that inhibits self-agency and diminishes the capacity to transform potentialities into accomplishments that distinguishes the condition of women from men scientists.

References

Abir-Am, Pnina (1989). *Science Policy for Women in Science: Lessons form Historical Case Studies.* North Atlantic Treaty Organization (NATO) Advanced Studies Institute in Science Policy, Il Cioccio, Italy.

Chimisso, Cristina and Gad Freudenthal (2003). 'A Mind of Her Own: Helene Metzger to Emile Meyerson, 1933', *ISIS*, 94(3): 477–91.

Etzkowitz, Henry (1971). The 'Male Nurse: Sexual Separation of Labor in Society', *Journal of Marriage and the Family*, August.

—— (1992). 'Individual Investigators and their Research Groups', *Minerva*, Spring.

Etzkowitz, Henry, Carol Kemelgor, and Brian Uzzi (2000). *Athena Unbound: The Advancement of Women in Science and Technology.* Cambridge: Cambridge University Press.

Gupta, Namrata (2002). 'Women in Academic Science: A Study of Social and Work Environment at Institutes of Higher Learning in Science and Technology in India', PhD dissertation, Department of Sociology, Indian Institute of Technology (IIT), Kanpur.

Hargittai, Istvan and Magdolna Hargittai (2003). *Candid Science Interview with Jacqueline Barton*. London: Imperial College Press.

Jacobs, Jerry (2001). 'Revolving Doors: Sex Segregation and Women's Careers', in David Grusky (ed.) *Social Stratification: Class, Race & Gender in Sociological Perspective*, 2nd edition, pp. 714–19. Boulder: Westview Press.

Kohler, Robert (1994). *Lords of the Fly: Drosophila Genetics and the Experimental Life.* Chicago: University of Chicago Press.

Lear, Linda (1997). *Rachael Carson: Witness for Nature.* New York: Henry Holt.

Lovitts, Barbara (1996). 'Leaving the Ivory Tower: A Sociological Analysis of the Causes of Departure from Doctoral Study', PhD dissertation, Department of Sociology, University of Maryland.

Marcuse, Herbert (1974). *Eros and Civilization*. Boston: Beacon Press.

Merton, Robert K. and Harriet Zuckerman (1973). 'Age and Scientific Productivity,' in Norman Storer (ed.), *The Sociology of Science*. Chicago: University of Chicago Press.

Smith-Doerr, Laurel (2004). *Women's Work: Gender Equality vs. Hierarchy in the Life Sciences.* Boulder: Lynne Rienner Publishers.

Tri-national Conference: UK/France/Canada (2003). 'The Role of the National Academies in Removing Gender Bias'. London: The Royal Society.

Wesseling, Elisabeth (2004). 'Judith Rich Harris: The Miss Marple of Developmental Psychology', *Science in Context*, 17(3) (Autumn): 293–314.

Wilson, Robin (1999). 'An MIT Professor's Suspicion of Bias Leads to a New Movement for Academic Women', *Chronicle of Higher Education*, 46(3 December): A16–18.

Section II

Illustrative Examples

CHAPTER SIX

Women and Science in the Netherlands

A Dutch Case?*

Mineke Bosch

> ...even in the long-industrialized European countries, the story has not been one of automatic growth and progress. Thus...in the Netherlands...there too the situation for women academics has deteriorated over the past two decades. Where in 1970 there were 2.7 per cent women professors, by 1980 this was down to 2.2 per cent and by 1988 to 2.1 per cent.
>
> *(Rose 1994, p. 103)*

In April 2000, the European Commission (EC) organised a conference in Brussels, titled 'Making Change Happen'. The aim was to discuss the EC's report titled *Science Policies in the European Union – Promoting Excellence through Mainstreaming Gender Equality* (EC, 2000) and to determine the follow-up strategies for the report's findings. With regard to the Netherlands, the report made it clear that in terms of a European (and global) comparative perspective, Dutch women in the field of science were not faring well. Although the report presented some good news about practices in the Netherlands, this was overshadowed by statistical negatives – one table in the European Commission report showed that the Netherlands came in last out of 24 nations with the lowest percentages of women in the rank of professor. Another table confirmed that the Netherlands had 'the leakiest pipeline' of

* An earlier version of this paper was published in *Science in context,* 15(4): pp. 483–527, (2002) Cambridge University Press.

all (EC, 2000, Table 2.1, p. 10 and Table 2.4 (the scissors diagram), p. 13; for the 'leaky pipeline', see p. 12).[1] While the Netherlands comes in last for full professors (1998 figures), it is tied with Austria and Belgium with 7 per cent associate professors, and higher than Austria, Belgium, and Ireland with 20 per cent assistant professors.[2]

On the fringes of the EU conference, there was a screening of a Dutch documentary called *Waste of Talent* that focused on the under-representation of women in science and research in the Netherlands, starting from the famous Swedish case of gender bias in the selection process of the Swedish Medical Research Council.[3] This gender bias had convincingly been documented by Christine Wennerås and Agnes Wold, who subsequently published their findings in the leading journal *Nature* (Wennerås and Wold, 1997). Their article alerted leaders in academic and research institutions about the problems that women face in science. The film also caught some remarkable opinions of the Dutch research establishment. A representative of the Royal Dutch Academy of Sciences (Koninklijke Nederlandse Akademie van Wetenschappen [KNAW]), for example, stated that women had to make choices – that women who really wanted to be professors had to realise that they could no longer sew their own clothes.

At this conference in Brussels, the idea of a 'Dutch case' (as the worst case) was born. The dismal statistics as well as the opinions recorded in the documentary suggested that something odd was going on behind Dutch dikes. How could it be that a country known for its traditions of tolerance and liberty in matters of religion and sexuality, a country in which gay men and women can legally marry, in which a rational and non-moralistic policy on drugs is pursued, a country proud of its republican system which advocates non-hierarchical difference over sharp social inequalities remain so backward when it comes to women and equal opportunities in science?

1 The three highest ranks in Dutch universities – professor, universitair hoofdocent (UHD), and universitair docent (UD) – correspond to the American ranks of full, associate, and assistant professor. In the Netherlands, however, these ranks cannot be attained by good performance' they refer to fixed positions and not to a tenure track. Formerly they were called professor, lector, and universitair (hoofd) medewerker.

2 The Netherlands is doing worse than Germany, a country which until far into the 1980s could be compared with the Netherlands (see Hawkins and Schultz, 1990).

3 The documentary *Verlies van talent* (Waste of Talent) was made by Marjan Tjaden for the TV series *Noorderlicht* (Northern Light) of the Broadcasting Association VPRO. It was shown on 6 February 1998.

This chapter will explore the possibility of 'a Dutch case' regarding the under-representation of women in science. Is Dutch science or are Dutch universities lagging behind other countries in respect to equal opportunities for women? The very first answer is both yes and no. On one hand, there is no unique Dutch case. For example, the impression – based on comparative figures – that in Finland and Great Britain women fare better in science is misleading. Because of the restructured ranking systems in Finland and the fusion of polytechnics with universities in Great Britain, the reported figures for these countries appear better, but in reality, they report different categories than in the Netherlands. On the other hand, there are factors at play within the Netherlands, as indeed there are in every country, which function to create a unique nation-specific scenario. What are these factors? And, can these factors be attributed to long-term historical factors such as the influence of religion, or conceptions of labour or the family? Or is the current situation simply a product of more recent developments such as university reform, or ineffective equal opportunities policies? To what degree is the situation part of a 'general Dutch case' (reflecting Dutch women's comparatively unstable social-economic position) and to what degree is it a case born out of Dutch academic traditions or the organisation of the sciences particular to this country? In this chapter, 'science' refers, in the continental sense, to science as the whole spectrum of sciences and the humanities (*wetenschap, Wissenschaft*), and women's participation in it primarily in the context of academia.

A Historical Perspective on 'the Dutch case'

Getting In: Female Students in the Netherlands

Aletta Jacobs (1854–1929): from first female academic to feminist politician
In the nineteenth century, when women everywhere in the Western world were knocking on university doors, in the north of the Netherlands, the Jewess Aletta Jacobs became the first woman to enroll as a medical student. On 20 April 1871, Jacob's name, in her own handwriting, appeared in *The Album Studiosorum* of the State University of Groningen (Jacobs, [1924] 1996; see also Bosch, M. and A. Kloosterman, 1990; Wilde, 1979, 1998). The story of her entrance is simple. While demonstrating to a family friend her proficiency in Latin, a skill she had learned from her father, Aletta revealed her unhappiness at the lack of educational opportunities available to girls in

her small provincial town. The friend, L. Ali Cohen, was a medical inspector and leading figure in the profession, and he suggested that she take the assistant apothecary's examination, an option that had just been opened to women in the previous year. After she passed the exam one year later, another friend of her father, a professor of medicine and the Rector of the State University of Groningen, S.S. Rosenstein, pointed out that some students had received permission to enter the university on the basis of the assistant apothecary's exam. Aletta Jacobs wrote to the prime minister, J. R. Thorbecke (an important liberal reformer), requesting exemption from the university entrance requirement and permission to attend university classes. Thorbecke granted this permission in a letter to her father, stating that if Mr Jacobs had no hesitations, Thorbecke would raise no objections either. However, Aletta would be required to renew her exemption request when she was ready to take her first official exam. A year later, when a rumor circulated that Thorbecke was seriously ill, Jacobs's father wrote to him, indicating the courses Aletta had already successfully completed and requested that she be granted the necessary dispensation to continue. A few days later, a black-edged letter containing the final dispensation arrived. It was one of Thorbecke's last official acts before he died. From then on, Aletta Jacobs studied in Groningen and Amsterdam and, after having defended her thesis 'On the Localization of Physiological and Pathological Symptoms in the Cerebrum', she received her degree on 8 March 1879.

Aletta Jacobs did not intend to pursue a laboratory or scientific career. She wanted to make use of her education to improve the position of women in the Netherlands. After receiving her medical degree, she went to London, where she met Elizabeth Garrett Anderson, the second woman to be officially registered as a doctor in the UK, and her sister Millicent Garrett Fawcett, the long-time leader of the National Union of Women's Suffrage Societies. Through introductions by the radical liberal Carel Gerritsen, whom she later married, Jacobs also became acquainted with the free-thought campaigners and Malthusians, such as Charles Bradlaugh, Charles and George Drysdale, and Annie Besant, who taught her various methods of birth control. In the 1880s, she corresponded with a German doctor, Mensinga, about a pessary that she began to prescribe as a contraceptive. It was originally known as 'Mensinga-pessary', but later became famous – especially in the United States – as the 'Dutch pessary', substantially because of Jacobs's efforts. She also had an interest in women's political rights, which is evident in her attempt to register to vote in 1883. Her request was turned down by the city council, a decision she fought, unsuccessfully, all the way to the High Court. In 1894, she helped find and direct the Dutch Woman Suffrage Association from 1903 until Dutch women received the right to vote in 1919.

Although Jacobs turned out to be more of a politician than a medical woman or feminist theoretician, her accomplishments served to secure her a prominent role in the Dutch Women's Suffrage Association. Certainly, her reputation as the first university-educated woman in the Netherlands was important. She also gave the women's movement in the Netherlands a broader philosophical base by translating Charlotte Perkins Gilman's *Women and Economics* (in 1900) and Olive Schreiner's *Woman and Labour* (in 1911) into Dutch.

Aletta Jacobs' admittance to university – certainly when compared to the experiences of the German and British women – seems to show Dutch tolerance (Boehm, 1958; Brittain, 1960; Burstyn, 1980; Soden, 1979). However, although all Dutch universities (three at the time) were state institutions and subject to the same regulations, Jacobs' admission proved to be the exception rather than the rule (Baranelli, 1990; Bertels *et al.*, 1986; Lucassen and Peeters, 1988). One year after Jacobs was admitted, two Russian women, Olga von Stoff and Fanny Berlinerblau (later Fanny Berlin), were denied admission to the medical faculty in Leiden university, despite the university's long tradition of admitting foreigners, particularly those facing discrimination elsewhere (Bosch, 1994: 93–94). Both women had started their medical education three years earlier in Zürich, but had to leave because of the Russian *ukase*, which forbade female students from continuing their studies in that 'revolutionary' city (Bonner, 1992). Although the Dutch universities did not explicitly refuse to admit women, it was a decade after Jacobs before another woman applied to a medical school. This was Catharine van Tussenbroek, 1852–1925 (Bosch, 1994: ch. 4; Schoon, 1989). Until 1898, only a handful of women became regular students (Jansing and Dasberg, 1978; Jensma and De Vries, 1997).

The year 1898 marked a turning point. In that year, Dutch women organised a large National Exhibition of Women's Labor, symbolically using the occasion of young Queen Wilhelmina's ascension to the throne to present their case (Grever and Waaldijk, 1998; Posthumus-van der Goot and De Waal, [1948] 1977). Although women's participation in science or

Table 6.1 *Female Students per Faculty per year (totals and percentages)*

	1898			1928			1948		
	Total	Female	%	Total	Female	%	Total	Female	%
Theology	393	1		678	42	6,2	779	71	9.1
Law	476	*5*		1,590	262	16,5	1,610	430	26.7
Law and Arts				337	15	4,5	375	49	13.1
Medicine	1,126	*29*	2.3	3,125	374	12	3,470	634	18.3
Math and Natural Science	402	48	11.9	1,862	508	27,3	1,437	418	29.1
Geography and Psychology				57	15	26,3	112	52	46.4
Philosophy and Literature	178	18	10.8	1,224	561	45,8	865	380	43.9
Economics				159	22	13,8	190	18	9.5
Vetinary Science				158	1	0,5	135	7	5.2
Technology and Agriculture					113 (*)		2,778	119	4.3
Total	2,716	1081	3.7	9,561	1,810	18.9	11,251	2,044	18.2

Source(*): Freie, 1948; Jensma and DeVries, 1997: 193, 204, 210

academic education was not a central issue at the exhibition, it did spark a widely publicised debate concerning women's higher education. A young gynecologist and professor at Amsterdam University, Hector Treub, who was sympathetic towards the cause of education for women, responded to the question 'Is woman fit for science, and science fit for woman?' with a resounding 'yes' (Bosch, 1994; Kloosterman, 1992; Treub and Winkler, 1898). His statement provoked several counter-attacks, reminiscent of earlier works by Harvard professor Edward Clarke and London University College professor Henry Maudsly (Burstyn, 1980; Clarke, 1873; Maudsley, 1874). One Dutch psychiatrist reduced the question of sexual equality to a simple social (or rather sexual) Darwinist scheme, and declared that in his opinion female scientists were 'monsters' (Cox, 1898). One of Treub's colleagues at Amsterdam, Cornelis Winkler, argued more politely that female academics, by ignoring that 'biology is destiny', threatened social fecundity through education-induced sterility (Treub and Winkler, 1898). One of the few feminist voices in this debate was that of Catharine van Tussenbroek, who became a leading gynecologist, combining clinical practice with research. She electrified the National Exhibition of Women's Labor with her opening speech on the lack of vital energy in young women caused by society's expectations that girls should wait for a husband instead of prepare for meaningful work (Tussenbroek, 1898a, 1898b; also Bosch, 1994; Grever and Waaldijk, 1998). After 1898, the number of Dutch female university students rapidly increased (see Table 6.1).

Never Just about Women: Discourses of Gender and Science, 1898–1948

The 1898 debate concerning women's abilities in the field of science has often been interpreted as the last serious effort to bar women from higher education. However, women were to stay in academic institutions, and 1898 marked the beginning of a new discourse about 'women's particular place' in science. It took place almost daily in diverse venues – speeches opening the academic year, addresses opening conferences, necrologies, vocational guidebooks, government documents setting policy for higher education, and in histories of disciplines and laboratories. Although the specific arguments changed over time and shifted according to specific contexts, the function of the debate remained the same, namely to define the place of women in science.[4] The result of this 'restrictive logic', to use Margaret Rossiter's term, has been women's limited access to science (Rossiter, 1984: xvii).

What complicates the picture in these debates is that they are never just about 'women' or explicitly about the social relations between the sexes in science. Ideas about women and science (and implicitly about men and science) can also contribute to creating hierarchies between sciences and within disciplines, tasks, and functions in terms of gender.[5] Directly or indirectly such discourses produce and reinforce horizontal and

[4] In the introduction to her work on women scientists in America, Rossiter uses the phrase 'women's historically subordinated "place" in science' (Rossiter, 1984: xv).

[5] A widely accepted definition of gender – going back to Joan Scott – is that languages of gender not onlyserve to give meaning to the relations between the sexes, but also serve to give meaning to power and construct power and power differences (Scott, 1988).

vertical segregation, which is often explained only in terms of 'women's preference'.[6]

Thus, in 1898, when women's access to higher education was at stake, the concepts of higher education, science, and what it meant to be a scientist were being redefined. The nineteenth century 'man in the lab', as Londa Schiebinger aptly labelled the heroic figure of the scientist who pursues his goal day and night, undistracted by private or bodily cares, was not as yet well established (Schiebinger, 1989).[7] At this time, this image was still overshadowed by the older image of the learned man whose training departed from the classics and whose authority depended on elite status rather than scientific merit. Eventually, the newer image would triumph and signify an ideal that few women could achieve. The image of the scientist as a heroic 'man in the lab' which was upheld in the debates on 'women and science' therefore had the double effect of redefining the meaning of 'science' to the narrower domain of 'natural science' and of defining 'natural science' as a 'manly profession'. After 1898, women were never again completely excluded from the broader realm of science. However, by reserving some disciplines (the humanities, languages, and pharmacy)[8] for women and in the process feminising these fields, the 'real' natural sciences became masculinised.

The reason Winkler intervened in the debate about women and science is only clear when seen in the context of his efforts to create a new division within the medical profession between general practitioner and scientific specialist. In the debate about women and science, he aptly used the gendered associations of theory (masculine) and practice (feminine), thereby excluding women from scientific careers, but

[6] The unequal distribution of women over the disciplines (or the question of how disciplines produce gender and vice versa) has until now hardly received any systematic attention. There is sensitivity to the differences between disciplines regarding gender, but not to how these differences come about, or what explains them. Exceptions are Honegger's study of the origin of the social sciences (Honegger, 1992) and Rossiter's study of the many differences within the natural sciences (Rossiter, 1997).

[7] How influential this figure still is, was more recently pointed out by Naomi Oreskes (Oreskes, 1996).

[8] Martha Kirejzyk has rightly pointed out that at the turn of the century the faculty of natural science attracted the largest number of women, due to the study of pharmacy. Pharmacy, however, had a much lower status than physics or mathematics, and led to an already well-known women's career outside the university, almost never to a career in science. Only 22 out of 596 women doctors earned a PhD (Kirejzcyk, 1993).

including them in nursing or general practice. This not only limited women's access to the medical profession, but also strengthened his case for these basic divisions within the profession (Bosch, 1994; Winkler, [1901] 1918). Similarly, the harsh criticism directed at 'girl-students' by the history professor at Leiden, P J. Blok in 1907 can be explained in the context of his efforts to professionalise historical science (Blok, 1907). In his opinion, only men were capable of pursuing 'the heroic and solitary quest for truth' because 'the female brain lacks impartiality. By nature it is superficial and becomes focused unnecessarily upon one single fact instead of the cohesion of facts' (cited in Grever, 1997; see also Bosch 1994; for an elaborate study of gender and the historical profession, see Grever, 1994). Like Winkler, Blok advocated that women be included in the infrastructure of the historical profession as 'assistant archivist' or 'assistant librarian', fields that were already feminised. That again had the effect of making the core business of historical research a decidedly male pursuit.

In the 1920s and 1930s, women's place in higher education was contested not so much in terms of women's capacities or brains, or specific fields and functions, but in terms of the general social and economic value of higher education. In a lecture on the relation between university and society, the chemist H.R. Kruyt argued that the Dutch universities produced 'learned men' while the Dutch polytechnics had been downgraded to simple professional schools (Bosch, 1994; Kruyt, 1931). The Dutch, he recommended, should look to the United States, where universities contributed directly to society.[9] He recommended further that natural scientists, industrialists, and other professionals should be appointed to university boards with the aim of demoting the old elite with their heavy commitment to the humanities and law, and promoting instead the new elite of natural scientists.

This plea for a university system that stressed the social value of the natural sciences over the humanities (a plea later repeated in C.P. Snow's *Two Cultures*) did not bode well for women. Not only were women drastically

[9] Other academic leaders at that time (among them the historian Huizinga) noted that there was not a marked division between general education, professional education, and graduate education in the American college and university system and the choice was less determined by social class. The mutual influence of science and society was always seen in the connection between alumnae and university (Berkel, 1990).

Table 6.2 *University Students in the Netherlands, 1898–1970*

	Total	Female	% Female
1898	2,716	101	3.7
1928	9,561	1,810	18.9
1940	11,251	2,044	18.2
1960	40,727	7,305	17.9
1970	102,219	29,99	19.8

Source: Bosch, 1999; Freie, 1948

under-represented in the natural and technical sciences in this period, they were under-represented precisely because of the widespread notion that a woman's education should be directed towards general knowledge to make her a better wife and mother, and not towards a profession or a scientific career. Although the Dutch never implemented explicit quotas for women, as was done in Oxford in 1927 and in Nazi Germany in 1933, the constant worry about overcrowded laboratories or female's failure to contribute to the labour market had the effect of limiting females to 20 per cent of the student body well into the 1960s (Brittain, 1960; Weyrather, 1981; for Dutch discourses on overcrowding, see Bosch, 1994: ch. 9; *De toekomst*, 1936; and Idenburg, 1934) (see Table 6.2).

The debate over women's place in science took another turn immediately after World War II, when a commission was set up in 1947 to reorganise Dutch higher education (Bosch, 1994: ch. 10; *Rapport van de staatscommissie*, 1949; Rupp, 1997). In response to the war, debates about the purposes of university education changed yet again. Instead of preparing students for a professional career, the university was now to be organised in order to provide a general humanist education, focusing on moral and ethical responsibility of autonomous persons in times of war. Theoretically, such an approach should have provided women with more opportunities. The few women invited to participate in this state commission, however, were placed on one subcommittee whose task was to consider 'women and higher education'. They were pressured to develop a special 'women's curriculum' focusing on psychology and pedagogy. Despite the popular arguments that women had a special contribution to make through their feminine difference, the women's committee of 1947 was able to prevent the formalisation of a special women's curriculum that would have marginalised academic women even after.

Dutch Women and the 'Second Golden Age': Horizontal Segregation

The gendered division between science and humanities occurred in Netherlands in the second half of the nineteenth century. Although many late twentieth century authors suggest that the relationship between women and natural sciences had been adversarial from their beginnings in the scientific revolution, one may wonder why it seemed so necessary in the nineteenth century debate about women and science to stress the masculinity of the natural scientific enterprise. Could it be that doubts existed regarding the 'manliness' of the scientific profession, and that this uncertainty spilled into the debates?

In her book, *The Scientific Lady*, Patricia Phillips claims that the relationship between women and the sciences flourished in seventeenth and eighteenth century Great Britain, and did not become strained until in the second half of the nineteenth century when an important secondary school reform took place (Phillips, 1990). The same is true for the Netherlands.[10] Although the Netherlands did not produce best-selling female science writers, such as Priscilla Wakefield or Jane Marcet, or female scientists of the stature of Mary Sommerville, the Dutch women did foster the 'new experimental philosophy' as useful and claimed that it would lead to a knowledge of divine creation (Bosch, 1998). The German-born Maria Sybilla Merian, renowned for mapping Surinam insects, remained exceptional (Schiebinger, 1989). More commonly in eighteenth century Holland, authors of popular science were men of God, as was the case of the vicar who penned the *Catechismus der Natuur* (Catechism of Nature) and taught his lessons to a ladies' group. At the end of the eighteenth century, the only Ladies' Physical Society in the world was founded in Middelburg, which should be interpreted as an early expression of middle-class gender segregation, rather than as a sign of feminist activism (Sturkenboom, 2001). As a matter of fact, girls' education in the nineteenth century commonly included science education, even though it was presented in a 'domestic' format, such as a dialogue among friends or as a board game (Bosch, 1998).

Any positive relationship between women and science cultivated in the Netherlands in an earlier age, however, had soured by the last quarter of the nineteenth century as a consequence of educational reform and the

[10] Phillips' and my own research throw into doubt presentist judgements that women's curricula in the beginning of the nineteenth century were exceptional or even 'prefeminist'.

rapid expansion of Dutch science. The period after 1863 is known as the 'Second Golden Age' because six Dutch scientists won Nobel Prizes by the first decade of the twentieth century (Berkel, 1985, 1998; Willink, 1980). The 'Second Golden Age' dawned with the reform acts of 1863 and 1876 that created the modern two-tiered educational system. The most important creation of the 1863 Secondary Education Act was the Higher Burgher School (HBS) and its goal to prepare middle-class boys for careers in business or the technical trades (Idenburg, 1960).[11] When it was first established, the HBS was not intended to lead to a university education, but, could lead to education at the Polytechnical School at Delft (today Technical University) where engineers were trained. The HBS curriculum was set up as an alternative to the gymnasium with its traditional classical education, and it emphasised the natural sciences, modern languages, and accounting. Many schools were equipped with excellent laboratories for physics and chemistry, good collections for geography and biology, and some even had experimental gardens. Teachers specialised in one or more fields and were paid well. However, eventually the HBS did begin to supply students to universities, especially in the fields of medicine and the natural sciences, granting them special privileges to gain entrance and promotion. In 1917, a special act formally granted HBS-educated students admission to a PhD programme, though only in the faculties of medicine and natural sciences (Groen, 1987).[12]

Social class and for that matter, gender difference – not individual merit or learning–still lay at the heart of the Dutch educational system. The HBS schools were for boys of the middle classes while boys from the academic and upper classes went to a gymnasium and subsequently to university. Nonetheless the university system underwent its own 'scientific revolution' in step with the HBS system. The Higher Education Act of 1876 changed the university's goal. Instead of providing classical training as an introduction in

[11] Recently the centrality of educational reforms in the historiography of the 'Second Golden Age' has been criticised as being 'too narrow and even erroneous' by Maas, who notes the importance of the parallel development of individual entrepreneurship and innovation in Dutch culture (Maas, 2001).

[12] In the nineteenth century, degrees were not considered so important, but the PhD or doctorate degree, which a *doctorandus* could earn by writing a doctoral thesis or dissertation was the final degree. In 1921 the candidate's degree, as well as the doctoral (not doctorate) degree got *effectus civilus*, and from that moment the doctoral degree (Drs), especially, can be seen as a final degree. The doctorate by then became a way to qualify for professional science.

the 'learned classes', the university's goal was to teach young men to become independent researchers and prepare them for careers in science and society. Even though the gymnasium, with its classical curriculum, remained the formal path to university admission, Latin was abolished as the language of scholarly communication. The 1876 Act encouraged specialisation and new disciplines in the faculties and sub-faculties and created new professorships in those disciplines. In addition, a considerable sum of money was invested in institutional infrastructure to build labs and purchase specialised equipment. These reforms, combined with the opportunities offered by the Netherlands' far-flung empire, led to a real boom in the natural sciences (Bosch, 2000; Bossenbroek, 1996).

But just as the late historian Joan Kelley asked, 'Did women have a Renaissance?' we can pose the question, 'Did Dutch women have a "Second Golden Age"?' (Kelly, 1984). Did they partake in the triumphs and the rewards? The answer is no. The 1863 law, reforming secondary education did not explicitly exclude girls from the HBS, yet it implicitly did so by stating in one by-law that the education of girls was the responsibility of the citizens themselves. This stimulated a lively debate about secondary education for girls (Essen, 1990; Pouwelse, 1993). What would be its purpose? Should it prepare women for marriage or work, and if for work, what kind of work? How should it be done, in co-educational or separate schools? And what should be on the programme? In the course of these debates, the HBS curriculum became the point of comparison with the ideal curriculum for girls, and as a result, came to be seen as the ideal masculine education. In the 1860s, the first secondary schools for girls that strongly emphasised the humanities were founded. Unlike the (male) HBS, these schools would not lead to university. Despite these gender inequalities in the newly developed realm of secondary education, a few girls did manage to enrol in the HBS as early as 1872 (among them one of Aletta Jacobs's sisters). Until 1906, however, women had to obtain permission from the central government to enrol as HBS students.

What emerged in secondary and higher education, then, were two gendered tracks within a system of co-education. At the secondary school level, privately funded girls schools did not prepare them for higher education. The natural sciences, due in large part to their connection to the HBS-curriculum's focus on practical issues and careers in industry or economics, became masculinised, while the humanities and a general education (and

Table 6.3 *Student M/F of Seitral Tipes of Secondai)' Education, 1898–1947*

	1898			1911			1930			1947		
	M	F	%	M	F	%	M	F	%	M	F	%
Gymnasium	2,265	155	7				7,833	2,407	24	11,533	3,557	24
*1900		230										
HBS	7,612	534	7	10,860	2,451	18.4	23,605	5,510	20	36,888	8,377	19
*1900		855										
Lyceum							8,487	3,525	29	31,683	13,348	30
Secondary School Girls		1,706						1,924			4,617	

Source: Freie, 1948; (*1900): 'De vrouw 1813–1913,' Afdeeling Onderwijs. Statistische en andere mededelingen.

even the formerly male-identified classical education) became feminised.[13] This can be seen in the increasing enrollment of women in the gymnasium (Table 6.3). It also explains why the social reformer, Helene Mercier, in her plea of 1876 for more women to enroll in the university, argued specifically for women's active participation in the moral teachings of 'the Old' (Socrates, Plato, and Aspasia). Her message emphasised that as keepers of classical learning, women could serve as a counterbalance to the growing dominance of 'materialistic sciences'. Mercier spoke for women and also against the growing prestige of the natural sciences and against attempts to remake medicine from an art into a science (Bosch, 1994, 1997; Wilde, 1985). Mercier was convinced that women should be able to make important contributions to 'science', but her definition of learning was based on the classics, which was fast becoming outmoded. The new masculinised 'science' was natural science, and women were not to share in it and the 'Second Golden Age' of Dutch culture that it came to represent.

Remaking and redefining science in the Netherlands in the last decades of the nineteenth century was supported by emerging notions of gender and questions about what type of learning was appropriate for each sex. Here, the basis was laid

[13] The growing attractiveness of Latin and Greek for women is not specifically Dutch. Solomon has pointed out the increasing importance of the liberal arts curriculum at women's colleges: 'In 1910 no one remembered that educators in the previous century thought women were incapable of learning Greek and Latin' (Solomon, 1985: 81).

for the horizontal gender segregation between the sciences and the humanities that still remained a major characteristic of academic life until today.

The Exception that Proves the Rule: The Relative Success of Dutch Female Biologists in Dutch Colonial Perspective

One exception to the strong masculinisation of the natural sciences in the late nineteenth century Holland was the field of biology. Like today, this was a field where women were successful and visible. Johanna Westerdijk, who became the first woman professor in the Netherlands in 1917, was a biologist. Likewise, the second female professor appointed two years later, Tine Tammes (Stamhuis, 1995; Wilde, 1989) was a biologist. Because of institutional restrictions, both women earned their doctorates by circuitous routes. Westerdijk, after completing her exams to become a biology teacher in Holland, went to Zurich to take a doctorate in 1905. Tammes received an honorary PhD from the University of Groningen in 1910. Two other female biologists who also received honorary doctorates were: Anna Weber-van Bosse from the University of Amsterdam in 1910 and Gerda Wilbrink from the University of Utrecht in 1922. Nearly half of all women who completed a university education in biology in this period earned a doctorate. The representation of women in the field is even higher when we take into account all the female biologists who trained for a teacher's exam and did not get an academic degree.

Johanna Westerdijk (1882–1963): Challenges and opportunities in colonial botany for Dutch women

One explanation for the success of female biologists can be found in the formidable person Johanna Westerdijk, who mocked the codes of high masculinity and femininity. Cigar smoking and (occasionally) trouser clad, she attracted considerable attention during a professional tour of agricultural research stations in the United States. Newspaper headlines read: 'Dutch Woman Scientist Pities American Sisters'; 'Westerdijk Says 'They Are Bound by Chains of Puritanism – They Lack Freedom Europeans Possess' '; 'Botanist Objects to Ban on Fair Sex Smoking'. Westerdijk criticised 'smokers' that excluded women, as 'ridiculous'; she was also critical of her male colleagues at a formal dinner who invited one 'white-haired' female botanist so that (according to Westerdijk) their reputations were saved and the other female biologists in town had not to be invited.[14]

Westerdijk grew up in a culturally privileged Amsterdam family. Her father was a doctor; her French mother a gifted (amateur) musician. Johanna turned to botany, inspired by the 'biological revival' that sent the Dutch people from all walks of life on botanical excursions throughout all parts of Netherlands. She enrolled at the University of Amsterdam, where the famous Hugo de Vries reigned, even though she could prepare only for a teacher's exam. In pursuing her PhD in Zürich, she was following in the footsteps of another Amsterdam woman, Mia Boissevain, the first Dutch woman to earn a doctorate in biology.

While still in Zurich, Westerdijk was invited to direct a small private Dutch laboratory for the study of plant diseases, the Willie Commelin Scholten Laboratory of Phytopathology (WCS-laboratory), founded in 1895.[15] The laboratory had been left financially devastated by the former director who took the lab's resources (a state-subsidised contract for crop protection and the journal for phytopathology) with him to a newly founded institute at the School of Agriculture at Wageningen (Löhnis, 1963; Kerling, 1969; Kerling et al., 1986). Luckily for the WCS-laboratory, the influential professor of botany at Utrecht University, F.A.F.C. Went, still believed there was room for additional scientific research in this field, especially dealing with colonial agriculture (Schoor, 1994). His staunch support for Westerdijk included asking her to curate the fungi collection of the *Association Internationale des Botanistes* (International Association of Botanists). This collection grew to form the foundation for a new institute, the *Centraal Bureau voor Schimmelcultures* (Central Bureau for Fungi-cultures), which attracted much

[14] Westerdijk's criticism reinforces Margaret Rossiter's analysis of the function of 'smokers' to exclude women from the networks of scientists:

> 'Smoking and organized 'smokers,' apparently an important and often deliberately intimidating part of the male professional culture at the turn of the century, greatly complicated the women's comprehension of whether or not they were actually welcome. Properly bred women apparently did not smoke or enter rooms where men were smoking before the 1920s.... Thus to include smoking on the program or to allow men to smoke after dinner was, whether men realised it or not, long an overt social message to the women not to attend'. (Rossiter, 1984: 92)

[15] The laboratory was founded by Mr and Mrs Scholten-Commelin in 1895 (the founding contract was signed in December 1894), in memory of their son, a biology student Willie Scholten. To his and his father's last name, the last name of his mother, Commelin, was added to the name of the laboratory. Several members of the famous Amsterdam Commelin family played major roles in the field of (colonial) botany. One of them was founder of the Amsterdam Botanical Garden.

interest from scientists and industrial researchers. The institute was privately supported for many years by a former tea planter.

The WCS-laboratory developed into an internationally recognised research institution. Between 1913 and 1915, Westerdijk visited the Dutch East Indies to study colonial phytopathology at various research stations. In 1919, she published a phytopathological system of classification in cooperation with the German biologist Otto Appel. This classification system paved the way for studying the anatomy of diseased plants, as well as the process of infection of healthy plants. It also focused on crop plants and ornamentals often neglected in pure botanical studies. In 1920, the laboratory moved from a small house in Amsterdam to a large and spacious mansion called Java in the 'green village' of Baarn, formerly owned by Dutch plantation owners. The gardens bordered the beautiful 'Cantonspark', which was part of the University of Utrecht's botanical gardens. In 1930, the coach house 'Madura' was converted into a laboratory annex and party hall using private funds donated by one of the lab's first PhD students and assistants, Marie Löhnis. Above the entrance was inscribed: *Werken en feesten vormt schone geesten* (Work and celebration make for creative spirits). Over the years, Westerdijk's lab hosted several international conferences and received visits from many foreign scientists.

In the meantime, on the initiative of Went, Westerdijk had been appointed *buitengewoon* professor (extraordinary professor) at the University of Utrecht in 1917 and at the University of Amsterdam in 1930.[16] In this capacity, she attracted a large number (56) of PhD students, almost half (23) of whom were women. Westerdijk was a strong supporter of female students and, through her hiring practices, turned her lab into one where women's representation was exceptionally high. She also used her influence as a professor to influence academic hiring – two of her South African female students became professors in their own country. Westerdijk was also active in the Dutch Association and International Federation of the University Women, which she chaired from 1931 to 1938. In 1951, she was the second woman in the Netherlands to become a member of the Royal Dutch Academy of Sciences (Noordenbos, 2000).[17]

[16] The vital distinction between full or regular professor (*ordinarius*) and extraordinary professor (*extraordinarius*) is explained further on.

[17] The first member of the Royal Dutch Academy of Sciences was Christine MacGillavry, a crystallographer. Crystallography seems to have been a relatively feminised field in the 1950s and 1960s.

Although she would have denied being a feminist, Westerdijk left a gender rebellious mark on the biology profession. She often ridiculed women's marriage expectations and made no secret of her preference for the unmarried state. Relying on her irresistible sense of humour, she established many rituals and traditions, and staged puppet shows or theatre plays that poked fun at what Westerdijk saw as the pompous behaviour of many academics. Cross-dressing was often part of these productions, with men acting as women and vice versa. She mocked the invasive way scientists investigate nature by personifying cabbages and potatoes, fungi and trees staging them in roles which turned upside down the usual division between subjects and objects of scientific research.[18]

The relative success of the Dutch female biologists cannot, of course, be fully explained by Westerdijk's biography. Other factors were the traditional involvement of women in botany and biology, the *scientia amabilis*, which Ann Shteir has discussed (Shteir, 1996), and, more importantly, the competition which raged between academic biologists and vocational biologists for the colonial research market at the turn of the century (Bosch, 2000). In this competition Went made use of old traditions and associations with learning to advance a scientific ideal of 'pure science' and 'fundamental research' done at universities as opposed to 'applied science' done at the vocational School of Agriculture at Wageningen (Schoor, 1994). In the process, Went also mobilised the women botanists in upper-class family networks who felt encouraged to enter a career in biology.

Women Scientists and the Pillarisation of Academia: Vertical Segregation

In 1948, on the occasion of Queen Wilhelmina's fiftieth Jubilee, Dutch women organised a large exhibition to display women's accomplishments. In conjunction with the exhibition, the Dutch Association of University Women prepared a quantitative study of women's representation in all scientific disciplines and in the labour market (Kolf, 1950). Marie van der Kolf's statistical analysis focused on women who had completed an academic education. By consequently dividing the women into categories of 'married', 'unmarried', and 'having been married', she could show that

[18] Several historians of science have analysed the scientific image of nature as female, the passive gender to be dominated, explored, ravished, unveiled, searched, and so on. Westerdijk in her fantasies gave an active role to 'nature' vis-à-vis the exploring biologists.

women were 'as good as men', and that it was only due to marriage that women did not work outside the home as much as men. The majority of unmarried women certainly did have a career. One chapter summarised women's scientific rewards in terms of academic positions and honorary doctorates, showing that there were eleven female professors in the Netherlands (plus two who had taken their degrees in Holland but now taught at foreign universities), fourteen lecturers (plus six teaching at foreign universities) and 84 other academic personnel (private lecturers, assistants, laboratory assistants, and so on). Overall, the tone of the book was optimistic. Even though barriers remained (especially outside academia in the realm of marriage and marriage law), unmarried women had proven their worth.[19] This was all certainly meant to be good news; she could not avoid the implicit assessment, however, that, for women, science and marriage (and motherhood) were incompatible.

The central metaphor in the book invoked 'frontier' language. Van der Kolf fancied women scientists as 'pioneers' who had entered 'virgin' territory, and had 'ploughed', 'sowed', and 'harvested' in the fields of science. Van der Kolf believed that women must be judged on their own merits, which explains why academic men were absent from the book (and the metaphors). Had she considered men, perhaps academic women would more properly have been figured 'immigrants' rather than 'pioneers' in science (see Table 6.4). Van der Kolf 's optimism was based on a definition of science as an autonomous process separate from the society. This definition became widely contested in later decades by historians and philosophers of science as well as by scientists themselves. The church historian, and sometime Rector of Utrecht University, Otto J. de Jong contributed to this debate by pointing out that between 1876 and 1931 the Ministers of Education, Culture and Science fundamentally 'pillarised' the four state universities, by appointing Board members as well as professors according to a system of 'equal representation' (Jong, 1982).[20]

'Pillarisation', or as it is also called, the 'politics of accommodation' or 'consociational democracy', is a central concept in Dutch history, referring to a specific Dutch variant of social segmentation in the age of state formation and the creation of a unified nation. The four (recognised) pillars (the orthodox protestants, Catholics, socialists, and a 'neutral' liberal group) represented

[19] Marriage laws in the Netherlands until 1956 did not grant married women legal status as persons in their own right.

[20] In his speech at the 346th *dies natalis* (anniversary) of Utrecht University in 1982.

Table 6.4 *Female Academic Scientists in the Netherlands 1948 (Universities and higher vocational-economical, technical, agricultural-education)*

	Professor	Lector Associate	Other	Total University	Drs.	Ph.D.	Dr.%
Theology					98	7	7.1
Law	1		2	3	1,381	160	11.5
Medicine	1	3	12	16	1,354	108	7.9
Mathematics and Natural Sience	3	(5+1*)	41	49	1,106	204	18.4
Biology*	1	2	22	25	239	100	41
Pharmacy*	1	2	4	7	596	22	3.6
UF2: Arts and Natural Science	2	1	3	6	134	22	16.4
Literature and Philosophy	4	(4+5*)	24	32	1,117	243	21.8
Economics					45	1	2
Veterinary Science					8	1	12.5
Technology and Agriculture		1	2		247	12	4.8
Total	11(+ 2**)	14(+6**)	84	109	4,109	598	14.5

* Only biology and pharmacy in this table are shown specifically, they are included in the overall figures for mathematical and nature science.
** Graduated in the Netherlands, appointed at foreign universities.
Source: Kolf 1950.

important segments of the Dutch population, which at the top were organised as a political party. State subsidies were divided according to pillars, and pillars completely dominated social and private life until far into the 1950s (Blom and Talsma, 2000; Rooy, 2001).[21] At the heart of the pillarisation process was the political struggle about free (special or religious) and state-subsidised education; at the periphery were all social and cultural activities organised along these divided interests. The famous example mentioned here is always the art of goat breeding which had its several organisations: protestant, catholic, and socialist. Later, the division of the broadcasting

[21] Blom and Talsma (2000) give an overview of the most recent literature on the subject, where gender as an analytical category has not yet been given serious attention.

system became symbolic for twentieth-century segmentation (recently also called ethnicitisation) of Dutch society. 'Mixed marriages' were a concept quite different from today, until far into the sixties.

Though pillarisation has been studied often locally and in relation to social class and nation formation, the above mentioned study by De Jong has looked at a special case of pillarisation, by asking the question: How did pillarisation affect the organisation of higher education and research in the Netherlands? In a way this is a revolutionary question, for it is a generally accepted opinion that although the nation was segmented the state kept its liberal (neutral) signature, and with the state also the state-run institutions such as the legal system, universities, and the state bureaucracy.[22]

According to De Jong, one prerequisite for the pillarisation of Dutch academia – the centralisation of the university system – was put in place in 1815, when the Higher Education Act was passed which ensured that the universities would not go back to the earlier regional status they had had in the republic. This meant that now three state universities were created, at Groningen, Utrecht, and Leiden, Amsterdam being added to the list in 1876, though not as a state but as a city university. Usually, the foundation of the private Free (orthodox Protestant!) University in Amsterdam in 1880, followed by the Catholic University at Nijmegen in 1925, are seen as the material outcome of pillarisation in higher education and research, the state universities representing the liberal pillar.[23] According to De Jong, however, the three state universities were run also segmented according to the main pillars in the long run. In the academy this meant that from the very first time a protestant government ruled, in the 1880s, political and religious affiliation played a role in the appointment process not just for the Board of Curators (the administrators of the university), but also for the academic personnel in so far as they were appointed by the 'crown' (state): the professors and senior lecturers. The result of this was the instalment of a system of equal representation of religious groups in the universities.

[22] De Rooij writes that the state could be seen as the 'liberal pillar' which had its 'constituency' in the public sphere: the universities, law courts, and public bureaucracy (Rooij, 2002: 145).

[23] The Free University and the Catholic University at Nijmegen were for the most part state-subsidised universities. The boards of these universities selected the professors to be appointed by the Crown. In Amsterdam, which was also largely state subsidised, professors were selected by the city council, even if the actual appointment was done by Royal Decision (the Minister of Education, Culture, and Science).

This means that social (religious) considerations and not scientific qualifications have often played a decisive role in the appointment procedures of scientists, even though these considerations were never mentioned publicly, or at least were not recorded in the dossiers that have survived. Discussion of any requirements for political, religious, and sexual conformity cannot be traced in the documents. They record only scientific qualifications, supplemented occasionally with a note on age or didactic requirements. De Jong concludes: 'Whatever has been said about political background, religious conviction, sexual preference, social behaviour, and whether these factors have been obstacles, has not been documented.' He goes on to say that the process of pillarisation happened at the expense of some social groups, such as women. In his words: '[N]o…female faculty and no persons to the left of liberal' were appointed during this period (Jong, 1982: 32).

This is quite a revolutionary conclusion to be drawn by a respectable church historian and a high representative of the academic system in the Netherlands. The myth of meritocracy in science is so strong that it often eludes criticism and prevents serious discussion of the relation between science and society, even historically. But De Jong's conclusions are generously supported by archival sources dealing with the appointment of women faculty at the Dutch universities in the same period under study. Particularly interesting in this respect are the files of 28 'crown appointments' (the female professors and lecturers [*lectoren*] appointed by the state) in the archives of the Ministry of Education, Culture, and Science. The files are all of women born before 1903 and appointed between 1883 and 1964: nine full or regular professors, seven special or irregular professors and twelve senior lecturers or assistants.

In line with De Jong's conclusion that between 1876 and 1939 women were not even considered for appointments to professorships in state universities is the fact that the first full female professor in the Netherlands, the pediatrician Cornelia de Lange, was installed in 1927, at the University of Amsterdam, which at that time was still a city university, appointments being done by the city council. Amsterdam was considered a 'red' city, and its council broke the ice by appointing both women and socialists (and even communists) to university positions already before the World War II. And not just De Lange, six of the nine full female professors in our files were at the University of Amsterdam. The files of the three other female professors who were appointed to a regular chair at one of the state universities show that this happened in all cases after 1945 and only after they had reached the age of retirement.[24] In all cases, the women had been lecturers for several decades. One of them succeeded her father-in-law, a full professor, in 1929, but with an appointment as a lecturer; she was appointed professor only in 1960 at the age of 66.[25]

The fact that Johanna Westerdijk in 1917 and a few other female professors before the World War II were appointed at state universities does not contradict De Jong's findings; rather it underlines the importance of the distinction between full and extraordinary professorships. In comparison with a regular (or full) professorship, 'extraordinary' or irregular professorships were poorly paid, part-time positions, reserved for people who earned their living elsewhere. This type of professorship was created to strengthen the ties between science and society, and to make room for areas of expertise that had not yet been a part of the university curriculum. Irregular professors had the *jus promovendum*, and were members of the faculty and senate, but they lacked the power to vote in these decision-making bodies. Most women, then, were appointed only to irregular positions at these universities even if they were excellent scholars, who in some cases had developed whole new fields of study or developed and run major institutes. The fact that they were women, however, limited their appointments to the status of '*extraordinarius*' or irregular professor. That there were excellent women scientists in the pre-war period who did not come into positions according to their merits is also shown by the many female scholars who ended long and distinguished careers still in positions of (senior) lecturers.

The conclusion to be drawn is that the archives reveal a strong pattern of vertical segregation, which kept women in subordinate positions well into the 1960s. This may in part be explained by a Dutch version of women's exclusion, namely as a by-product of pillarisation as the Dutch system of dealing with social and religious difference. Though outwardly the three state universities were secular or religiously neutral institutions this neutrality was assured only by a subtle system of equal representation of religious difference, which did not allow for the systematic inclusion of other differences, such as gender. As we have seen, the few women who held faculty positions were either

[24] Professors were the only group of professionals who were allowed to remain in their jobs until the age of seventy.

[25] These statistics are drawn from the following archive: Algemeen Rijksarchief, 2e afdeling, 's-Gravenhage: Ministerie van Onderwijs, Kunsten en Wetenschappen: Personeelsdossiers Rijkspersoneel t/m geboortejaar 1903. Also Bosch, 1994.

appointed at the City University of Amsterdam, or were irregular professors without the power to make decisions. Given the strength of pillarisation and its long endurance, this may well have contributed to the Dutch case (as the worst case).

Not just 'first women in' but 'exceptions to the rule': The case of Antonia Korvezee

All women who succeeded in getting an appointment were exceptions to De Jong's rule of 'no female faculty'. In each case, particular decisions were taken concerning a particular woman. Thus, in some cases, such as Westerdijk's, an influential man could play a supportive role, but influential men could also block a woman's career. In some cases, the state ministry played a positive role, encouraging faculty boards to appoint a woman; in other cases the ministry used its influence in a negative way, not even considering women for positions as professors. This whimsicality is painfully illustrated by the case of Antonia Korvezee (Jong, 1988).

Antonia Korvezee (1899–1978) is remembered mainly as the first woman to become professor at the Technical University at Delft. This presentation tends to hide the complex and painful process of women's integration into science behind a rosy image of progress and increasing democratisation. Those who remember her this way, give her both too much and too little credit. Too much because she was appointed only as irregular professor, and too little because to celebrate her as the first female professor at Delft downplays her scientific contributions to the field of theoretical chemistry and nuclear research.

Korvezee started her career at the Technical University in 1922 as an assistant in the department of analytical chemistry. Two years later, she switched to physical chemistry to study with F.E.C. Scheffer and took her PhD in 1930. She then went to Paris to work in Marie Curie's laboratory where she stayed for two years, partly on a French stipend, partly at her own expense. In 1938, she was appointed head assistant at Delft, where she was also admitted as an unpaid private lecturer to teach in the new field of radioactivity. In this period, her name was put forward twice for a professorship, each time unsuccessfully. In 1948, her department proposed to the Minister of Science that she be appointed senior lecturer. This appointment was supported by pointing out that her appointment as a head assistant had not 'taken stock of the special scientific merits of Miss Korvezee, nor of her longtime experience that determines the value of the tasks she performs.' She was passed over for a professorship a third time in 1953 when her teacher Scheffer was retired. As compensation, her department requested that the ministry appoint her to an 'extraordinary' professorship. She was said to be too 'theoretical' and lacking the managerial skill required for a regular full professorship. This time the minister's reaction was irritated and sharp – although sharper in the margins of the letters where he recorded his reactions than in the final draft sent to Delft. In his letter to the Delft department, the minister stated that an irregular professorship would not be proportional to her extraordinary merits. The following correspondence between University and Ministry ended, however, in a decision to have her appointed to an irregular position. It made her the first female professor in Delft, indeed. However, as a consequence of the appointment, she had to leave the lab where she worked for thirty years and move to another. Her working days rapidly drew to a close after the move. At the end of the 1950s, she was overtaken by fatigue (probably from radiation poisoning) from which she never recovered. The under-recognition of Antonia Korvezee in historical studies of Dutch science continues today (Willink, 1998); she is remembered nonetheless in an Antonia Korvezee Equal Opportunities Award that was established by the Technical University in 1989.

Are there Historical Arguments for a Dutch Case?

The question now is whether there are historical arguments for a 'Dutch case' in the sense that there was something unique in Dutch culture or science that has put in place barriers not found in other countries. Looking at the patterns of women's participation in the sciences in the world, the Dutch case does not seem particularly unique. Just as other countries have produced their own Clarkes and Maudsleys, the Netherlands spawned Cox and Winkler to preach ruin in the face of women's aspirations to higher education. Just as elsewhere, Dutch women scientists endured horizontal and vertical segregation, and just as elsewhere the two cultures split along the lines of gender. And even if there was no formal quota system for women students in the Netherlands, their 'invasion' was stopped at 20 per cent.

But if the gender gap may have been somewhat wider and the numbers of women making a career in science somewhat lower than elsewhere, a first explanation is always seen in Dutch women's unique weak position in the labour market in general as compared to her European and American sisters.[26] Sometimes Dutch neutrality during World War I is mentioned as its cause, or the Netherlands' slow industrialisation, but most often a specific Dutch brand of Protestantism is blamed, which in its strong emphasis on motherhood ideals ties women to the home more than in other countries.[27] This explanation, however, seems to arise from a dichotomy between religion and modernity, and tends to ignore the importance of medical and biological theories which were invented as an inseparable part of Enlightenment thinking, and which legitimated the separation of the public and the private spheres in the nineteenth and twentieth century. Indeed, what about the role of doctors, psychologists, and pedagogues in constructing an over-demanding motherhood, not just in the Netherlands but in the whole Western world? Moreover, compared to other European countries, the Netherlands seems not to have been exceptionally protective towards women or mothers in its labour legislation (Bock, 2002). Just as in other countries, laws were passed – sometimes earlier, sometimes later – regulating (paid) maternity leave, night work, or equal pay. And just as in other European countries, a breadwinner income came to be the rule, which was further refined in the income tax and welfare system. According to Plantenga (1993), the main reason for Dutch women's low labour participation was an economy that allowed many Dutch couples to live up to the ideals, thereby strengthening the ideology of women as not available for or capable of serious labour. We should add that this economic affluence of the Netherlands was due to its colonies, which before World War II contributed more than 13 per cent to the national income (Rooy, 2002: 212).

This is not to say that religion did not play a role, but a different one than presumed, namely as pillarisation. As we have seen, religion comes in as an important organiser not only of social institutions, but quite unexpectedly, also of state institutions, such as state universities.[28] The pillarisation of academia for a long time seems to have influenced patterns of appointment, preventing women from being included on an equal basis.

[26] Comparative historical figures of women in science in European countries and the United States, let alone countries in other parts of the world, do not exist.

[27] Plantenga checked all these explanations: slow industrialisation, poverty, or neutrality during the World War I. She concludes that it was the good economic situation, a high productivity rate, and pillarisation that account for Dutch women's low participation in labour (Plantenga, 1993).

[28] Plantenga also points to pillarisation as a factor in Dutch women's labour position, claiming that the vertical segmentation of society was strengthened by lessening the horizontal segmentation of social class. Therefore, within the pillars a consensus could more easily be forged on the gendered division of labour.

Another, rather paradoxical outcome of the pillarisation of the university system may be the absence of a fully differentiated university system which would have allowed women to have the privilege of different routes towards higher education or a career in science. Amsterdam became the only exception to the rule that women were not normally involved in the selection of university positions. Related to this may be the absence of women's colleges which emerged in many other countries. As Rossiter (1982) has shown, women's schools and women's colleges have been important paths through which women would enter higher education and science. Moreover, in co-educational schools there were several patterns of segregation, hierarchical segregation being more rigid than territorial segregation. In 'women's institution' women had at least a fair chance to become professor. In fact, only Westerdijks laboratory seems to have functioned as an almost all female institution, and as such has produced more women scientists then anywhere else in the Netherlands.

Dutch Women in Science: A Contemporary Perspective on 'the Dutch case'

Anneke Levelt Sengers: a narrow escape from the Netherlands

In 1963, the physicist Anneke Levelt Sengers left for the United States, together with her husband, Jan Sengers, a chemist. Looking back later, Levelt said she was convinced that a career in science would have been impossible had she stayed in the Netherlands because of the complete lack of understanding for working mothers (and the absence of childcare) and because of the general negative attitude towards women scientists: 'In the Netherlands I heard the doors shut softly in my face when I was married and pregnant.'[29] Her concerns seem warranted. She grew up in an unusual family – her father was a chemist and her mother a physicist – but it was really thanks to her own efforts that Anneke Levelt took up academic studies. Her mother was completely absorbed in raising her ten children, and when family finances were strained by WWII, it was self-evident that her brothers would take up academic studies and not Anneke. This challenged her to apply for state funding. Upon finishing her doctoral exam, she decided to do a PhD in the Van der Waals Laboratory under the guidance of Professor Michels, whose

[29] The information on Anneke Levelt Sengers is based on one of four interviews which Elma Drayer did for the AWT backgound study (Drayer, 1999).

wife had earned a doctorate at his lab and a cum laude promotion, which had almost been a media event in 1937 since she was a mother of four. As Levelt's position at the lab did not become stable, she left and spent a postdoctoral year in the United States. When she returned with four published articles, no one except her future husband seemed to notice, and they decided to leave the Netherlands. Shortly before the couple left, her husband was offered a job as lecturer at the University of Nijmegen – it was, in fact, Levelt's position.

At the National Institute of Standards and Technology in Gaithersburg, Maryland, however, there were no objections to her wish to combine motherhood and work, if she adhered to normal standards of productivity. Of course it was not an easy life with four children and a career, but Levelt did become a successful scientist who received high awards for her work on the critical phenomena of liquids (steam). She eventually became a member of the US National Academy of Sciences, the Academy of Engineering in the USA, and one of the very few female corresponding members of the Royal Dutch Academy of the Sciences.

The Success of Women's Studies: Another Dutch Case

Five years after Anneke Levelt left for the United States, Joke Smit, a senior lecturer of English at the University of Amsterdam published the article '*Het onbehagen bij de vrouw*' (The Discomfort of Women) (Kool-Smit, 1968). Like Betty Friedan's *Feminine Mystique*, it was a seminal article that eloquently protested against society's expectations that women become housewives, a group that Smit ironically referred to as a 'herd of Hoover cattle'. It sparked off the second wave of feminism in the Netherlands. Within a few years a mass movement came into being consisting of many different organizations and groups (Meijer, 1996; Ribberink, 1998).

As part of the new activism in 1974, women's groups at several Dutch universities assembled to discuss women's minority position in the universities, which was first and foremost attributed to 'sexist science'. Following these discussions, women's studies groups in many disciplines were formed which pressed for research and teaching from a feminist perspective. In Amsterdam, this resulted in the appointment of the first lecturer of women's studies in the political and social sciences faculty (department) in 1977, although for many years to come many women were contributing to the new learning as students or as volunteers. At the end of the 1970s and the beginning of the 1980s the first women's studies journals were founded and a national organisation was formed to provide a platform for adherents of women's studies, the Netherlands Association of Women's Studies which exists today as the Nederlands Genootschap Vrouwenstudies (NGV).[30]

From then on women's studies (also known as women's and gender studies or gender studies *tout court*) in the Netherlands have developed impressively. From its start, women's studies attracted government funding, much of it due to its claim to solve the problem of under-representation of women in academia. Saskia Grotenhuis, who played a central role in the organisation of Dutch women's studies, wrote on the occasion of the fifth anniversary celebration of the Dutch *Tijdschrift voor Vrouwenstudies* (*Journal of Women Studies*, recently renamed *Journal of Gender Studies*): 'The marginal position of women in the universities and in science, the inequalities between women and men, were an important reason for the development of women's studies in the 1970s' (Grotenhuis, 1988). Indeed, when equal opportunity policies (in the Netherlands called 'emancipation policies') became an issue in the Ministry of Education, Culture, and Science at the end of the 1970s, the first action by the minister was to support the appointment of women's studies' teachers and coordinators at almost all universities.[31] In 1985, an advisory group Voorlopige Begeleidingsgroep voor Emancipatie-Onderzoek (VBEO) was installed by the government to stimulate women's studies in the academy and scientific research, to define priorities for women's studies research and to help divide a sum of F 4,800,000 (about E 2,000,000) to promote women's studies research across the disciplines through grants.[32] At about the same time women's studies got a formal representation in the Netherlands' Organisation for Scientific Research (Nederlandse Organisatie voor Wetenschappelijk Onderzoek [NWO]) in order to promote the funding of women's studies.[33]

These measures resulted in a stream of PhD theses in women's studies at the end of the 1980s. At the same time chairs and departments in women's studies were founded, resulting in a comparatively strong institutionalised field. In 1995 there were about twenty full professors in women's studies,

[30] The most important journals were: *Socialistisch-Feministische Teksten* (1978), the *Tijdschrift voor Vrouwenstudies* (1980) (Journal of Women's Studies, today *Journal of Gender Studies*) and the *Jaarboek voor Vrouwengeschiedenis* (Yearbook of Women's History, that still exists today).

[31] In 1979, the first white paper on gender equality in education came out. The first measure on equality in higher education was to introduce women's studies at the universities to enhance women students' motivation. The minister was prepared to finance a women's studies coordinator on a 50 per cent matching basis, and one chair (for all universities) in the field of emancipation or equal opportunities oriented research (Bosch, 1999: ch. 1; Mottier, 2002).

[32] For the division of this money the *Stimuleringsgroep Vrouwenstudies- en Emancipatie Onderzoek* (STEO) was founded.

[33] *Dwarsverbandcommissie Vrouwenemancipatie-onderzoek bij* NWO, later *Werkverband Vrouwenstudies en Emancipatie Onderzoek* (WVEO).

thirteen associate professors, and 89 assistant professors though not all in a full time position (Jansen, 1996; Lasthuizen, 1995). The jewel in the crown is the Dutch Research School for Women's Studies (Nederlandse Onderzoekschool Vrouwenstudies [NOV]), a national scientific and educational network that was founded and recognised by the Royal Academy of the Sciences in 1995.[34] In this school, women's studies researchers cooperate in five interdisciplinary research programmes. Also the NOV offers a PhD programme in women's studies. Due to the strong international orientation of Dutch women's studies, the Women's Studies International Europe (WISE) was organised in 1990 and the foundation of the *European Journal of Women's Studies* set up its administrative centres at the University of Utrecht.

At the beginning of the 1990s the success of women's studies was widely advertised nationally and internationally. In an inventory of about 300 women' studies scholars the fact that 150 of them had fixed positions provoked the remark: 'For a young scientific field, developed in a period of budget cuts, not a bad performance' (Lasthuizen, 1995: 5). An interview entitled 'The Dutch Case' with Margo Brouns, who did the first large scale evaluation of women's studies, appeared in the first issue of the *European Journal of Women's Studies* (Brouns, 1988/1990; Davis and Grünell, 1994). Brouns attributed the success of women's studies to a mixture of government-supported institutionalisation and a kind of Dutch tolerance and pluralism in forming academic disciplines. Willy Jansen, who pondered the question in 1996, attributed the successful development of women's studies to a stubborn adherence to the principles and policy of integration and autonomy, or 'the double track'. She claimed that even though the pressures on integration have been and are always strong, women's studies as a field can only survive on an independent basis, even if integration is an important goal (Jansen, 1996).

Reorganisations and Cutbacks

'The success of women's studies has a shadow side as well.' Kathy Davis wrote these words focusing mainly on the internal divisions and conflicts she perceived within women's studies in the Netherlands. However, when

[34] The Secretary of the NOV is the University of Utrecht (UU); its director, Professor Rosi Braidotti, who chairs one of the largest women's studies units in the Netherlands, in the humanities faculty, UU (see www.let.uu.nl/nov/).

Table 6.5 *University Students in the Netherlands 1970–2001*

Year	Total	Female	% Female
1970	102,219	29,99	19.8
1975	120,134	45,83	25
1985	166,636	61,38	36.8
1995	176,096	80,83	45.9
2001			49*

Source: Bosch, 1999; (*) Mottier, 2002

Table 6.6 *Percentage of Women (in full time equivalent) in the Three Top University Ranks in 1970, 1980, 1988, 1990, 1995 and 2000*

Rank	1970	1980	1988	1990	1995	2000
Professor	2.7	2.2	2.1	2.6	4.2	6.3
Associate Prof.	9.4	8.9	4.6	5.7	7.1	10.7
Assistant Prof.	11.8	15	15.1	15.5	18.2	22.3

Source: Hawkins and Noordenbos, 1990; Noordenbos, 1994–95; WOPI, 1995, 2000

she added, 'Women's studies in the Netherlands may be flourishing, but the position of women in the academy couldn't be worse,' it was clear that the 'shadow side' also referred to something else (Davis, 1996). This truth gained ground in the beginning of the 1990s, posing a kind of counter narrative to the success story of women's and gender studies. Even though the number of women students had increased tremendously (though not yet in an evenly distributed way), and although women's studies had taken a firm stance throughout the institutions, the position of women academics in general had even deteriorated as compared to twenty years earlier. The news became also internationally known through several publications in the beginning of the 1990s (Hawkins and Schultz, 1990; Hicks, 1991; Lie et al., 1994; Rose, 1994: 102) (see Tables 6.5 and 6.6).

According to many, this loss in an age of feminist agitation and emancipation policies is due to reorganisation and cutbacks in Dutch higher education and science in the 1970s and 1980s (Beekes, 1991; Bosch et al., 1999; *Ministerie van Onderwijs, Cultuur en Wetenschappen* [Ministry of Education, Culture, and Science] 1979, 1981, 1982, 1985, 1986; Mottier, 2002). Even if universities had been willing to make changes with respect to women academics, so it is reasoned, they could not succeed because of the many intrusions from the central government in The Hague. From the beginning of the 1980s, there were major changes in the academic and scientific landscape. One important change was the shift in funding research due to the desire for more democratic control and better quality. Increasingly, research funding – which had earlier been provided and divided primarily by the universities – had to be earned through competitive application and peer review evaluation by the NWO.

Also a system of evaluation (or visitation) of university faculties was installed. Another important reorganisation was an effort to restrict the number of professorships and a more general restructuring of the ranks. Consequently, the academic career track was changed from an individual promotion path (or tenure track) to a system of fixed positions (formation). Individuals were not automatically promoted anymore on the basis of merit and/or seniority but had to apply each time a position became available (which was not very often).

Especially this last reorganisation and the underlying budget cuts were responsible for blocking progress with regard to women in academia. At the end of the 1980s, it became clear that although the participation of women in academics had grown slightly, the percentage of women in higher positions of associate and full professor had actually dropped. Hawkins and Noordenbos in an evaluative study in 1990 argued that the whole operation started from the gender-blind assumption that something had to be done to reduce the number of professors, who held 50 per cent of the positions. That women held only about 2 per cent of the chairs was overlooked. They also drew attention to the fact that after the operation 85.6 per cent of the women in academia held the rank of assistant professor; for men this was 53.4 per cent. Men, therefore, had a much greater chance than women to become professor or senior lecturer (Hawkins and Noordenbos, 1990).

The Growing Gap between Bureaucrats and Scientists

Cutbacks and reorganisations in the academic system helps to explain the poor showing of women in science, yet this is not the whole story. If equality for women had been an issue of importance for science administrators in the rank and file of the university, that would have influenced how reforms were carried out. In the 1980s, however, if the university management considered gender inequality at all, it was defined as a 'woman's problem', either a social problem outside the purview of the university or a biological problem. The centre of focus was women's capacity of motherhood and all that entailed: lesser ambitions and the belief that it was impossible for women to give full time to their jobs. Even if in the course of the 1980s '*emancipatie-commissies*' (equal opportunities committees) were formed at many universities – most often linked to the democratic structures of the university, and on a voluntary basis – these were supposed to deal with childcare, parental leave, or measures to empower women individually. At the end of the decade, the Ministry of Education, Culture, and Science advised universities to install a system of affirmative action and defined target figures. At the same time the first university employed equal-opportunity officers to support activity in the realm of gender-conscious human resources management.

The national policy of university reform increased the power of university administrators and decreased the traditional academic liberties enjoyed by the faculty. These developments were pushed further with the move towards further privatisation of the universities after 1988, which was concluded by the Law on Modern University Administration (MUB) in 1997. Scientists increasingly began to strengthen their claim to manage and select their own scientific personnel in order to protect themselves against too much managerial intrusion. This they did by keeping a tight rein on their professional right to define scientific quality as an internal matter or a matter of peers, and shielding recruitment and selection procedures from the hands of bureaucrats. While managers and administrators increasingly adhered to the advantages of professional human resources management to support the move towards autonomy, scientists preferred the mechanisms of the old-boys' network. Actually, as professor of history, Chris Lorenz remarked, 'the professors did not even protest the gradual invasion of the managers because they knew their recruitment and selection methods were anything but fair, but they wanted to keep things as they were' (Lorenz, 1993).[35] The development of a more socially responsible science that emerged after the War and accelerated in the 1970s resulted in conceptions of a university that depended for its accountability on a growing bureaucratic apparatus in the next two decades. In the 1980s, discourse about 'critical science' was exchanged again for discourse about the principles of academic freedom that stressed the scientists' need for autonomy in defining scientific quality.

[35] This essay was signed by thirteen other scientists who commented upon the developments in the media turning it into a kind of manifesto.

Women lost out twice in this situation. Recruitment through the old-boys' network had always favoured applicants already in the system or new applicants who looked and performed very much like the recruiters. Women scientists, in not looking or behaving like the 'old boys' found this to have a negative influence on recruiters' perception of their abilities (Vianen, 1997). Second, equal-opportunity policies and affirmative-action measures were resented as an intrusion. As a result, these policies became marginalised to personnel departments within the central bureaus of universities (Braat, 1991). Equal-opportunity policy-makers were kept at a distance from the faculties and were allowed to deal mostly with social measures, such as daycare centres or women's personal career plans. These policy-makers were not concerned, however, with the core activities of the university, that is, research and teaching, and the criteria used to assess quality. The gap between bureaucrats and scientists explains why even at the progressive University of Amsterdam, affirmative-action measures (such as the rule that in recruitment procedures all women who applied should be considered, or that a woman who was qualified for a job should be appointed even if there was a 'better man') collapsed under collective protest from professors. An in-depth analysis of five controversial recruitment procedures under this affirmative-action recruitment regime, in order of the Ministry of Social Affairs, shows how affirmative action and 'women' *tout court* came to symbolise the very antithesis of scientific quality, and how in comparing the qualifications of male and female applicants, the female applicants were generally seen as inadequate because in all cases the male's profile in the course of the procedure became the standard profile (Verhaar, 1991).[36]

Equal Opportunities, Gender Studies and Female Scientists: Divided Interests and Perspectives

Except for the above-mentioned structural developments in the 1980s and 1990s, the failure to increase the number of women scientists in academia in the 1970s and 1980s was also due to the divided interests, perspectives, and powers of the equal-opportunity policy-makers, the gender-studies scholars, and the women scientists. Indeed, the situation holding women at bay in American sciences described by Londa Schiebinger in an article in *Osiris*

36 The University of Amsterdam (UAV) did go on to develop effective emancipation policies. According to Portegijs, the Amsterdam University's active stimulation programme, based upon a special fund, has been the exception to the rule of ineffective university policies (Portegijs, 1998). The percentage of female professors at the UVA was 8.3 on 31 December 2000 (WOPI, 1999).

seems to hold for the Netherlands as well: 'One reason [for the continuing inequalities women in science] is the tragic divide between three increasingly self-contained groups of experts: those who fashion and fund intervention programs designed to increase the number of women in science; historians and philosophers of women and gender in science; and scientists' (Schiebinger, 1997: 202). In the same issue, Evelyn Fox Keller describes hostility and an almost-unbridgeable gap between gender and science experts on the one hand and women scientists on the other (a gap related to the differences between feminists and other women), and says her task is 'to go beyond assumptions either of a natural commonality or of an entrenched opposition between women scientists and feminist analysts of science' (Keller, 1997: 16).

That the gains for academic women in the Netherlands were contained within women's and gender studies units and even ran counter to the situation of women scientists in other disciplines began to be noticed in the second half of the 1990s, especially among the gender studies scholars. Thus, Willy Jansen, an anthropologist and professor in women's studies, observed in an overview on women's studies in the Netherlands in *Women's Studies Quarterly* in 1996:

> As in other countries, Dutch universities suffered from the tightening of budgets for higher education. During reorganizations, university staff was laid off on the basis of 'last come first out.' This rule hit women harder than men. Departments had tried to hire more women over the last decade, but these newcomers suffered most.... As a result, the already small number of female teaching and research staff was approaching zero in many departments. The establishment of women's studies centers and the creation of women's studies chairs (often part-time and temporary) enabled universities to save face and keep their best employees. Also women's studies units used the diminishing female staff in general as an argument to gain or maintain their place at the university. This partly explains the curious situation mentioned above that the Netherlands belong to the countries with the lowest percentage of female university staff in general, but nevertheless has a fair number of women's studies posts. For instance, of all the professors at the university, women make up only 3.6 per cent. But of the 148 female professors in the Netherlands in 1995, 17 are in women's studies, that is 11.5 per cent. Not bad for a new discipline, but we would prefer a fair representation of women in the highest echelons in all discipline, not just women's studies.
>
> (Jansen, 1996: 332–33)

Jansen implicitly states that women's studies profited from the cutbacks at the cost of women scientists. Kathy Davis is more critical in formulating the same process: '[W]omen's studies has increasingly taken on a token function for a university more interested in cutting costs than eliminating gender inequalities' (Davis, 1996: 188). Davis also gives an underlying explanation for this development as the general lack of interest on the part of women's studies departments:

> Academic feminists in the Netherlands have been wary about having women's studies equated with the overall position of women in the university, arguing and quite rightly – that women's studies has a different focus. Feminist scholarship is concerned with the analysis of 'gender' rather than with emancipation issues. Moreover, many academic feminists have their hands full carving out a niche for women's studies without having to sit on hiring committees or tackle the affirmative action policies of their universities. In addition, some feminist scholars are critical of affirmative action for theoretical reasons: it doesn't 'get to the root of the problem' or it 'reifies rather than deconstructs gender dichotomies.'
>
> *(Davis, 1996: 187)*

Her view is sustained by Mottier who in her national report on the Netherlands – underlying the recently published *Helsinki Report* (Rees, 2002) – remarks that in the 1980s only about ten gender studies researchers (out of more than 300) were interested in questions concerning women, gender, and science (Mottier, 2002). Saskia Grotenhuis, ten years earlier had argued that women's studies were interested in 'all women' and not only academic women who formed a very small elitist group (Grotenhuis, 1988). Moreover, the concept of equality became a contested concept and 'equality or egalitarian feminism' was considered by some to be a lesser kind of feminism that lived by superficial 'sex counting' and 'fixing women's problems'.

The distance between emancipation policies and related research has been an unnoticed but persistent feature of women's studies in the Netherlands. By favouring a conviction that women's inequalities in science were epistemological rather than social and cultural, and demonstrating a certain disdain for the applied science of diagnosing and empirically investigating the relation between women, gender, and science, or formulating and implementing gender expertise in academic policy. Dutch women's studies fuelled the idea that equal opportunities policies lay outside the core of university business. In so doing women's studies assessed the separation between science and society as well as the gap between scientists and bureaucrats. It was one aspect of their strategy to gain respectability in the academic or scientific world.

Because of this situation, equal opportunities officers have often missed the insights from women's and gender studies regarding the complex relations between women, gender, and science. Lacking the necessary support and power, their policies and programmes for too long focused too much on 'women as the problem in need of fixing', too little on the central role of science as a social process, and producing rather than reflecting gender relations. Too often they were 'enclosed' in the bureaucratic section, most often in the personnel management department in the universities, far from the universities' core business.

Far into the 1990s, female scientists who were not in gender studies have more often than not turned their backs on the active politics of changing science and the position of women in science by gender studies specialists or equal opportunities officers. Their effort was to remain as much as possible invisible 'as a woman' even if they sometimes profited from the effects of feminist activities. In contrast to the feminist professor Mieke Bal who declared she would rather be appointed 'as a woman' than not to be appointed 'as a woman', many female academics turned up their noses at positive action measures or affirmative-action recruitment programmes (Drayer, 1999). It is only recently that female professors have organised the association, the *Landelijk Netwerk Vrouwelijke Hoogleraren* (LNVH).[37]

Contemporary Arguments for the Dutch Case: Pillarisation Again?

The divisions between women or groups of women are noted because, although they are not unique, they may serve to explain the Dutch case – the relative large under-representation of women – as indeed again a Dutch case of pillarisation. This suggestion is inspired by Margo Brouns' observations: 'By subsidizing opposing religious or political groups, the state has traditionally ensured that all parties have their needs met – at least to some extent – thereby ensuring a kind of peaceful coexistence. This social tradition paved the way for the institutionalisation of women's studies, which became in a certain sense just one more 'pillar' – albeit a small and relatively weak one' (Davis and Grünell, 1994: 102). Though in the 1980s the Netherlands was already far into the process of de-pillarisation, we can agree with the idea that the success of Dutch women's studies could be explained by the old system and mechanisms of pillarisation; but, we should add, so can the failure of integrating women scientists in academia, as well as the rather unsuccessful integration of gender studies in the 'regular' disciplines. The institutionalised segmentation of

[37] The network has a strong commitment to equal opportunities policies. The first action taken was a letter to the Minister to promote the appointment of women professors.

society on the basis of political or religious conviction is capable of recognising and awarding – albeit to a limited extent – the feminism of women's studies (which is based upon an ideology and a 'constituency'), but not 'women' or 'women scientists' as a group, and not gender as a category of political division. Pillarisation may also explain the typical Dutch situation of women in science because of the role of elites. One aspect of pillarisation is the central role of the elites of the pillars, representing the pillar as a whole. One could argue that women's studies in a way came to be seen as the elite of the 'women's pillar', at least in academia, and so became representative of 'women scientists' all over the university. Women's studies, therefore, could claim that what was good for them was good for 'all women' in the university.

If we compare Norway and Sweden, we can see how important the cooperation of women's groups and women's interests is (Lykke, 2001). But there is of course much more to learn from good practices in other European countries and the United States. For instance, Germany is doing visibly better than the Netherlands on the issue of 'women in science' in the last few years. To learn from Germany's experience, we should examine how its equal-opportunity policies were organised and by whom, how women's studies developed, and the role of the federal state and universities in implementing principles and programmes. Such an analysis would also give a more definite and well-founded answer to the question of the 'Dutchness' of the Dutch case.

Turning the Tide?[38]

In 1997, almost a hundred years after the public debate concerning women in higher education and almost fifty years after Dutch academic women documented their accomplishments in a book of statistics, the academic world was shocked into action by the Swedish study, 'Nepotism and Sexism in Peer Review'. Although the study proved the existence of a gender bias in the peer review process of the Swedish Medical Research Council in only one particular year, it suggested a wider pattern of discrimination in the sciences. The study seemed to distill the essence of what had been suggested by numerous 'anecdotal' answers to the question: 'Women Scientists: Why so Few?' (Rossi, 1965).

[38] The metaphor of the 'rising tide' was used in Great Britain in the title of an influential report (Committee on Women and Science, Engineering and Technology, 1994).

Table 6.7 *Skewed Pattern of Women in University Disciplines, 2001/2002*

Disciplines	Percentages
Mathematics	28
Informatics	13
Physics	18
Chemistry	28
Biology	48
Pharmacy	56
Architecture	31
Mechanical engineering	6
Electrical engineering	5
Medicine	60
Dentistry	50
Economics	27
Law	55
Psychology	76
Pedagogy	89
Political science	41
Theology	48
French	79
English	76
German	68
History	42

Source: Ministry of Education, Culture and Science; Mottier, 2002

In the Netherlands, the *Nature* article caught the attention of the media as well as the NWO, the Advisory Council for Science and Technology Policies (AWT), and the Ministry of Education, Culture, and Science. The NWO immediately did an analysis of the success rates of men and women in its programmes and concluded that on the basis of the available figures, no conclusion could be drawn about possible gender bias in the peer review system. However, the findings did indicate that very few women applied for funding. NWO cooperated also in a more in-depth study into the peer- review process ordered by the Ministry of Science and Education and undertaken by the Netherlands Association for Women's Studies (NGV). This study revealed no hard evidence of gender bias, but it did show that the qualifications of men and women tended to be evaluated differently (Brouns and Scholten,

Table 6.8 *Membership of University Boards, Committees of Deans and Leading Positions in the KNAW and NWO*

	Total	Women	% Women
University Boards	40	2	5
Committees of Deans	135	3	2.3
KNAW	43		
NWO	65	5	7.7

Source: Bosch, 1999

1999; see EC 2000: 34–35). An independent study demonstrated that few women applied for grants at the NWO because of the low numbers of female professors and the unequal distribution of women across the disciplines. It further showed that the under-representation of women is greatest in those disciplines where funding is greatest – in the natural sciences (see Table 6.7) (Noordenbos, 1999).

The Swedish study also fuelled interest in the 'Law on Equal Representation of Women in Leading Positions in Education', which was passed in 1997 after years of discussion and lobbying by the national platform of equal opportunities policy-makers in science and higher education (LOEKWO) and others. Although the law aimed at women in secondary school administration, the minister likewise charged universities to set target figures for women in university positions and make plans for reaching those targets. The minister also distributed the results of a study that documented women's lack of advancement in the universities. The study clearly demonstrated that due to the immobility of academic personnel, women largely did not get promoted. It concluded that if nothing changed, the under-representation of women would worsen. One of its recommendations was to return to the system of a tenure track for academic personnel based on performance (Portegijs, 1998).

This report and various statistical publications aptly captured the situation of women in faculty positions (cf. Balen and Fisher, 1998). By the year 1995, women represented 45 per cent of all university students in the Netherlands (up from 17 per cent in the 1950s). In the 1990s, the percentage of women earning doctorates was 23.65 (up from 4.1 in the 1970s). This growth at the undergraduate and graduate levels was not reflected in academic staff. Among Dutch professors, women were still heavily under-represented with 5 per cent of 2,445 full-time positions in 1998. For associate professors (fixed positions not automatically leading to professorships), the same was true with 7.5 per cent women. Female assistant professors fared only a little bit better with 20 per cent (13.8 per cent of them in temporary positions) against 80 per cent men (8.7 per cent in temporary jobs) (Bosch, 1999). The figures for women in university boards as deans or in administrative positions in the KNAW or NWO drew attention to the adverse effects of the modernisation process on the position of women (see Table 6.8). Only then did it become clear that the Netherlands' record lagged far behind that of other countries and that the percentage of Dutch women professors was lowest in the world except for that of Botswana which had zero women professors.[39]

The heightened attention to the subject of women in science generated more reports. The Advisory Council for Science and Technology Policy (AWT) commissioned a background study to the LOEKWO, posing the question 'why so few?' (Bosch, 1999).

The history of the development of Dutch science over the past twenty years and general gender theory provided some answers. The heart of the book was a plea for universities to develop broader, multidimensional definitions of scientific quality more in step with the multiple functions and forms of today's research, referring to Schiebinger's conception of 'sustainable science' and Brouns' and Harbers' conception of science as taking place in the agora instead of on Olympus (Brouns and Harbers, 1994; Hoving, 1999; Schiebinger, 1997).[40] The study ended with a review of equal opportunities and diversity policies in academic and research institutions in other European countries and the United States and made five recommendations:

(a) To create extra chairs for women (like in Sweden).
(b) To enlist gender equality goals in the system of contract compliance between state and universities (like in Germany).
(c) To mainstream gender equality goals in all policies at all levels of universities and scientific institutions (like in the European Union and Germany).
(d) To integrate gender and diversity perspectives in all teaching and research plans.

[39] This information is from a table in Lie et al., 1994. We have never liked this comparison for its roots in racist civilisation theory.

[40] The book attempted to broaden the scope of diversity policies to stress the interconnectedness of gender with other axes of difference, such as ethnicity. In European, as well as in National policies regarding gender and science, there is still a very restricted view of gender, which implies and perhaps strengthens a definition of science as a 'white' phenomenon.

(e) To create a national advisory and expert centre on women, gender, and science to coordinate and stimulate research, and to develop and implement policies (Bosch, 1999: ch. 5).

The AWT study was presented at a major conference on women and science at the Royal Academy of Arts and Sciences (KNAW) in Amsterdam, 12 October 1999, that was organised by the NWO, the The Association of Universities in the Netherlands (VSNU), and the AWT. In several lectures, references were made to the Swedish study and there was a report on inequalities for women scientists at the Massachusetts Institute of Technology (MIT, 1999; AWT et al., 1999). Both studies were the result not of actions by equal opportunity policy makers or gender researchers, but rather by women scientists, and perhaps for that reason made such an impression. Both documents stressed that gender inequalities result not from one large organised cause but from a series of small and almost imperceptible differences in the treatment of men and women. Both documents evoked worldwide reactions on the website of *Nature*.

A sense of breakthrough was in the air. The president of the NWO, Reinder van Duinen, referred to his own changed understanding of the gender issue in academia as a 'paradigm shift' (AWT et al., 1999: 93). In his opening speech, the minister of education, culture and science, Loek Hermans, offered praise and financial support for a programme to promote thirty female assistant professors to associate professors. This programme, named after the female Greek philosopher Aspasia, was to be implemented jointly by the Netherlands Organization for Scientific Research and the universities. Under this programme, female assistant professors (or junior lecturers) could apply for funding of a four-year PhD research project, or a two-year post-doctoral research project. The plan foresaw thirty Aspasia promotions in two groups of fifteen, the first call being in 2000 and the second in 2002.

A few months after the meeting at the KNAW headquarters, the Advisory Council for Science and Technology Policy brought out a well argued advice to the government under the title (in translation) Am biguous Science: Under-use of women's potention as an existential problem for academia. (AWT-Advies, 2000). Two of its recommendations were to enlarge the Aspasia Program from 15 to 30 grants, and to found an expert centre which could function as a powerhouse for research and policy making in respect of women and science, both of which are (being) realised. At about the same time the association of Dutch universities (VSNU) came into action by organising a few expert meeting. Last but not least a National Network of Women Professors (LNVH), that had been functioning informally during the 1990s, was formalised in 2001 so as to be able to act as interlocutor in matters regarding women in science.

Also around 2000 in the European Commission the awareness of women's under-representation in science rose. The international report that was mentioned in the opening section of this paper was made, and a gender impact assessment was done of the Fifth Framework Program, which was the main research funding programme of The European Union. This resulted in several prescriptions for researchers to pay attention to the gender balance of research teams and the gender dimension of the content of their research plans in the Sixth Framework Program. In 2005 the European Platform of Women Scientists (EPWS) was founded with support of the European Commission. (http://www.epws.org/)

In the meantime the NWO started to ask the network of women professors in the Netherlands (LNVH) to recommend female scientists for their prestigious awards. As a result, Spinoza awards were granted to the Australian Ann Cutler in 1999, E.F. van Dishoeck in 2000, and D.I. Boomsma in 2001.41 Both the KNAW and NWO have appointed more women to their several boards and committees. In the meantime, the Aspasia Program has proved to be an overwhelming success. At the first call, instead of 15, 68 women were promoted, bringing the percentage of female associate professors from 8.5 to 11 per cent. The large number of highly qualified applications alerted universities to the under-use of women's potential and gained enthusiastic support from the NWO-board (Bosch and Potting, 2001).[41]

The increased interest in the multiple aspects of women in science can be seen in the case of the age criteria for applying at the NWO and KNAW research programmes which came before the National Committee for Equal Treatment by the NGV.[42] The NGV added to its goal of promoting women's and gender studies, also women in science generally. Research on

[41] At the first call 31 women received grants from the NWO and they were promoted to the position of UHD. In addition, universities promoted 37 women who were qualified as excellent, but could not be funded. Sometimes the universities funded the research project as well. On the second call, 40 women received an Aspasia grant from the NWO. It is expected that universities will also promote women who received good evaluations but did not get grants.

[42] The committee ruled that age criteria indirectly discriminated against women and should therefore be dropped or implemented on a differential basis. In a way the rule was a mixed blessing since the committee based its judgement on fertility statistics only, thereby implying that women's under-representation in science had a 'natural' cause (Steen, 1998).

issues relating to women, gender, and science increased notably, and a few dissertations appeared on the subject (Balen, 2001; Dekker, 2000). In April 2001 the research school NOV organised a conference on the 'Dutch case' and the National Network of Women Professors (LNVH) held a conference dedicated to women professors.[43]

Whether these developments will turn the tide is still to be seen. Except for the visible increase of women as associate professors (due to Aspasia) there is still no progress. The most recent *Arbeidsmarktmonitor* (Labour-market monitor) for the academic sector has come up with figures that show that the gap between men and women with respect to the chances they have to become professors has become even wider (*Arbeidsmarkmonitor*, 2001: 16–20).[44] This was calculated on the basis of a comparison between student generations (m/f) and related age groups of professors (m/f). While men between 55 and 59 (who finished their academic studies between 1970 and 1974) had a 3.7 better chance than women to become professors, men between 40 and 44 have an even better chance: 5.5. The position of women's studies also proves less stable than the strong institutional position seemed to indicate. Although the National Research School passed the (international) visitation, and was recognised by the KNAW for another five years, locally the picture is much less promising. Many centres are based upon 'special regulations' and full integration of gender in disciplinary research and teaching has not been attained.[45]

Even the turnover of academic personnel expected in the near future due to the asymmetrical age curve is lamented without taking into account the many talented women who represent a potential which is still underused.[46] An exception to this forms the report *Talent voor de Toekomst, Toekomst voor Talent* (A Future for Talent, Talent for the Future) which fully integrated a gender dimension (Vucht Tijssen, 2000). Many of its recommendations echoed earlier ones such as to drop the system of fixed formation and instead turn

[43] The conference, '*Iedereen hoogleraar*?' (Everybody a Professor?), was held at the KNAW headquarter in Amsterdam, May 30 2002.

[44] In the report a reservation was made about the possibility that men become professors at a younger age. This will probably be an important explanation for their findings.

[45] This conclusion can be drawn from the contributions to the conference '*De IJkpunten geijkt*' which was an assessment of the results of the large historical project financed by the NWO between 1992 and 2002 on 'Dutch culture in a European context' (Bosch and Hellevoort, 2002).

[46] A typical example of the neglect of the availability of talented women in the face of the great turnover of academic personnel is the interview with the newly appointed president of the KNAW, Pim Levelt, a brother of Anneke Levelt, in the national daily paper *NRC*, 1–2 June 2002: '*Talent wordt nu verspild*' (Talent is wasted now). He focused only on 'young scientists,' mentioning women scientists only in passing.

to a career pattern based on quality and experience. Also now for the third time – it was recommended to install an expert centre on Women in Science. The failure of this last initiative to materialise is, however, an ominous sign. More than a year after the minister assigned the VSNU to make plans for such an institution in cooperation with the NWO and the KNAW and based upon the knowledge and experience in the field, a plan was put on the table that lacked vision as well as flesh and bones. Thereupon, the minister – all too easily – dropped the plan.

Towards a Delta Plan for Total Equality in Dutch Academia and Science

'Do something structural about the Dutch case,' declared the newly appointed professor at the Faculty of Mathematics and Natural Science at the University of Groningen, Jeanine Olsen.[47] Taking into account the good practices of countries such as Germany, Sweden, and 'Europe' itself, one could name a few basic needs to confront the Dutch case. First, to develop a structural, multi-level national policy plan – a real Dutch Delta Plan – which would combine a stimulation plan for women professors (such as in Sweden), which is at all levels committed to gender mainstreaming (such as in European research) and which plays a central role in the management control cycle (or contract compliance) between universities and Ministry (such as in Germany). The plan should have a firm financial basis and make use of financial steering mechanisms (cf. Bosch, 1998).

The commitment of this Delta Plan to gender mainstreaming at all levels of the scientific and academic process means the inclusion of all aspects of women, gender, and science into all institutions of science and education. The plan should target the integration of gender perspectives in research and education; the development of gender studies as an autonomous field of research; as well as equality issues (or issues of difference) in recruitment and selection of women scientists, the culture of science, the representation of science and scientists. The commitment to gender mainstreaming should also mean that the strategies and recommendations are always firmly connected to the different contexts of science. Last but not least, the plan

[47] Interview with Prof. Dr Jeanine Olsen (*UK*, personeelskatern, April 2001). Since 2002 the University of Groningen runs a very successful tenure track program for excellent women, the Rosalind Franklin Fellowship program.

should build into its basic principles a recognition of the interconnection between gender and other axes of difference such as age and ethnicity. In Germany for this inclusive definition of gender mainstreaming the term 'total E-quality' has been developed and connected to an award system for universities.

To succeed, such a plan would require a powerful structure to deal with the issue of women in science, for instance, a strong women and science unit connected to the VSNU, or a national steering committee in close contact with the most powerful players in the field, KNAW, NWO and the ministry, as well as in 'Europe'.[48] Leaders in academic policy and research, gender (and science) researchers, equal opportunities policy makers, and women scientists should consult regularly and commit to an orientation on the recognition of 'difference' and social responsibility instead of on 'pillars' and a non-committal tolerance in science and academia. In the mean time Dutch women scientists' motto should be clear: '*Luctor et emergo*!' (I struggle and come up).[49]

Epilogue

The new Monitor on (Dutch) Women Professors that came out in October 2009, brings good news, indeed, but it carries also older messages. Let us begin with the good news. The participation of women in science in the Netherlands is still growing. Since 2003 the percentage of women professors increased annually with 3 per cent to a percentage of 11.7 per cent at the end of 2008. Most universities, moreover, have an explicit policy to promote the participation of women in science. Notwithstanding this good news, too often the policy plans remain good intentions that do not turn into deeds. Still a lot of talented women scientists get lost in the academic labyrinth. Moreover, the European target for women professors, 25 per cent in 2010 is far from reached, and the same holds for the figure of 15 per cent that the Ministry of Education and Research deemed a more realistic target. The

[48] The Helsinki Report claims to have a Women and Science Unit in the Dutch Ministry of Education, Culture, and Science. While a few dedicated civil servants are dealing with the issue, it is little compared to the Women and Science Unit in the Federal Ministry in Germany, or in Great Britain (Mottier, 2002; cf. EC (Osborn a.o.) 2000: 97).

[49] This is the motto of the 'water province' of Zeeland, which in its weapon shows a lion struggling to come up out of the water. Zeeland harbours the core of the 'Delta works' which were started after the great flood of February 1953.

growth of women professors in the Netherlands (2.8 per cent) is slower than many of the European countries that in 2003 also had about 10 per cent women professors, and also lower than the European average (3 per cent). Nowadays, the glass ceiling is not anymore between post graduates and assistant professors, but between assistant professors and associate professors, with variations among the disciplines. The glass ceiling is thicker in disciplines with relatively many women students such as Language and Culture, Behaviour and Society, Law, Agriculture, and Health. For the first time in the Netherlands also the salaries have been assessed for gender effects. The outcome was perhaps not surprising, but to realise its full meaning was rather a shock: in all the university ranks women earn less than men, and the differences that occur in the end can be enormous. A last topic to be covered by the monitor was women's participation in decision-making functions. Of the 41 board members of universities in the Netherlands only three were women in 2009. And while women make up 24 per cent of the directors of education, among the directors of research, the more prestigious position, women are represented with only 6.3 per cent. Female faculty deans make up only 5.2 per cent of all the deans at all Dutch universities. Besides the monitor, in 2009 an interesting analysis of academic appointments was published by Marieke van den Brink, under the title *Behind the Scenes of Science: Gender Practices in the Recruitment and Selection of Professors in the Netherlands* (PhD Dissertation, Raboud University, Nijmegen, 2009: ISBN 978-90-9023959). Among the many interesting results of this dissertation was the outcome that with no woman in an appointment committee for a regular chair women applicants had only a 7 per cent chance to be nominated, with one woman in the committee their chance rose to 14 per cent, but with three or more women in the committee women applicants had a chance of 22 per cent to be selected. Another shocking outcome was that in spite of the call for open and transparent recruitment more than 60 per cent of the appointments were not advertised publicly at all. The citations she recorded from the many interviews she held with academic leaders were sometimes even breath taking, others were almost endearing for their naivety. 'Men have been working with men for ages. That is natural. It is easy.' To conclude, it is fair to say that there is progress, but that we can still argue that there is a Dutch case in the sense that Dutch universities are still at the far end of women's participation in science and research in Europe and worldwide. Thus, the Dutch case still can not be closed.

References

Arbeidsmarktmonitor academische sector (2001). *Werkgelegenheid in de wetenschap in onderzoeken en cijfers.* Een uitgave van SoFoKles in samenwerking met Research voor Beleid en de Vereniging van Samenwerkende Universiteiten VSNU.

AWT, KNAW, NWO, VSNU (1999). *Verslag van de conferentie 'Vrouwen in de wetenschap'* d.d. 12 oktober,. S.a., s.l.

AWT-Advies: Adviesraad voor Wetenschaps-en Technologiebeleid (2000). *Halfslachtige wetenschap. Onderbenutting van vrouwelijk potentieel als existentieel probleem voor academia.* Den Haag: SDU.

Balen, B. van (2001). *Vrouwen in de wetenschappelijke arena. Sociale sluiting in de universiteit.* Amsterdam: Het Spinhuis.

Balen, B. van and A. Fischer (eds.) (1998). *De universiteit als modern mannenklooster.* Amsterdam: Het Spinhuis.

Baranelli, M. e.a. (1990). *Geachte promovenda, 110 jaar promoverende vrouwen aan de universiteit van Amsterdam.* Amsterdam: SUA.

Beekes, A. (1991). *De hordenloop. Ontwikkelingen in de achterstand van vrouwelijke op mannelijke academici aan Nederlandse universiteiten in de periode 1960–1985.* Utrecht: Faculteit Sociale wetenschappen.

Berkel, K. van (1985). *In het voetspoor van Stevin. Geschiedenis van de natuurwetenschap in Nederland, 1580–1940.* Meppel: Boom.

—— (1990). "De amerikanisering van de Europese universiteit. Het begin van de discussie in Nederland (1900–1940)", in: *Amerika in Europese ogen: facetten van de Europese beeldvorming van het moderne Amerika.*, edited by K. van Berkel. Den Haag: 135–160.

—— (1998). *Citaten uit het boek der natuur. Opstellen over Nederlandse wetenschapsgeschiedenis.* Amsterdam: Bert Bakker.

Bertels, P. et al, eds. (1986). *Het naadje van de blauwkous. Over een eeuw vrouwen in de wetenschap aan de RUU.* Een uitgave bij de gelijknamige tentoonstelling van 9 maart – 20 april 1986. Buro Studium Generale/Universiteitsmuseum. Utrecht.

Blok, P.J. (1907). "Vrouwelijke studenten." *Onze Eeuw* 7(3):447–426. (Ook in: *Het Kind, tijdschrift voor ouders en opvoeders* 9:69–70, 77–78, 86–87, 95–96, 103.

Blom, J. C. H. and J. Talsma (eds.) (2000). *De verzuiling voorbij. Gosdienst, stand en natie in de lange negentiende eeuw.* Amsterdam: Het Spinhuis.

Bock, G. (2002). *Women in European History.* Oxford: Blackwell Publishers.

Boehm, L. (1958). 'Von den Anfängen des kademischen Frauenstudiums in Deutschland. Zugleich einKapitel aus der Geschichte der Ludwig-Maximilians-Universität München', *Historisches Jahrbuch*, 77: 298–327.

Bonner, T.N. (1992). *To the Ends of the Earth. Women's Search for Education in Medicine.* Cambridge: Harvard University Press.

Bosch, M. (1994). *Het geslacht van de wetenschap. Vrouwen en hoger onderwijs, 1878–1948.* PhD dissertation, Erasmus Universiteit Rotterdam. Amsterdam: SUA.

—— (1997). 'Kies exact! In historisch perspectief: veranderende visies op meisjesonderwijs en de exacte vakken, 1650–1880', *Gewina*, 20: 184–210.

—— (1998). 'Naar een deltaplan voor *total E-quality* in de universiteit?' *Tijdschrift voor Genderstudies,* 1(4): 46–54.

—— (ed.) (1999) *In het hart van de wetenschap. Naar total E-quality en diversiteit in de universiteit. Achtergrondstudie van het AWT,* Den Haag: SDU.

—— (2000). 'Wetenschap voor het vaderland: een wereld zonder vrouwen?' in M. Grever and F. Dieteren (eds.), *Een vaderland voor vrouwen (*A Fatherland for Women). *The 'Nationale tentoonstelling van Vrouwenarbeid' in retrospect,* pp. 167–82. Amsterdam: Stichting Beheer IISG.

Bosch, M. and A. Kloosterman (1990). *Politics and Friendship. Letters from the International Woman Suffrage Alliance, 1902–1942.* Columbus: Ohio State University Press.

Bosch, M., I. Hoving, and G. Wekker (1999). *In het hart van de wetenschap. Naar Total E-quality en diversiteit in de universiteit.* Background study Adviesraad voor Weten-schaps-en Technologiebeleid. Den Haag: SDU.

Bosch, M. and M. Hellevoort (eds.) (2003). "De IJkpunten geijkt." *Tijdschrift voor Sociale Geschiedenis* March.

Bosch, M. and M. Potting (2001). '*Vrouwen moeten door dat plafond heen.' Evaluatie van het Aspasia- programma, eerste ronde: ervaringen, resultaten, effecten.* In order of LOEKWO. Maastricht. (Evaluation study of Aspasia Program)

Bossenbroek, M. (1996). *Holland op zijn breedst. Indië en Zuid-Afrika in de Nederlandse cultuur omstreeks 1900.* Amsterdam: Bert Bakker.

Braat, J. (1991). 'Leonardo da Vinci en de kroonprinsen. Positieve actie voor vrouwen aan de Universiteit van Amsterdam'. M.A thesis, UvA, Amsterdam.

Brittain, V. (1960). *The Women at Oxford: A Fragment of History.* London: George Harrap.

Brouns, M. (1988). *Veertien jaar vrouwenstudies in Nederland; een overzicht.* Zoetermeer: RION. English translation: 1990. *The Development of Women's Studies. A Report from the Netherlands.* Dordrecht: Foris.

—— (2001). 'Wetenschap tussen Olympus en Agora. De vele gezichten van gender in dewetenschappen', *Tijdschrift voor Genderstudies,* 4(4): 24–33.

Brouns, M. and H. Harbers (1994). *Kwaliteit in meervoud. Reflectie op kwaliteiten van vrouwenstudies in Nederland.* Den Haag: VUGA.

Brouns, M. and M. Scholten (1999). *De kwaliteit van het oordeel. Een onderzoek naar beoordelings-procedures van KNAW en NWO.* Uitgebracht door het NGV. Utrecht: NGV.

Burstyn, J.N. (1980). *Victorian Education and the Ideal of Womanhood.* London/Totowa, NJ: Croom Helm/Barnes and Noble.

Clarke, E. H. (1873). *Sex in Education.* Boston: Osgood.

Committee on Women and Science, Engineering and Technology (1994). *The Rising Tide: A Report on Women and Science, Engineering and Technology.* London: HMSO.

Cox, W. H. (1898). *Over de aequivalentie van man en vrouw. Een biologische beschouwing.* Print from Psyciatrische en Neurologische Bladen.

Davis, K. (1996). 'What's a Nice Girl Like You Doing in a Place Like This? The Ambivalences of Professional Feminism', in L. Stamley (ed.), *Borderlands: Feminism in the Academy,* pp. 184–96. London: Sage Publications.

Davis, K. and M. Grünell (1994). 'The Dutch Case: An Interview with Margo Brouns', *The Journal of Women's Studies*, 1: 100–06.

De toekomst der academisch gegradueerden (1936). *Rapport van de commissie ter bestudeering van de toenemende bevolking der universiteiten en hoogescholen en de werkgelegenheid voor academisch gevormden.* Groningen/Batavia: J. B. Wolters.

Dekker, R. (2000). *De wetenschappelijke match. Persoon-cultuur fit en loopbanen van vrouwelijke en mannelijke wetenschappers.* Amsterdam: Thela Thesis.

Drayer, E. (1999). 'Anneke Levelt Sengers'; 'Mieke Bal', interviews. In Bosch 1999, 39–43; 105–109.

Eijl, C. van (1994). *Het werkzame verschil. Vrouwen in de slag om de arbeid 1898–1940.* Proefschrift UU. Hilversum: Verloren.

Eindadvies van de VBEO (1983). Voorlopige Begeleidingsgroep voor Emancipatie-Onderzoek, Deel II `Een programmering van emancipatie-onderzoek in deelprogramma's'. 's Gravenhage: VBEO.

Essen, M. van (1990). *Opvoeding met een dubbel doel. Twee eeuwen meisjesonderwijs in Nederland.* Amsterdam: SUA.

EC (European Commission) (2000). *Science Policies in the European Union. Promoting Excellence through Mainstreaming Gender Equality.* A report from the ETAN Expert Working Group on Women and Science, Research Directorate-General, European Commission. Luxembourg: Office for Official Publications of the European Communities.

Freie, M.J. (1948). 'In opvoeding en onderwijs', in M.G. Schenk (ed.), *Vrouwen van Nederland 1898–1948. De vrouw tijdens de regering van Koningin Wilhelmina*, pp. 89–107. Amsterdam: Scheltens & Giltay.

Friedan, Betty (1963). *The Feminine Mystique*, W.W. Norton and Company, New York.

Gilman, Charlotte Perkins (1900). *Women and economics: a study of the economic relation between men and women as a factor in social evolution.* Small, Maynard & Co, Boston.

Grever, M. (1994). *Strijd tegen de stilte. Johanna Naber (1859–1941) en de vrouwenstem in geschiedenis.* (Proefschrift KUN) Hilversum: Verloren.

—— (1997). 'Controlling Memories: Gender and the Construction of Scientific History', *Annali dell'Instituto storico italo-germanico in Trento,* 12(1996): 385–400.

Grever, M. and B. Waaldijk (1998). *Feministische openbaarheid. De Nationale Tentoonstelling van Vrouwenarbeid in 1898.* Amsterdam: IISG.

Groen, M.G. (1987). *Het wetenschappelijk onderwijs in Nederland van 1815 tot 1980. Een onderwijskundig overzicht. Deel I: Wetgeving, civiel effect, godgeleerdheid, rechtsgeleerdheid, indologie, geneeskunde.* Z.p. [Eindhoven].

Grotenhuis, S. (1988). 'Over geleerde vrouwen en geleerde feministen', *Tijdschrift voor Vrouwenstudies*, 24(6/4): 395–412.

Hawkins, A. and G. Noordenbos (1990). 'Blokkades in het doorstromen van vrouwen naar hogere functies aan de universiteit', *Universiteit en Hogeschool*, 36(5): 269–279.

Hawkins, A. and D. Schultz (1990). 'Women: the Academic Proletariat in West Germany and The Netherlands', in S. Lie and V. O'Leary (eds.), *Storming the Tower: Women in the Academic World*, pp. 47–57. London/New York: Nicholson/GP Publishers.

Herwerden, C.A.B. van (1948). *Marianne van Herwerden, 16 Februari 1874–26 Januari 1934*. Rotterdam: W. L. and L. Brusse.

Hicks, E.K. (1991). 'Women at the Top in Science and Technology Fields. Profile of Women Academics at Dutch Universities', in V. Stolte-Heiskanen (ed.), *Women in Science. Token Women or Gender Equality*? pp. 173–192. Providence/Oxford: Berg Publishers.

Hoving, I. (1999). "Lof der diversiteit." In Bosch 1999, 111–139.

Idenburg, Ph.J. (1934). *Studie, Crisis, Studiecrisis*. Openbare les, gehouden bij de aanvaarding van het privaat-docentschap in de geschiedenis, theorie en statistiek van het schoolwezen aan de Universiteit van Amsterdam op 24 April. Amsterdam.

—— (1960). *Schets van het Nederlandse schoolwezen*. Groningen: J. B. Wolters.

Jacobs, Aletta Henriette. (1879). 'Over Localisatie van Physiologische en Pathologische verschijnselen in de groote hersenen'. Dissertation, University Groningen. Groningen: Oppenheim.

—— ([1924]1996). *Memories: My Life as an International Leader in Health, Suffrage and Peace*. Edited by H. Feinberg and translated by A. Wright. New York: Feminist Press.

Jansen, W. (1996). 'Women's Studies in the Netherlands', *Women's Studies Quarterly*, 1,2: 327–42.

Jansing, J.W.G. and L. Dasberg (1978). 'Het sociaal-culturele leven in Nederland, 1875–1895, 1895–1914: onderwijs', in *Algemene Geschiedenis van Nederland* (AGN). Deel 13 (Nieuwste Tijd): 129–44, 361–72.

Jensma, G. and H. de Vries (1997). *Veranderingen in het hoger onderwijs tussen 1815 en 1940*. Hilversum: Verloren.

Jong, F. de (1988). "'Die oude Aloë toch...' Prof. Dr. Ir. Antonia E. Korvezee (1899–1978), de eerste vrouwelijke hoogleraar aan de Technische Hogeschool Delft," 227–34. In: *Geleerde vrouwen. Jaarboek voor Vrouwengeschiedenis 9*. Nijmegen: SUN.

Jong, O. J. de (1982). 'Benoemingsbeleid aan de Rijksuniversiteiten (1876–1931)'. Rede bij de viering van de 346ste Dies Natalis der Rijksuniversiteit te Utrecht op 19 maart 1982 in de Janskerk. z.p. [Utrecht].

Keller, E. F. (1997). "Developmental Biology as a Feminist Cause?" *Osiris*, 12:16–28.

Kelly, J. (1984). "Did Women Have a Renaissance?" In *Women, History and Theory. The Essays of Joan Kelly*, 19–50. Chicago: University of Chicago Press.

Kerling, L.P.C. (1969). *Phytopathologisch Laboratorium 'Willie Commelin Scholten,' 18 december 1894–18 december 1969*. Baarn: Mededeling.

Kerling, L. P. C., J. G. ten Houten, and G. de Bruin-Brink (1986). 'Johanna Westerdijk: Pioneer Leader in Plant Pathology', *Annual Review of Phytopathology*, 24: 33–41.

Kirejzcyk, M. (1988). "Marianne van Herwerden (1874–1934). De eerste vrouwelijke lector aan de Rijksuniversiteit Utrecht." *Geleerde vrouwen* 146–151.

—— (1993). 'Vrouwen kozen exact; studie en beroepsuitoefening rond de eeuwwisseling', *Gewina*, 16: 234–247.

Kloosterman, G.J. (1992). 'Hector Treub (1856–1920)', in J. H. C. Blom a.o. (ed.), *Een brandpunt van geleerdheid in de hoofdstad*, pp. 119–40. Amsterdam: AUP.

Kolf, M.C. van der (1950). *Zeventig jaar Vrouwenstudie*. Rotterdam: Brusse.

Kool-Smit, J. (1968). 'Het onbehagen bij de vrouw', *De Gids* 130: 267–81.

Kruyt, H.R. (1931). *Hooge school en samenleving*. Amsterdam.

Lasthuizen, K. (1995). *Onderzoekers vrouwenstudies in Nederland: een overzicht*. Utrecht: NederlandsGenootschap Vrouwenstudies.

Lie, S. S., L. Malik, and D. Harris (eds.) (1994). *The Gender Gap in Higher Education: World Yearbook of Education*. London: Kogan Page.

Löhnis, M.P. (1963). *Johanna Westerdijk Een markante persoonlijkheid. Met medewerking van enkele oud- leerlingen en vrienden*. Wageningen: Private publication.

Lorenz, C. (1993). *Van het universitaire front geen nieuws*. Baarn: Ambo.

Lucassen, H. and Th. Peeters (1988). 'Hora Est. Gepromoveerde vrouwen aan de Katholieke Universiteit Nijmegen, 1923–1988'. Centrum voor Vrouwenstudies van de Katholieke Universiteit Nijmegen. Nijmegen. Unpublished paper.

Lykke, N. (2001). 'Women in Science: Sweden versus Denmark. A Case Story about Neighbouring Countries with Different Gender Profiles'. Unpublished paper.

Maas, A. (2001). 'Tachtigers in de wetenschap. Een nieuwe kijk op het ontstaan van de Tweede Gouden eeuw in de Nederlandse natuurwetenschap', *Tijdschrift voor Geschiedenis*, 114(3): 354–76.

MIT (Massachusetts Institute of Technology) (1999). *A study on the Status of Women Faculty in Science at MIT*. Massachusetts: MIT. Available at http://web.mit.edu/fnl/women/women.htlm.

Maudsley, H. (1874). 'Sex in Mind and Education'. *Fortnightly Review*, 15: 466–83.

Meijer, I.C. (1996). *Het persoonlijke wordt politiek. Feministische bewustwording in Nederland, 1965–1980*. Amstertdam: Het Spinhuis.

Mercier, H. (1911). "Wegwijzers." In *Verbonden schakels*, by H. Mercier, 13–36. Amsterdam: Maatschappij voor goede en goedkope lectuur.

Ministerie van Onderwijs, Cultuur en Wetenschappen (Ministry of Culture, Education and Science) (1979). *Beleidsnota Universitair Onderzoek*. 's- Gravenhage: sdu.

—— (1981). *Kernnota inzake beleidsvoornemens betreffende het universitair wetenschappelijk personeel*. 's-Gravenhage: sdu.

—— (1982). *Beleidsnota taakverdeling en concentratie in het weten-schappelijk onderwijs; standpunt van de minister*. 's-Gravenhage: sdu.

—— (1985). *Hoger Onderwijs: Autonomie en Kwaliteit* (hoak). 's-Gravenhage: sdu.

—— (1986). *Het proces van taakverdeling en concentratie in het wetenschappelijk onderwijs*. 's-Gravenhage: sdu.

Mottier, I. (2002). 'Women and Science: Review of the situation in the Netherlands'. *Helsinki Report*. Available at http://www.cordis.lu/improving/women/reports.htm.

Noordenbos, G. (1993–94). '"Dat was niet te voorzien': consequenties van universitaire reorganisatie voor de man/vrouw-verdeling'. *Universiteit en Hogeschool* (Special issue 'In dienst van Alma Mater' of 'Hoe aantrekkelijk is de universiteit als werkgever'). 40(3): 107–17.

—— (1994–1995). "'Gender gap in Academe': de positie van vrouwen aan de universiteit in internationaal perspectief." *U and H*, 41(3): 105–120.

—— (1999). 'Genderasymmetrie in de aanvraag van onderzoekssubsidies.' *Tijdschrift voorGenderstudies*, 2(4): 36–45.

—— (2000). s.a. *Vrouwen in de Academies van Wetenschappen. Van uitsluiting tot uitzondering.* Zutphen: Walburg Pers.

Oreskes, N. (1996). 'Objectivity or Heroism? On the Invisibility of Women in Science', *Osiris*, 10: 87–113.

Phillips, P. (1990). *The Scientific Lady: A Social History of Women's Scientific Interests, 1520–1918.* London: Weidenfeld and Nicholson.

Plantenga, J. (1993). *Een afwijkend patroon. Honderd jaar vrouwenarbeid in Nederland en (West) Duitsland.* Amsterdam: SUA.

Portegijs, W. (1998). 'Eerdaags evenredig? Belemmeringen en beleid ten aanzien van de doorstroom van vrouwen naar hogere wetenschappelijke functies'. *Beleidsgerichte studies Hoger Onderwijs en Wetenschapsbeleid.* Den Haag: SDU.

Posthumus van der Goot, W. H. and A. de Waal (eds.) ([1948] 1977). *Van Moeder op Dochter. De maatschappelijke positie van de vrouw in Nederland vanaf de franse tijd.* Nijmegen: SUN.

Pouwelse, W.J. (1993). *Haar verstand dienstbaar aan het hart. Middelbaar onderwijs voor meisjes, debatten, acties en beleid.* Tilburg: Het Spinhuis.

Rapport van de Staatscommissie tot Reorganisatie van het Hoger Onderwijs (1949). Voorzitter: H.J. Reinink. Ingesteld bij KB van 11 april 1946. 's-Gravenhage.

Rees, T. (2002). *National Policies on Women and Science in Europe.* The Helsinki Group on Women and Science. EC, Directorate-General for Research, Women and Science. Luxembourg, Office for Official Publications of the European Communities.

Ribberink, A. (1998). *Leidsvrouwen en zaakwaarneemsters. Een geschiedenis van de Aktiegroep Man Vrouw Maatschappij (MVM), 1968–1973.* Hilversum: Verloren.

Rooy, P. de (2001). 'Voorbij de verzuiling?' *BMGN*, 116(1): 45–57.

—— (2002). *Republiek van rivaliteiten. Nederland sind 1813.* Amsterdam: Mets en Schilt.

Rose, H. (1994). *Love, Power and Knowledge: Towards a Feminist Transformation of the Sciences.* Cambridge: Polity Press.

Rossi, A. S. (1965). 'Women in Science: Why So Few?' *Science*, 148: 1196–1202.

Rossiter, M.W. (1982). *Women scientists in America: Struggles and strategies to 1940.* Baltimore, MD: Johns Hopkins University Press.

—— (1997). 'Which Science? Which Women?' *Osiris*, 12: 169–85.

Rupp, J.C.C. (1997). *Van oude en nieuwe universiteiten. De verdringing van Duitse door Amerikaanse invloeden op de wetenschapsbeoefening en het hoger onderwijs in nederland, 1945–1995.* Den Haag: SDU.

Schiebinger, L. (1989). *The Mind Has No Sex? Women in the Origins of Modern Science.* Cambridge: Harvard University Press.

—— (1997). 'Creating Sustainable Science', *Osiris,* 12: 201–16.

Schoon, L. (1989). *Het ideaal van de nieuwe vrouw. Catharine van Tussenbroek (1852–1925).* Doctoraal- scriptie Politieke Theorie en Nieuwste Geschiedenis, UvA. Amsterdam.

Schoor, W. van der (1994). 'Biologie en landbouw. F. A. F. C. Went en de Indische proefstations', *Gewina,* 17: 145–61.

Schreiner, Olive (1911). *Woman and Labour.* T. Fisher Unwin, London.

Scott, J. W. (1988). *Gender and the Politics of History.* New York: Columbia University Press.

Shteir, A.B. (1996). *Cultivating Women, Cultivating Science: Flora's Daughters and Botany in England 1760 to 1860.* Baltimore: Johns Hopkins University Press.

Soden, K. von (1979). 'Zur Geschichte des Frauenstudiums', in K. von Soden and G. Zipfel (eds.), *70 Jahre Frauenstudium. Frauen in der Wissenschaft,* pp. 9–42. Köln: Pahle-Rugenstein.

Solomon, B.M. (1985). *In the Company of Educated Women: A History of Women and Higher Education in America.* New Haven: Yale University Press.

Snow, C.P. (1958). The Two Cultures Cambridge University Press, Cambridge.

Stamhuis, I. H. (1995). 'A Female Contribution to Early Genetics: Tine Tammes and Mendel's Laws for Continuous Characters', *Journal of the History of Biology,* 28: 495–531.

Steen, M. van der (1998). 'Op leeftijd. Wetenschapsbeleid en seksediscriminatie', *Tijdschrift voor Genderstudies,* 1: 50–53.

Steinmetz, S. R. (1934). "De overvulling onzer universiteiten." *Paedagogische Studiën* 15:280–291.

Stevens, I., I. van Lamoen. 2001. *Manual on Gender Mainstreaming at Universities.* Leuven-Apeldoorn: Garant.

Sturkenboom, D. (2001). 'Une société savante exclusivement féminine. Présage des temps modernes ou vestige de Áncien Régime?' *Annales historiques de la Révolution française,* 4: 117–27.

Treub, H. and C. Winkler (1898). *De vrouw en de studie.* Voordrachten en debatten gehouden in de vergadering van 3 Maart en 10 November 1898 van de Vèreeniging ter Behartiging van de Belangen der Vrouw' te Rotterdam. Haarlem.

Tussenbroek, C. van (1898a). *Over het tekort aan levensenergie bij onze jonge vrouwen en meisjes.* Rede uitgesproken door Dr. Catharine van Tussenbroek op de Nationale Tentoonstelling van Vrouwe- narbeid. S.l. s.a.

—— (1898b). *Over de aequivalentie van man en vrouw. Weerlegging van de biologische beschouwing van Dr. W.H. Cox uit Deventer.* Print from *Belang en Recht.*

Verhaar, O. (1991). *Prima inter pares. Over de voorkeursbehandeling van vrouwen. Analyse van de argumentatiestrategieën van voor- en tegenstanders in de openbare discussie over het voorkeursbeleid.* 's- Gravenhage: Vuga.

Vianen, A. E. M. van (1997). *Het selectie interview. Over de rol van stereotypen.* Academisch Proefschrift. Leiden: RUL.

Vucht Tijssen, B.E. van (2000). *Talent voor de Toekomst. Toekomst voor Talent. Plan van aanpak voor het wetenschapsbeleid.* Utrecht.

Wachelder, J.C.M. (1992). *Universiteit tussen vorming en opleiding. De modernisering van de Nederlandse universiteiten in de negentiende eeuw.* Hilversum: Verloren.

Wekker, G. (1999). 'De belofte van gender- en vrouwenstudies', in Bosch 1999, 149–68.

Wennerås, C. and A. Wold (1997). 'Nepotism and Sexism in Peer-review', *Nature* 387: 341–43.

Weyrather, I. (1981). "'Die Frau im Lebensraum des Mannes.' Studentinnen in der Weimarer Republik." *Beiträge 5 zur feministischen Theorie und Praxis. Dokumentation des 3. Historikerinnentreffens in Bielefeld*, April 1981, 25-39. München: Verlag Frauenoffensive.

Wilde, I. de (1979). *Aletta Jacobs in Groningen.* Groningen: Studium Generale/ Universiteitsmuseum/ RUG.

—— (1985). "'Er is een heilig móeten, waartegen geen bezwaar is bestand.' De betekenis van Hélène Mercier (1839–1910) voor de vrouwenbeweging." *De eerste feministische golf. Jaarboek voor vrouwengeschiedenis* 6, 59–77. Nijmegen: SUN.

—— (1989). "Jantina Tammes (1871–1947), Nederlands eerste hoogleraar in de erfelijkheidsleer." Om niet aan onwetendheid en barbarij t4e bezwijken. Groningse geleerden 1614–1989. Edited by G. A. van Gemert, J. Schuller tot Peursum-Meijer, A. J. Vanderjagt, 187–206. Hilversum: Verloren.

—— (1998). *Nieuwe deelgenoten in de wetenschap. Vrouwelijke studenten en docenten aan de rijksuniversiteit Groningen, 1871-1919.* Proefschrift UVA. Assen: Van Gorcum.

Willink, B. (1980). 'Een inleiding tot de Tweede Gouden Eeuw. De wetten van 1863 en 1876 en de wedergeboorte van de Nederlandse natuurwetenschap'. *Hollands Maandblad* 22: 391–92 (juni/juli 1980): 3–9.

—— (1998). *De tweede gouden eeuw. Nederland en de Nobelprijzen voor natuurwetenschappen 1870–1940.* Amsterdam: Bert Bakker.

Winkler, C. ([1901] 1918). 'Het hooger onderwijs in de geneeskunde', In *Opera Omnia*, III, *Cornelis Winkler*, pp. 515–39. Haarlem: Tjeenk Willink.

CHAPTER SEVEN

Japanese Women Scientists

Trends and Strategies*

Mariko Ogawa

THE aim of this chapter is to consider the status of Japanese women scientists. First, the historical conditions of Japanese women scientists have been surveyed, and then their present situation is examined and described. Finally, new strategies for women in science and technology (S&T) in the twenty-first century are introduced.

It is easy to understand why there were few women scientists in the past. The simple fact is that women were excluded from the universities and academic associations. The question is, why are there still so few women scientists now that such discrimination has ended? Institutional equality with men does not seem to be enough for women to be fully active as scientists, even though they have far better circumstances than before; they can enter graduate schools and set themselves up in science if they wish. Why then, in spite of this, do so few women become scientists or engineers?

* This chapter is based largely on Chapter 3 of *Feminism and Science/Technology* (Tokyo: Iwanami, 2001) (in Japanese) by the author. Some new literature and information were added. Figures 7.1–7.4 and 7.5–7.8 were made by the author based on the data from the Japanese Ministry of Education. Figure 7.9 was made by the author based on data from The Academic Profession by E. Boyer, P. Altbach, and M. Whitelaw, The Carnegie Foundation for the Advancement of Teaching, 1994. Figure 7.10 was made by modifying Figure 7.6 (Share of women researchers employed in the business and public sector, 1997–2002 or nearest years, p. 9) of *Background Paper* in Joint OECD-French Government Workshop on 'Women in Scientific Careers: Unleashing the Potential', 16–17 November 2005.

The fact that women's careers may be interrupted by marriage, childbirth, and childrearing is not a sufficient reason for there being such few women scientists. These difficulties are just as common in other professions. The 'leaking pipeline' and 'glass ceiling' that women meet in the pursuit of their careers are not problems specific to women scientists. All this suggests that we should consider the nature of scientific knowledge as well as the social problems associated with it (Sonnet, 1995: 8–13).[1]

The present state of women scientists, problems they have to cope with, and the policies to increase their numbers, has been analysed using various data from many countries. In the developed countries, there are some common trends such as a tendency for students to avoid science courses. There is an urgent necessity to draw more women into S&T in order to weather the typical difficulties in the human resource of developed societies, such as their having a lower birth rate and more elderly people. Above all, Japan, which has fallen particularly behind in the assignment of women to scientific and technological fields, is under pressure to deal with this problem (Kuwahara, 1999; Yokoo, 1993).

From the Meiji Era to the End of World War

It is difficult to imagine women scientists in Japan which had not yet absorbed modern science from Europe. However, a search for a practitioner of Western medicine rather than a scientist as such finds the name of Ine Siebold (1827–1903), whose father was a naturalised Dutch military doctor of German origin. Philip Franz von Siebold came to Japan in 1823 and studied Japanese plants and animals, geography, history, and the Japanese language. After his return to Germany, his daughter Ine (Japanese surname: Kusumoto) was brought up by her Japanese mother and grew up to be an obstetrician with the latest medical knowledge. Other female doctors may have existed, but there are no details before the Meiji Era (Akiko and Satoko, 1996: 28).

After the Meiji Restorations, the Japanese government imported Western civilisation rapidly and hired a great many foreign science teachers to develop modern science in Japan. In the course of these developments, in 1874, it introduced an examination system for a license to practice Western medicine. Those women who hoped to become doctors could attempt it and enter the medical world if they passed the examination. This system ensured women a degree of equality in professional status. In 1885, several women attempted

[1] As to the nature of scientific knowledge, see Keller (1985) and Oreskes (1996).

this examination, but only one woman, Ginko Ogino, passed. She became the first woman doctor. This was not a bad start, particularly if the situation in Japan was compared with that in Britain, the most advanced country of its day. The first woman doctor in England was Garret Anderson, who started her job at St. Mary's Dispensary for Women in 1866. After Ginko Ogino, more women attempted the examination and subsequently became doctors. One exceptional Japanese woman doctor who followed an unusual career was Kyoko Okami. She graduated from Pennsylvania Women's Medical College and then came back to Japan, where she was appointed chief doctor of obstetrics in the Jikei Dispensary (later Jikei University) in 1889.

In the ten years following Ogino's qualification, over 50 women became doctors. One of them, Sonoko Maezono, founded the Japan Medical Women's Association in 1902, and another, Yayoi Yoshioka, founded the Tokyo Women's Medical School in 1900. The former association was organised to promote further research by women doctors in collaboration, and elevated their social position. It has played an important role as a unique association, distinct from normal academic associations. The latter school, the Tokyo Women's Medical School (later Tokyo Women's Medical College) was particularly important in a time when few opportunities in medical education were offered to women.[2] There were over 700 women doctors in early 1920s Japan, who formed a highly professional community.

As for medicine itself, the Imperial Women's Medical College (later the medical department of Tōhō University) opened in 1925 and the Osaka Women's Medical High School (later the Ōsaka Women's Medical College, now Kansai Medical University) in 1928. Gradually, female medical professionals secured their field. When higher education failed to open to women, some made efforts to establish women's colleges. The earliest examples were Tsuda Jyuku University, founded by Umeko Tsuda, and the Tokyo Women Medical School, by Yayoi Yoshioka.

Unlike the medical field, in which women doctors were able to build their careers, the emergence of women scientists was not until the late 1920s, and other organisations for woman in higher education were founded even later. In 1875, the Tokyo Women's Normal School was opened, and changed its name to the Tokyo Women's Higher Normal School in 1908. At the same time the Nara Women's Higher Normal School was also opened. In 1905, a graduate course was established at the Tokyo Women's Normal School, basically in order to produce teachers, not researchers. So, if women wanted to continue

2 As for women doctors in early Meiji Japan, see Patessio and Ogawa (2005).

their research, they had to enter and study at those national universities which were only just beginning to admit women.

It was the Tohoku Imperial University that first opened its doors to women. In 1913, Chika Kuroda, Raku Makita, and Ume Tange were the first to enter this university. After 1925, the Kyushu Imperial University, Hiroshima Humanities and Sciences University, and others first admitted women. During the 1920s and 1930s, eight women acquired doctorates in science – four doctors of pure science (Kono Yasui, Chika Kuroda, Sechi Kato, Fumiyo Ogawa); three doctors of agriculture (Michiyo Tsujimura, Yasu Honma, Yasu Namikoshi); and one doctor of pharmacy (Hideru Suzuki). After graduating from the Tohoku Imperial University, Ume Tange went to the United States to study and gained a PhD there in 1927. Between 1931 and 1937, the number of women who acquired doctorates reached fourteen. The first was Kanaeko Nishimura, and most of them were from the Tokyo Women's Medical School, founded by Yayoi Yoshioka. From the viewpoint of research, chemistry and biology as well as medicine were dominant, especially the analysis of pigments and other constituents of food and plants. Toshiko Yuasa, who went to Paris to study atomic physics after 1940, was unusual for a Japanese woman scientist of this period. She returned just before the end of the World War, but she left Japan soon after the end of the war and never returned (Ito, 2004).

In the 1940s, more universities opened their doors to women, as the influence of war aggravated the shortage of scientists and engineers. Against this background, women were encouraged to major in science or engineering, and the Ministry of Education actively set up new courses in scientific disciplines, and new science orientated women's high schools. In 1941, the Imperial Women's Science High School (now Toho University), the next year the Kyoto Prefectural Women's High School (now Kyoto Prefectural University), and some other mathematics and science courses were newly founded. However, at the end of the war, the scientific disciplines for women that had been hurriedly installed were mostly abolished because of their failure to satisfy the standard by which universities under the new system of education had been established. Even in the case of the Tsudajuku High School (now Tsudajuku University), only the mathematics course was left, both the physics and chemistry courses were abolished.

During wartime, educating medical doctors was an urgent problem, and new women's medical schools were founded in Nagoya, Gifu, Fukushima, Akita, Kyoto, Sapporo, and Seoul (then called Keijyo). However, with the exception of the Akita Women's Medical School, most of them eventually became co-educational medical universities.

Japanese Women Scientists in the Postwar Period

When a new university system started in 1948, equal opportunities for higher education for women also began. However, the numbers of female students in science and engineering courses remained very low. Originally the proportion of female students for all courses was less than 15 per cent, and the proportions for science courses or medical courses were only a few per cent; in particular, the proportion for engineering courses was less than 1 per cent (Figure 7.4).

In spite of the number of woman scientists being so few, the Society of Japanese Women Scientists was founded rather early, in 1958, to cultivate mutual friendships and promote the elevation of women scientists' status. In the highly developed period of the Japan of the 1960s, the so-called science and technology boom, the proportion of female students who went to university doubled to just over 2.5 per cent. On the other hand, in this period people became strongly conscious of men and women's distinct roles; the man working outside the home and the woman working inside. Ironically, the period of this boom in Japan was a period of strong sex divisions (Kuwahara, 2001: 208).

An opinion in 1962 from a professor of Japanese literature, Yasutaka Teruoka, that 'girl-students will destroy the state (Joshidaisei boukoku-ron)', fueled argument over female students who consequently got married and stayed at home even if they had been well educated at the tax-payer's expense. Some emphasised that the state could not collect so much tax, while others lamented that there was a crisis when the discipline of literature became like a finishing school. These arguments were typical of this tendency. In this period, the proportion of female students who went to university or college was about 7.4 per cent. However, the proportion of female students in the discipline of literature reached 37 per cent on average over the whole nation. Disciplines in which female students majored showed a remarkable bias. Female students who gained a bachelor of science were 25 out of 1130, about 2.2 per cent.

After International Women's Year in 1975, a remarkable change was seen in female student proportions in science and engineering departments at least initially. In this period, the female student proportion went into double figures and continued to grow to over 25 per cent in the late 1990s. By 2000, female students accounted for 35 per cent in universities and 25 per cent in graduate schools. In 1975, a committee for women scientists was convened by the Science Council of Japan (henceforth SCJ) and a paper submitted demanding the improvement of women scientists' status. In 1996, the Japanese Association for the Improvement of Conditions of Women Scientists (JAICOWS) was also founded and a proposal was submitted on improving the circumstances of women scientists. The contents of this proposal were very concrete; for example, the elimination of resistant attitudes regarding sex divisions; the enrichment of social support for childbirth, childrearing, and care of the sick and the aged; a guarantee that woman scientists could retain their maiden names; and a reconsideration from a woman scientist's standpoint of the application system for Grants-in-Aid for Scientific Research (JAICOWS, 1996).

In 1980, Dr Katsuko Saruhashi, who was a researcher in meteorology and earth science, was elected as the first woman to the SCJ. When she retired from the Research Institute of Meteorology at the Meteorological Agency the same year, she founded the Saruhashi Prize in order to encourage women scientists and publicise their activities. This prize remained the only one available to woman scientists until recently, and has been highly respected. As a result, Japan has been made aware of its wealth of excellent women scientists (Kozai et al., 2001).

The number of female scientists has increased by three to five times, parallel with the increase in female students who began majoring in science and technology in the 1980s and 1990s. The number of women scientists finally reached about 10 per cent by the end of the 1990s. (Details on this will be given in the section on Japanese female researchers in science and technology below.) The emergence of women presidents in the Japanese Society of Seismology and the Japanese Society of Physics, together with the achievement of a Japanese woman astronaut, resulted in increased awareness among people about women scientists in Japan. However, there are still very few women among professors and advanced researchers and most women scientists remain at a low status.

The Present State: Social Perspectives and Statistics

Social and Statistical analyses reveal a gender bias in the general body of scientists. According to *The Report on Female Researchers in Japan* published by the National Institute of Science and Technology Policy, there is a strong correlation between female students who major in S&T and women scientists and technology researchers (Yokoo). If this correlation is reliable, we can expect the following. If more female students major in S&T, more female researchers will emerge in S&T. So, the report suggested that it was necessary to encourage female students to take the courses in science and engineering, in which there are currently so few female students.

Self-evident as this appears, however, the American historian Londa Schiebinger maintained in her 1999 book that this pipeline model linking female students with female researchers was not necessarily correct, as it did not grasp the essence of the problem (Schiebinger, 1999: ch.3: The Pipeline). The model assumes that female researchers assimilate to the male-oriented stereotype of the scientist. It does not involve creative ideas as to how women scientists might manage to carry on working in conjunction with childbearing and childrearing. *The Report on Female Researchers in Japan* shows that the correlation between female students and women scientists is very poor for areas such as pharmacology, biology, and chemistry in which women have been working for a long time. The long history of female participation in these areas has extended the age range of female researchers and has produced many researchers who have now left their jobs. On the other hand, the high correlation in the engineering departments suggests that, here at least, there are few researchers who have left their jobs because of their youth. Young female researchers have not yet experienced the difficulties of coping with both job and family. The report comments that there is a striking limit to a process which assumes that female researchers lead the same sort of life as male researchers.

Japan, as a comparatively regressive country for women to go into science and engineering, has not yet analysed the nature of the pipeline model. Further detailed analysis and support systems for childbirth and childrearing, and care of the sick and the aged, are needed. Under current conditions, the present situation of female students and female researchers is given below.

Japanese Female Students in Science and Technology

First, the proportion of female students who enter universities or colleges – in this case the population comprises all girls who graduate from junior high schools. Until 1965, the proportion of female students had been under 5 per cent, that is, only five out of 100 girls who graduated from junior high school entered university or college. However, the proportion has more than doubled, from 17.3 per cent to 38.5 per cent, during the past fifteen years (Figure 7.1). Figure 7.2 shows the proportion of female students who enter university as opposed to college. The proportion of female students entering college has remained in the 20 per cent range since the late 1970s, but in 1996, this was exceeded by the proportion of female students who entered university. Now almost 40 per cent of girls study at university (Figure 7.2).

As we can see in Figure 7.3, the proportion of female students to all students who have just entered university has increased since the World War II and has

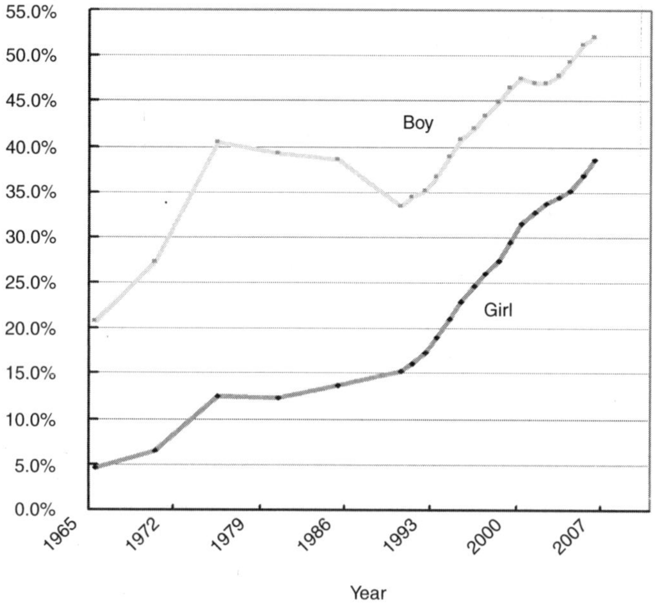

Figure 7.1 Proportion of Pupils Completing Compulsory Education Who Go on to Enter University

Source: Constructed by author on the basis of data by Ministry of Education, Culture, Sports, Science and Technology (MEXT), Japan.

now reached almost 40 per cent. Also, the ratio of graduate students is growing rapidly, as the graph shows. The proportion of female doctoral students is gradually approaching the proportion of female masters students, and in both courses female students comprise 25 per cent. This tendency for female students to pursue higher educational qualifications will probably continue, largely because parents who now have fewer children tend to expect their daughters to have as successful an academic career as their sons. In the context of university education (discussed in the following section) it is important to appreciate the great role that women's colleges played in the growth of female's university education before the 1990s. They have also been important work places for female researchers. Until 1990, 40 per cent to 48 per cent of female researchers worked at women's colleges, and they comprised 30 to 40 per cent of college teachers. Figure 7.4 shows the proportion of female students in various disciplines during the past thirty years.

A growth in this proportion is seen in all disciplines, although the humanities and education have shown only a gradual increase. Conspicuous

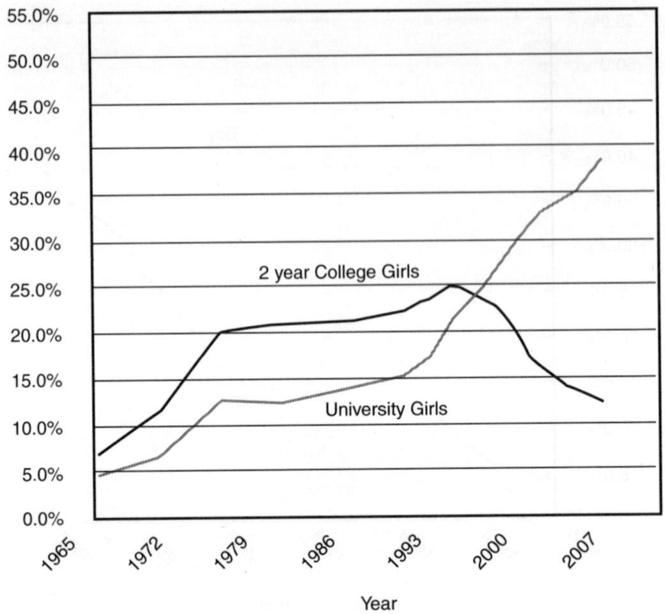

Figure 7.2 Proportion of Pupils Completing Compulsory Education Who Go on to Enter University or Two Year College

Source: Constructed by author on the basis of data by Ministry of Education, Culture, Sports, Science and Technology (MEXT), Japan.

growth has occurred in agriculture. The percentage of woman in engineering has increased dramatically. There is a substantial increase in the number of female students in engineering from 484 female students in 1970 to 11,309 in 2000; this is a 23-fold increase. In the same period in agriculture there is about eight-fold (from 842 to 6,703) increase. As the total number of students who major in engineering is about 100,000, it is regrettable that the existence of 11,000 female students is not reflected in the proportion of female students in engineering.

Next, the statistics in related disciplines are considered in detail. In science disciplines, biology and chemistry have high female student proportions. Generally speaking, girls seem to be poorer at mathematics, which is regarded as a key to taking a science or engineering major. However, the data shows that female students have remained at 20 per cent over the past thirty years in mathematics departments. This fact deserves more consideration. Figure 7.5 shows that the number of female students is smaller in physics rather than in mathematics.

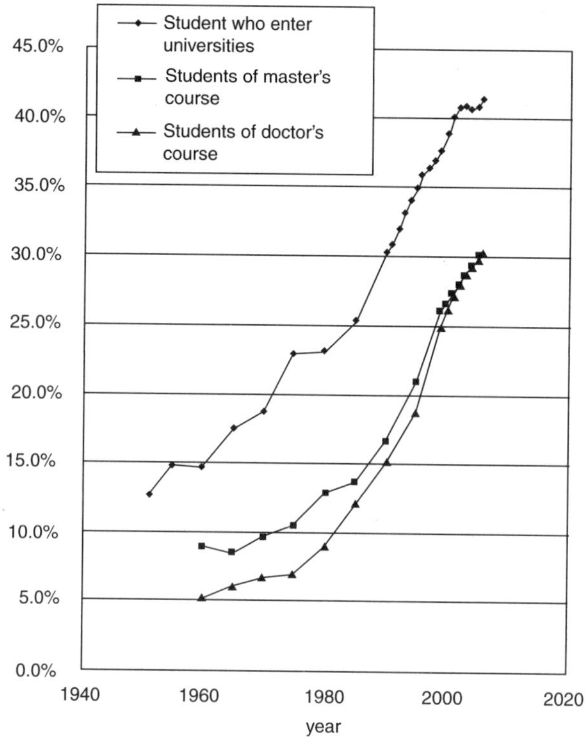

Figure 7.3 Proportion of Female Students

Source: Constructed by author on the basis of data by Ministry of Education, Culture, Sports, Science and Technology (MEXT), Japan.

However, since the late 1990s, the proportion of female students in engineering has been more than 10 per cent, and this may lead to an expectation of more activity from female students in future. There is evidence that women's conduct becomes less restricted once the proportion of women reaches 10–15 per cent, and according to the U.S. physicist Mildred Dresselhaus, a new phase in gender relations may be commencing (Dresselhaus, 1986).

The health disciplines, which include medicine, dentistry, pharmacy, and nursing, are in a different situation. As the science of nursing has recently changed from being a two-year course to being a four-year course, the proportion of female students has been pushed up. This is one reason for an increase in the female student proportion in health disciplines. Figure 7.6 shows proportion of women in the medicine, pharmacy, and health.

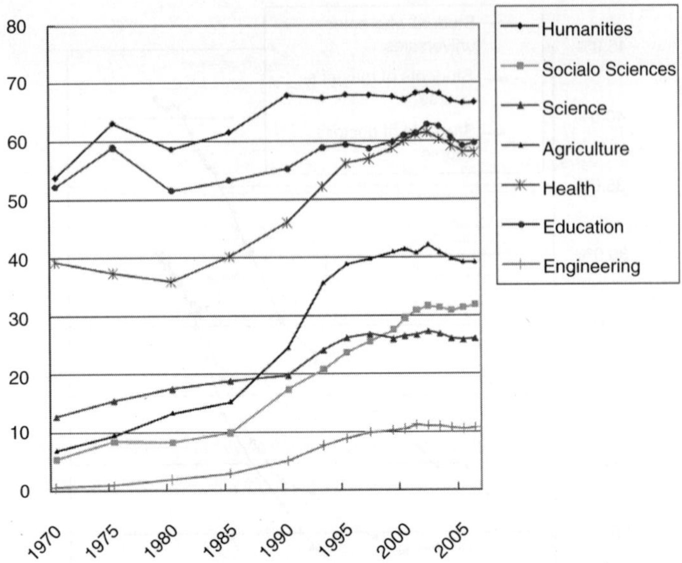

Figure 7.4 Proportion of Female Students in Various Disciplines
Source: Constructed by author on the basis of data by Ministry of Education, Culture, Sports, Science and Technology (MEXT), Japan.

Pharmacy has been very popular among female students because while it bestows a prestigious license, it is less demanding in terms of time and personal commitments than medicine. Pharmacy has sustained a female student proportion of almost 60 per cent for about thirty years, but in 2000, it dropped below that level.

From 2006, pharmacy training will take the form of a new six-year system as for medicine. What effect will this two-year extension of training have on the numbers of women planning careers in pharmacy? Another question concerns the fact that graduation from departments of science and engineering does not involve any national licenses. In contrast, female students in pharmacy and medicine can enjoy the advantage of a national license, which enables women to come back to their careers after childbirth and childrearing.

Medicine, which has a long history as a professional field for women, has shown a steady growth in the proportion of women since the 1970s. However, as in pharmacy, it has declined slightly recently. The proportion of women in engineering disciplines also exhibits a variety of trends. There are a high proportion of women in chemical, architectural, and environmental engineering, but civil, electrical, electronic, and mechanical engineering are not popular with female students.

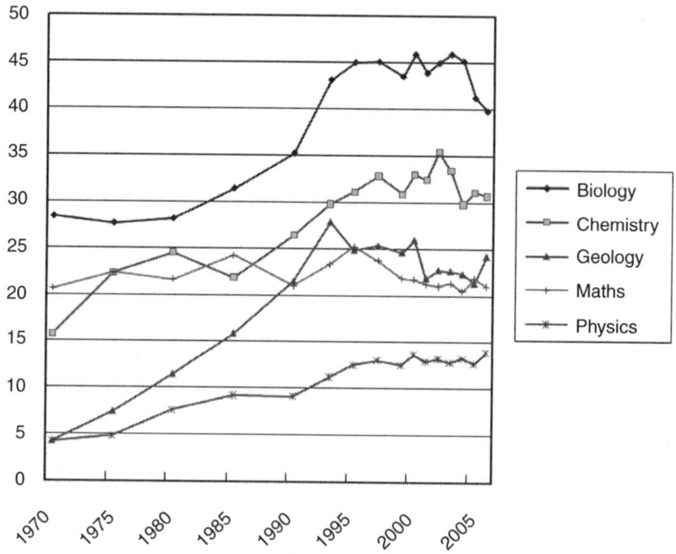

Figure 7.5 Proportion of Female Students in Science
Source: Constructed by author on the basis of data by Ministry of Education, Culture, Sports, Science and Technology (MEXT), Japan.

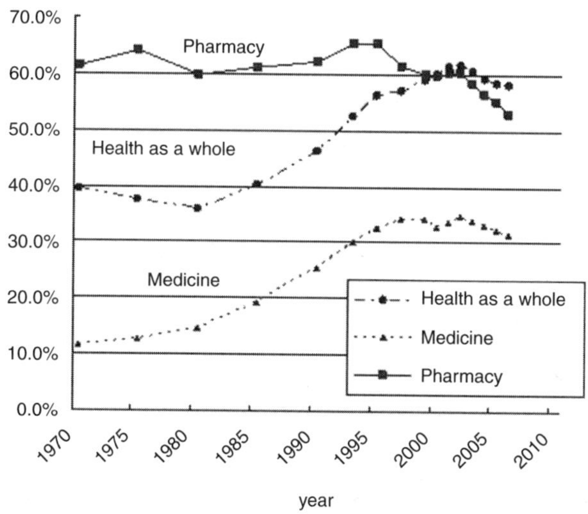

Figure 7.6 Proportion of Female Students in Medicine and Related Discipline
Source: Constructed by author on the basis of data by Ministry of Education, Culture, Sports, Science and Technology (MEXT), Japan.

As a whole, female students are increasing in practical disciplines in which they can obtain licenses, such as medicine, pharmacy, and architecture in engineering, but they remain few in physics and electrical, electronic, and mechanical engineering. The overall number of students in Japan is gradually decreasing, due to the falling birthrate. For this reason, some typical traditional women's universities have opened new science departments in order to secure enough students to justify their existence. In 2005, two famous private women's universities opened pharmacy departments. Recently, some women's universities have opened new departments for nationally licensed technicians. Career paths for female students in science and technology are gradually increasing, but those for postgraduates are still narrow.

Japanese Female Researchers in Science and Technology

Next, we will consider the present situation of female researchers in S&T. Differences in the proportion of female researchers depends on the institutions to which they belong. A total of almost 60 per cent of female researchers work for universities (including junior colleges and technical colleges), while the remaining roughly 30 per cent belong to private enterprises. On the other hand, almost 60 per cent of male researchers work for private enterprises, while 30 per cent belong to universities. The proportions of women researchers are 6.6 per cent in companies, 13.3 per cent of full-time researchers at national institutions, and 20.4 per cent of all researchers working for universities (MEXT, 2004; The Cabinet Office, 2004). In these three groups, the proportion of female researchers has increased year after year. In the five years since 1993, the average growth of the total number of researchers has been 2.5 per cent, while in contrast, the average growth of the number of female researchers has been 6.7 per cent. The most detail is available from universities where statistical material is abundant.

It is in engineering and science departments that a spurt in the proportion of female researchers has occurred. One reason is that these disciplines had too few women in the past. In 2002, the total number of researchers in the national, prefectural, municipal, and private universities reached 20,000 for the first time, and the number of male researchers reached 130,000. In 2006, the total number of female researchers in all universities was 28,600 and men and women together reached almost 164,500, with the proportion of women 16.4 per cent. However, a more detailed appraisal of these figures reveals a gender bias in these statistics (MEXT, 2006).

The proportion of female presidents is 7.6 per cent, of vice-presidents 5.5 per cent, of professors 10.6 per cent, of associate professors 17.7 per cent, of lecturers 25.2 per cent, and of assistants 25.2 per cent in 2006. The unexpectedly high number of female presidents is a result of the larger number of women in prefectural and municipal universities, which are more likely to appoint female presidents. Figure 7.7 shows temporal changes in the number of men and women in each position. For men, professor is the most common position, while for women assistant position is the most common.

Figure 7.7 covers all disciplines. In a diagram that showed only science and engineering, the assistant rate would increase further. In spite of the increase in overall numbers, there has been least growth among woman professors. However, in general, a slow increase in the proportion of women professors can be noted since 1985 (Figure 7.8).

Seen from an international viewpoint, Japanese gender equality has clearly fallen behind in the world. Figure 7.9 shows the comparative proportion of woman professors in a few selected countries.

The world average based on 14 countries' proportion of woman professors is 24 per cent, but Japan has the lowest at 7.9 per cent (including colleges). The corresponding proportions for science and engineering in Japan are 1.2 and 0.7 per cent, respectively.

How large is the proportion of Japanese female researchers in comparison with those in foreign countries? Take, for example, the proportion of female researchers in crystallography, a discipline distinguished among the physical sciences by its high proportion of female researchers. It may be significant that this discipline includes the collection and refining of data rather than the creative application of it, the latter being undertaken by complementary disciplines. The top ten highest proportions were in countries of the then Communist bloc and developing countries. This is also a common characteristic in other science disciplines. The highest placed of the Western countries was France, at 21 per cent ranked 11 out of 49. The USA, UK, and then West Germany ranked well below the middle, at 11 per cent, 10 per cent, and 5 per cent, respectively. Japan was ranked 46th, at 2 per cent, just above Bolivia, Ghana, and Kenya where there were no women crystallographers and where the total number of crystallographers is fewer than twelve overall. In fact, the total number of crystallographers in Japan was 516, after the USA with the largest number (1622), Russia with the second largest (657), and the UK with the third (651). While in the total number of crystallographers, Japan was ranked 4th, in

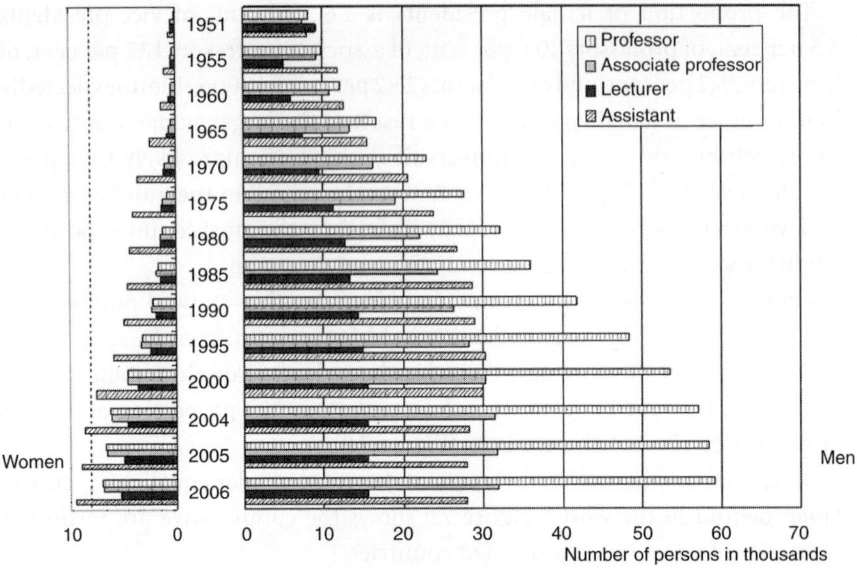

Figure 7.7 Distribution of University Staff by Gender

Source: Constructed by author on the basis of data by Ministry of Education, Culture, Sports, Science and Technology (MEXT), Japan.

terms of the proportion of women, it was ranked 46th. This is a realistic image of Japan.

In countries, where the total number of crystallographers was small, the proportion of women crystallographers was apparently high. The comparison of such proportions is not straightforward. However, in general, the poor situation of Japanese women crystallographers is clear, especially in comparison to USA and UK which had more than 10 per cent of women crystallographers and much bigger total number than that of Japan (Julian, 1990).

Ranking tables that indicate only the proportion of woman, without giving raw data, can be deceptive. For example, *Science*, in 1994 gave the proportion of female researchers in astronomy and physics, and has been often cited (*Science*, 1994: 1468, 1472). Yet it gave no indication of total numbers. Making allowance for this fact, the top three proportions of women in astronomy are in Georgia, Argentina, and Bulgaria, while the bottom three are in Japan, New Zealand, and Korea. The top three in physics are in Hungary, the Philippines, and Russia, while the bottom four are in Japan, the UK, the USA, and Korea. Generally speaking, the proportion of female physicists is

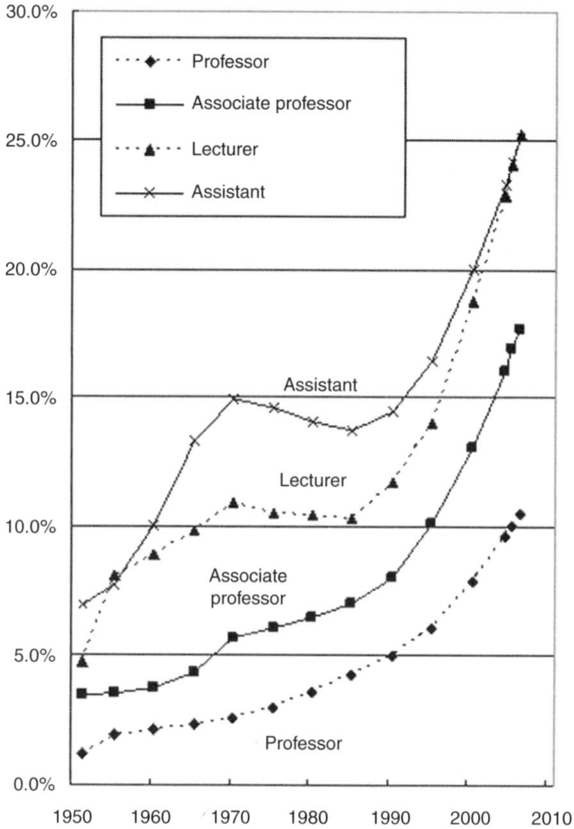

Figure 7.8 Proportion of Female University Staff

Source: Constructed by author on the basis of data by Ministry of Education, Culture, Sports, Science and Technology (MEXT), Japan.

lower in developed countries. This is related to the high prestige of physics in those societies. In particular, the prestige of physics in the USA was at an unparalleled high in the period of the Cold War. In crystallography, physics, and astronomy, French female researchers have made headway, but in all of these disciplines, Japanese female researchers remain at a strikingly low proportion.

In this regard, it is important to consider the ETAN report (by the European Technology Assessment Network on Women and Science, 2007). This provides statistics not only on numbers of female students in particular educational categories, but also follows their subsequent life-courses. Such

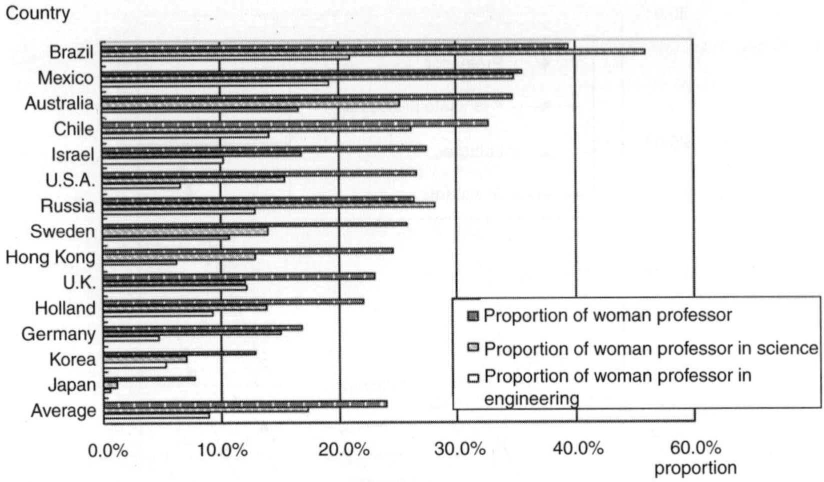

Figure 7.9 Proportion of Women Professors

Source: E.L Boyer, P. G Altbach, and Whitelaw, M.J (1994) *The Academic Profession:* An International Perspective, The Carnegie Foundation for the Advancement of Teaching, Princeton, NJ

research is crucial as it shows that, with the exception of computer science (in which women with families are more likely to be able to work at home), the proportion of women educated in scientific disciplines tends to decrease in the course of their career (the so-called leaky pipeline effect). Moreover, the pattern of decrease is not uniform for different disciplines. For example, in biology and chemistry, the proportion of women studying is strikingly high, while the proportion of women actually working in these disciplines is strikingly low. From the clinical philosophical viewpoint, tracing the proportions of these subsequently working in the various disciplines is important. Unfortunately, few studies comparable to the ETAN report exist for Japan.

These facts demonstrate how Japanese female researchers' status is poor. Historically, Japanese women scientists were not so far behind their contemporaries in the late nineteenth century, but problems emerged after the Second World War. The tremendous technological and economic development after the war was achieved under conditions of an extreme division of labour between the sexes. The opportunity of the true development of feminism from the 1960s in Japan was diverted by unprecedented economic development. The rate of Japan's economic growth at that time was over 10 per cent. However, despite the difficulties, Japanese women have gradually become independent.

The change is related to the decrease in birthrate. Since the 1980s, the birthrate has gradually decreased. Parents with only one or two children have tended to give as high an education to their daughters as to their sons. Women's access to higher education has resulted in the broadening of their professional talents in Japan. However, the country lacks adequate legislation to support women's education and their careers.

The UK, Canada, and Australia have reformed the principles of mathematics education for female students (Fennema and Leder, 1990; Kenway and Willis, 1998; Parker et al., 1996). In these countries, mathematics is regarded as the most fundamental subject for science education. If these governments did not cultivate the scientific talent of female students, that is, of half the population, it would be a national loss. If there is no intrinsic difference between the brains of the sexes, leaving female brains asleep means losing half of a country's treasury of ideas. The world has changed from heavy industry to light industry, and the interface between science and society has become more important and complicated than before. We may expect women scientists to add a fresh perspective to those of male scientists after former times. Japan must promote exploitation of this rich new source of ideas by various policies: affirmative action, quota systems, and various support structures. In 1999, United Nations Educational, Scientific and Cultural Organization (UNESCO) decided that the most important priority of science and technology was to promote more participation by women scientists. Japan must devise a drastic new gender policy in order to catch up from the long delay due to which Japan has fallen behind since the 1960s.

The New Strategies[3]

In recent years, new strategies have emerged. Some incentives for increasing numbers of female students and researchers in S&T are unprecedented in Japan. These stem from the activities of Japan Inter-Society Liaison Association Committee for Promoting Equal Participation of Men and Women in Science and Engineering (EPMEWSE) and the SCJ, and two new programmes of the Ministry of Education, Culture, Sports, Science, and Technology (MEXT). These incentives will contribute to a decrease in the horizontal segregation of female students in S&T and female scientists and engineers.

[3] This section basically depends on the author's presentation in the Workshop of *Organisation for Economic Co-operation and Development* (OECD) in Paris in 2005. See Ogawa (2006).

The EPMEWSE was founded in 2002 with only 12 member societies, but has now expanded to include 31 full member societies and 21 observer societies as of 2007. This is a unique meta-committee. Year after year, EPMEWSE's activities have become more influential.[4] The expansion of the application frame for research grants means that eligibility will be extended to part-time lecturers. This is because female scientists and engineers still find it difficult to obtain permanent positions. This committee has developed active lobbying, so it exerts great influence both on the election of members of the SCJ and on the policies of MEXT.

A science summer school for female high school students was held for the first time in 2005 under the joint auspices of the Physical Society of Japan, EPMEWSE, the National Women's Education Center, and the special committee of SCJ, and under the sponsorship of MEXT and the Cabinet Office. Hundred high school students and teachers took part. All participants hope to take part in similar events in future. It takes a long time to change something at this early stage of education, but it is the steadiest path to reform. In 2007, in addition to the summer school, the spring school was held in Kobe, which covered west Japan area, in contrast with the summer school in the eastern area.

In the SCJ election of September 2005, the number of female members increased dramatically. Before the election, EPMEWSE submitted a request for more female members to be elected from member societies. In 2000, the greatest expectation for female researchers was only that the percentage of female members would reach 10 per cent within the next ten years. However, this figure has reached 20 per cent in the last five years.

The Science and Technology Basic Plan for the third term 2006–10 has started. MEXT's two positive initiatives for female students and female researchers are of vital importance. One is to provide information on various career paths in S&T for female students at junior high school, high school, and university as well as for their parents; it also provides exchange opportunities between students and researchers in S&T. The second initiative is to balance research and childcare.

There are few female role models for scientists and engineers in Japan. Those that exist are in the public sector, not in the private sector (Figure 7.10).

[4] EPMEWSE summarised a large amount of data in 2004, which were brought with questionnaires to the members of the 39 academic societies. As many as 19,291 members sent their responses. EPMEWSE (2004).

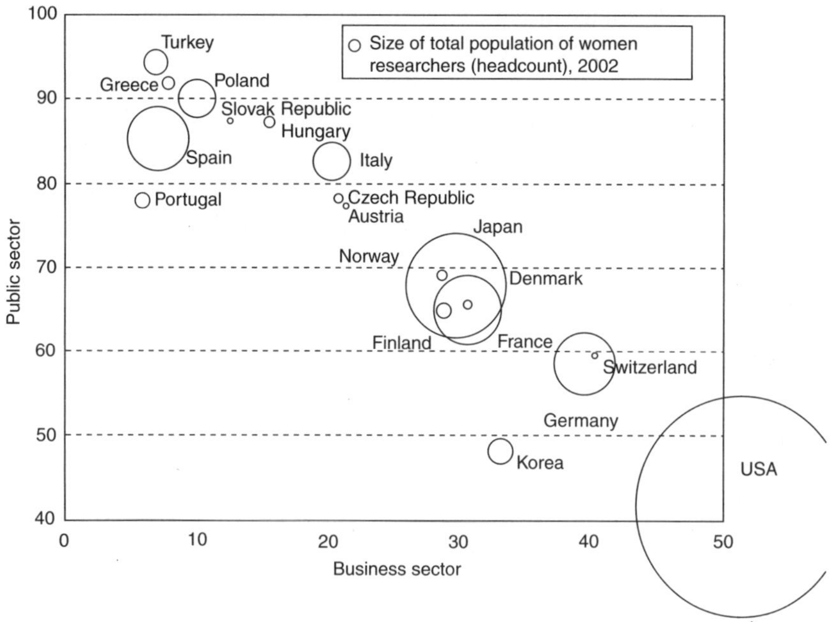

Figure 7.10 Business Sector vs Private Sector

Source: Main Science and Technology Indicators (MSTI, 2005), Women in Science: Unleashing the Potential, OECD (Organization for Economic Co-operation and Development), Paris, France.

Note: Germany and USA added by the author.

It is important to extend career paths for woman into the private sector (Ministère chargéde l'Enseignement Supérieur st de la Recherche, 2005: 23, Figure 5 on p. 9). This is especially true because, unusually, in Japan it is often the private sector that leads the way in scientific and technological innovation and discovery: witness recent Japanese Nobel Prize winners. A new program from MEXT aims to make junior high school and high school students aware of such role models in order to encourage them to choose majors in S&T; similar presentations will be made to university students to encourage various S&T-related job choices.

The interruption of research by childbirth and childcare is a big problem for female researchers in Japan. After childrearing, the barriers against returning to work are so high that many of them often give up their careers. However, a new programme from MEXT will supply these researchers with grants to encourage their return to work. MEXT is planning to allot 1.56 million euros to the programme.

As it is, the percentage of female researchers in Japan remains at 11.6 per cent. However, MEXT's Science and Technology Basic Plan for 2006–10 has the target of raising the percentage of female researchers to 25 per cent because the shortage of young researchers will become a serious problem in the near future. Female researchers will be indispensable for research and development in Japan in the twenty-first century. Now policies introduced by MEXT will encourage female students and female researchers. Japanese society is putting gender mainstreaming in S&T into gear. This will eventually bring about profound changes in dissolving horizontal segregation and even in the way we do science.

References

Akiko, Tsugawa and Kanomi Satoko (1996). *Hiraku* (Pioneering). Tokyo: Domesu.

Comparisons across cultures. Women in science (1994) *Science*, Vol. 263, Issue 5152, pp. 1467–1496.

Dresselhaus, Mildred (1986). 'Women Graduate Students', *Physics Today*, 39 (June). Issue 6, pp. 74–75.

European Technology Assessment Network on Women and Science (2000). *Science Policies in the European Union: Promoting Excellence through Mainstreaming Gender Equality.* Belgium: European Communities.

EPMEWSE (Japan Inter-Society Liaison Association Committee for Promoting Equal Participation of Men and Women in Science and Engineering) (2004). *Diverse Vision of Scientists and Engineers in the 21st Century: For the Promotion of Gender Equality.* Japan: EPMEWSE.

Fennema, E. and G. Leder (eds.) (1990). *Mathematics and Gender.* New York: Teachers College Press.

Holton, Gerald and Sonnert, Gerhard(1995) *Who Succeeds in Science?* New Brunswick: Rutgers University Press.

Ito, Kenji (2004). 'Gender and Physics in Early 20th Century Japan: Yuasa Toshiko's Case', *Historia Scientiarum*, 14(2): 118–36.

JAICOWS (Japanese Association for the Improvement of Conditions of Women Scientists) (ed.) (1996). *Jyoseikenkyusya no kanousei wo saguru* (Search for the Possibilities of Woman Scientists). Tokyo: Domesu Publisher.

Julian, Maureen M. (1990). 'Women in Crystallography', in G. Kass-Simon and Patricia Farnes (eds.), *Women of Science: Righting the Record*, pp. 335–83. Bloomington & Indianapolis: Indiana University Press.

Keller, Evelyn Fox (1985). *Reflections on Gender and Science.* New Haven and London: Yale University Press.

Kenway, J. and S. Willis (1998). *Answering Back.* London: Routledge.

Kozai, K., S. Kawashima, T. Tominaga, T. Hisatome, and K. Saruhashi (eds.) (2001). *My Life: Twenty Japanese Women Scientists.* Tokyo: Uchida Rokakubo.

Kuwahara, Masako (1999). 'Jyender to Kagakugijyutu' (Gender and Science/Technology), in Shigeru Nakayama, Kunio Goto, and Hitoshi Yoshioka (eds.), *Tsushi Nihon no Kagakugijyutsu* (The Social History of Science and Technology in Contemporary Japan), Vol. 5-II, pp. 718–36. Tokyo: Gakuyōshobō.

Kuwahara, Motoko (2001). 'Japanese Women in Science and Technology', *Minerva*, 39,2: 203–16.

MEXT (Ministry of Education, Culture, Sports, Science and Technology) (2004). *The White Paper on Science and Technology in Japan* (in Japanese). Tokyo, Japan.

—— (2006). *Gakko Kihontyosa*(School Basic Survey). Ministry of Education, Culture, Sports, Science and Technology (MEXT), Tokyo, Japan.

Ministère chargéde l'Enseignement Supérieur st de la Recherche (Ministry of Higher Education and Research) (2005). *Background Paper - Document D'Information: Joint OECD-French Government Workshop on 'Women in Scientific Careers: Unleashing the Potential'*. 16–17 November.

Ogawa, Mariko (2006). 'Women's Careers in Science and Technology in Japan', *Women in Scientific Careers: Unleashing the Potential*, pp. 87–94. Organisation for Economic Co-operation and Development (OECD) Publishing.

Oreskes, Naomi (1996). 'Objectivity or Heroism? On the Invisibility of Women in Science', *Osiris*, 11: 87–113.

Parker, L., L. Rennie, and B. Fraser (1996). *Gender, Science and Mathematics*. Dordrecht, The Netherlands: Kluwer Academic Publisher.

Patessio, Mara and Mariko Ogawa (2005). 'To Become a Woman Doctor in Early Meiji Japan (1868–1890): Women's Struggles and Ambitions', *Historia Scientiarum*, 15(2): 159–76.

Schiebinger, Londa (1999). *Has Feminism Changed Science?* Cambridge, Massachusetts: Harvard University Press.

The Cabinet Office (2004). *FY2004 Annual Report on the State of Formation of a Gender-Equal Society*. The Cabinet Office. Ministry of Education, Culture, Sports, Science and Technology (MEXT), Tokyo, Japan.

Yokoo, Yoshiko (1993). *Jyoseikenkyusha no genjyo ni kansuru kisotyosa* (The Report on Female Researchers in Japan) National Institute of Science and Technology Policy (NISTEP).

Yokoo, Yoshiko (1993). *Jyoseikenkyusha no genjyo ni kansuru kisotyosa* (The Report on Female Researchers in Japan), NISTEP Report No. 30, National Institute of Science and Technology Policy (NISTEP) Science and Technology Agency.

CHAPTER EIGHT

Saudi Women

Their Role in Science and Education

Samira I. Islam

SCIENCE education in most Muslim and Arab countries begins between the age of six and seven and is taught as an integrated compulsory subject to both boys as well as girls. The major science disciplines are then studied separately in the last two or three years of the high-school education (Sedgwick, 2001). Fewer girls than boys are enrolled in high school science curricula. There are various reasons related to gender stereotyping such as misleading perceptions that science and technology are subjects more suitable for boys. Another reason is the failure of the curricula to relate science and technology to the everyday life of women. Thus, there is a self-inhibition among school girls that affects not only the number of young women entering universities to study science and technology subjects, but it also results in the reluctance of talented women in introducing their own values and visions into a working world dominated by men.

Arab countries vary greatly in their culture, traditions, and social system. There is a wide range of attitudes towards educating women at the university level, for example in Egypt, women have attended university since the 1920s, whereas in other countries a university education for women is a recent phenomenon. Saudi women were admitted to formal university education in 1973. Although women have the right to a university education, those in more traditional, rural areas often do not exercise that right, whether for social, economic, or family reasons.

The percentage of females enrolled in science and technology university courses ranges from 70 per cent in the United Arab Emirates to 8 per cent in Djibouti (UNESCO, 1993). In Egypt, 35 per cent of science undergraduates are women. In certain disciplines, such as public health at the Lebanese universities, chemistry in universities in Iraq, and pharmacy in Syrian universities, there are more female than male undergraduate students. Women tend preferably to enroll in life sciences and chemistry, with far fewer studying physics, mathematics, and engineering. This seems more to be the result of female students choosing these subjects than some active discrimination by the education system. This trend is also seen among American and European women. For example, in some European countries, women constitute 40 per cent of natural science undergraduates, 28 per cent of mathematics and computer undergraduates, and 20 per cent of engineering undergraduates (European Commission, 1999; Lewis and Blenkinsopp, 1995).

In Arab countries, although accurate statistics are scarce, career opportunities in universities, research institutes, and scientific organisations are considerably more limited for female science graduates than those for men, particularly in senior positions (Hassan, 2003). The available published literature mainly discusses how technological changes have affected the household duties of women or the work environment where automation has led to their displacement. In Egypt and Saudi Arabia, more than 40 per cent of the faculty is women in disciplines such as pharmacy and dentistry. In the sciences, 25 per cent of the faculty are women, but this decreases to less than 10 per cent in departments of engineering and technology. These patterns are similar to those seen in US universities, where women constitute 50 per cent of health sciences faculty, 23.8 per cent of biological sciences faculty, and 6.1 per cent of engineering faculty. Although there is an increasing pool of highly qualified women scientists in some Arab countries, few women in universities and science institutes are holding top posts (Lewis and Blenkinsopp, 1995).

Saudi Women's Advancement in the Field of Science

This study aims to document Saudi women's advancement in the field of science and their contribution to the development of Saudi Arabia. Data documented in this study refers, unless otherwise stated, to the academic year 2002–03 which will be referred to as 2003. This data was sourced from the Ministries of Education, Higher Education, and Civil Service of the Kingdom of Saudi Arabia during the years 2002–03. Other sources are identified in the list of references. The term science is used to express all scientific and technical disciplines. Before examining the situation of women in science in Saudi Arabia, it is desirable to discuss the educational system of the Kingdom.

Education in Saudi Arabia

The history of higher education in Saudi Arabia is an important example of educational progress in general and women's education in particular. The modern education system in Saudi Arabia began in 1926, and the creation of the Riyadh University (now known as the King Saud University) in 1957, with an enrollment of 21 students and a staff of nine, marked the advent of the modern university education in the country. The education of boys' in Saudi Arabia started long before the establishment of the Ministry of Education, which took place in 1952 and led to the increase of education of men at several levels. Under the inspiration of the late then Crown Prince Faisal and his wife, Effat, formal schooling for girls started in 1960 (Islam, 2001). The Ministry of Higher Education, which oversees the Kingdom's colleges and universities, was established in 1975.

In Saudi Arabia, physically, the educational organisations are segregated by sex and divided into three separately administered systems: (a) general education for boys; (b) education for girls; and (c) traditional Islamic education for boys. Administratively, two organisations oversee most education institution in the Kingdom: (a) The Ministry of Education, which supervises the education of boys, special education programmes for the handicapped, adult education, and junior colleges for men and (b) The General Presidency for Girls' Education, an organisation staffed by *Ulama* and working in close cooperation with the Ministry of Education, oversees the general education of girls, kindergartens, nurseries for boys and girls, and women literacy programmes, as well as colleges of education and junior colleges for girls. In the year 2002, the Ministry of Education and the General Presidency for Girls' Education became one educational body.

Entering a new era of rapid development of science infrastructure and economy, the Kingdom embarked on a long-term master plan to enable the Saudi educational system to provide the highly trained manpower needed for the increasingly sophisticated economy. This led to a dramatic increase in student enrollment into university studies. Seven thousand students enrolled in 1970; 272,615 in 1997–98 of which 50.7 per cent were women; 444,800 in 2001–02 of which 55.8 per cent were women; and in the academic year 2002–03 the total students enrolled at Saudi universities was 525,344 out of which 305,988 (58.2 per cent) were women (Central Department of Statistics, 2002).

There are 81 Colleges of Education for Girls (CEG), which grant B.A. and B.Sc. degrees to qualify as teachers. Fourteen of these CEG grant Baccalaureate in Science subjects. In 2002–03, a total of 5437 women science teachers graduated from these fourteen CEG science colleges. There are eight major government universities, which have 39 colleges (faculties) that branch in towns throughout the Kingdom. Six of these universities teach science subjects (Table 8.1); however only five of them admit women students (Table 8.2). In 2002–03, 9,456 B.Sc. male and female students graduated from the different Science discipline (Table 8.3), while the total enrolled students were 54,512 of which 25,876 (48 per cent) were women (Central Department of Statistics, 2003).

The Workforce: Teaching, Health, and Research

Though higher education sector has seen enormous growth in Saudi Arabia, gender based segregation is one issue that probably continues to be unresolved and contested (Onsman 2011). The development of King Abdullah University of Science Technology (KAUST), first mixed-gender university campus, and Princess Nourah University, however, exemplifies the efforts in this direction. This section will describe the status of women students and teachers in university colleges in Saudi Arabia using various data sets. Table 8.1 describes Saudi Universities Teaching Science Subjects; Table 8.2 shows Years of Establishment of the Science Faculties and Women's access to these; Table 8.3 reveals Number of Students Enrolled and Saudi Staff; and Table 8.3b provides the percentage of Students and Faculty Members in Universities and Colleges.

Table 8.1 *Saudi Universities Teaching Science Subjects (during 2002–03)*

Universities	Location	No. of students in science colleges		% Female
		Male	Female	
Umm Al-Qura University	Makkah	4100	5119	55.5
King Saud University	Riyadh	10950	10519	48.9
King Abdulaziz University	Jeddah	5627	5355	48.8
King Fahd for Petroleum and Minerals University	Dammam	6896	–	0
King Faisal University	Al Ahsa	4728	4767	50.2
King Khaled University	Abha	3231	116	3.48

Source: Ministries of Education, Higher Education and Civil Service of the Kingdom of Saudi Arabia, 2002–2003.

Table 8.2 *Years of Establishment of the Science Disciplines' Faculties and Women's Admission in the Saudi Universities*

Universities' Faculties	King Saud		King Abdulaziz		King Faisal		King Khaled		Umm Al- Qura	
	Men	Women	Men	Women	Men	Women	Men	Women	Men	Women
Medicine	1969	1974	1975	1975	1975	1976	1980	2003	1996	1999
Dentistry	1995	1998	1987	1987	1995	–	–	–	–	–
Pharmacy	1960	1980	2002	2002	2002	–	2003	–	–	–
Nursing	1976	1976	–	1975	–	1988	1999	–	–	–
Medical Laboratory Technology	–	–	1982	2002	1989	–	2002	2002	–	–
Veterinary	1981	–	–	–	1975	–	–	–	–	–
Basic Science	1958	1985	1974	1975	1994		1976	2003	1981	
Education	1966		1977	1977	1979		1976		1952	1973
Computer Science	1984	1984	1985	1985	–	–	2003	2002	1981	–
Agricultural Science	1965	–	1974	–	1975	–	–	–	–	–

Source: Ministries of Education, Higher Education and Civil Service of the Kingdom of Saudi Arabia, 2002–2003.

Table 8.3 *Number of Enrolled Students and Saudi Staff (during 2002–03)*

Universities	Graduate	Total Students	Total Saudi Staff
Umm Al-Qura (Makkah)	526	9,219	369
King Saud (Riyadh)	3,961	21,469	327
King Abdulaziz (Jeddah)	2,594	10,982	799
King Faisal (Ahsa)	1,506	9,495	0
King Khaled	869	3,347	0

Source: Ministries of Education, Higher Education and Civil Service of the Kingdom of Saudi Arabia, 2002–2003.

Table 8.3b *Colleges, Students, and Faculty Members in Universities and Women's Colleges*

		Percentage of female students			
University	No. of colleges	Bachelor	Postgrad.	Total	Faculty
King Saud	14	37.0	34.8	36.9	22.5
Islamic/Madina	5	0.0	0.0	0.0	0.0
King Fahd	6	0.0	0.0	0.0	0.0
King Abdul Aziz	11	41.8	59.2	42.3	34.5
Imam Moh'd Bin Saud	5	22.6	22.2	2.6	9.7
King Faisal	10	50.1	31.9	49.5	22.1
Ummul Qura	9	43.7	28.0	42.5	20.9
King Khaled	8	0.9	–	0.9	0.1
Total	68	32.9	32.0	32.8	19.0
Women's colleges		100	100	100	90.5
distribution of females in universities		24.6	74.1	25.3	29.2
distribution of females in colleges		75.4	25.9	74.7	70.8

Source: Ministries of Education, Higher Education and Civil Service of the Kingdom of Saudi Arabia, 2002–2003.

Facilities for Research and Postgraduate Studies and Awards

Research Facilities

Several research institutions and programmes are well established in the Kingdom either as independent entities or attached to universities. The following are modules of such establishment.

Postgraduate Degrees

In the year 1999–2000, 372 graduates were granted Masters Degree in Science subjects of which 39 per cent were women. PhD degrees were granted to 52 graduates of which 79 per cent were women. This high percentage is due to the preference of males for studying abroad, especially for post-graduate work, while the number of women is more limited for such opportunities. Additionally, the fourteen teacher training science colleges, which only enroll women, offer postgraduate degrees as well.

Postgraduate Studies for Women Granted by UK Universities
The external students Joint Supervision Programme, is a collaborative support established between some United Kingdom (UK) universities and King Abdulaziz University (KAAU) for granting women postgraduate degrees from a UK university who provide the main supervisor. The KAAU staff members, although they are actually supervising the graduate students who perform the research work in their departments, act as external co-supervisor thus avoiding the need to travel abroad (Table 8.4).

King Fahd Medical Research Center is a part of Medicine and Allied Sciences of KAAU. It has 21 laboratory units, seven of which are headed by women (Table 8.5).

Funds for Research

Since the establishment of King Abdulaziz University Research Funds Program in 1982, one thousand seventy one (1071) Research projects

Table 8.4 *Ph.D Degrees Granted to Women through the U.K. Joint Supervision Programme*

Specialisation	Number	Percentage
Science	11	45.8
Administration	3	12.5
Arts	2	8.3
Medicine	7	29.2
Home Economics	1	4.2
Total (1990–2004)	24	100

Source: Ministries of Education, Higher Education and Civil Service of the Kingdom of Saudi Arabia, 2002–2003.

Table 8.5 *Number of Men and Women in Medical Profession*

	Men	Total	% Female
Consultative Doctor	745	226	23.30
Specialist	1,707	921	35.10
General.Practioner (G.P).	1,877	1,394	29.80
Diagnostician doctor	652	611	48.30
Pharmacist	632	289	31.30
Technician	14,048	7,336	34.30
Health Helper	139	1,495	8.50

Source: Ministries of Education, Higher Education and Civil Service of the Kingdom of Saudi Arabia, 2002–2003.

Table 8.6 *KACST-funded Programmes up to the Year 2003*

Programmes	Number of funded projects	Percentage
General Grand	566	34.8
Small Projects	335	20.6
Humanities	22	1.35
Graduate Students	612	37.6
National	86	5.29
Production Sector	3	0.18
Consultation	1	0.06
Joint External	1	0.06
Total	1626	100

were funded, of which 6.1 per cent projects had women as principal investigators. King Abdulaziz City for Science and Technology (KACST) is the highest government administrative sector in-charge of research all over the country (Directorate of Information, 2002) . From the time of its establishment in 1976 until the year 2003, more than 1,626 projects were funded with a budget exceeding 175 million US dollars (Table 8.6). Health and environment research projects had 45 per cent of the budget. The KACST's programmes are categorised according to the size of their budgets into: National, Grand, Small, and Graduate Students projects. Although Saudi women are actively involved in research, yet no woman gained a national projects support, and only 2.9 per cent of the grand projects, 12.7 per cent of the small projects, and 55 per cent of the students'

Table 8.7 *Women as Heads of Units at King Fahd Medical Research Center*

Name	Laboratory
Prof. Samira I. Islam	Drug Monitoring Unit
Dr. Soad AL Jaouni	Haematology Research Unit
Dr. Suhad Matoog	Nutrition research Unit
Dr. Faten Khorshed	Tissue Culture Unit
Dr. Samr Al Saggaf	Histological Slides Preparation Unit
Dr. Sawsan Jalalah	Electron Microscope Unit
Dr. Houda Abo Aurki	Animal House Unit

Table 8.7b *KACST Research Grants Awarded up to the Year 2002*

Grant type	Establishment date	No. of projects	Principal Investigators		% Female
			Male	Female	
National	1981	13	13	–	0
Grand	1979	68	66	2	2
Small	1996	118	103	15	12.7
Graduate students	1991	131	59	72	55

projects had women as principal investigators (Table 8.7) (Directorate of Information, 2002).

Awards

King Faisal International Prize (KFIP) is one the most distinguished world's pre-eminent scientific and humanitarian award. It was established in 1977 and its categories are – Science, Medicine, Arabic Literature, Islamic studies, and Service to Islam. There have been 134 Laureates from 33 countries worldwide. Out of these Laureates, three of which are women (King Faisal Foundation), none of them is Saudi, while fourteen Saudi men were granted the KFIP, none of them were in the science field.

Discussion

Science is a human activity and is influenced by the cultural norms and values. Science and technology exert major influence and are determining factors in the human development, yet, it remains a predominantly male domain.

Why? Could it be due to the under-representation of women, especially at the leadership levels or due to the overall social status of women, which has created stereotypes for the role of women in society?

The primary education for Saudi women was first established in 1960 and the first formal admission for girls into Saudi universities was in 1973. Women currently comprise 46 per cent of the secondary school students and 58 per cent of the university and the higher education students. However, women are still limited from studying certain subjects (UNDP-POGA, n.d.). When compared with the Egyptian women, admitted into the Egyptian universities in 1920s and with the European women who had the privilege of access to the university much earlier, it is noteworthy that 51.7 per cent of the Saudi undergraduates in the Science colleges are women, while in Egypt it is 35 per cent and in some European countries it is only 40 per cent (Hassan, 2003; Sedgwick, 2001).

The number of Saudi women working in the education sector exceed any other sectors (Table 8.8) (King Faisal Foundation Public Relation Department, n.d.). Saudi women scientists excel in the fields of teaching, health and research. These maybe due to the tendency of women to disseminate their knowledge at large which do not only influence their lives but also the lives of the children and society in general.

The health sector has always been given preferential emphasis. Although more women are required in this field, yet their participation and the presence require more appreciation and recognition. The increase in the percentage of women in the education field over that of women in the health sector may be due to the fact that, except in very rare circumstances, teaching of women at all levels is conducted solely by women while healthcare of women is performed by either sex.

King Abdulaziz University staff members in the basic science faculty only for the year 2003 were 439, out of which 14.2 per cent were women. Yet the

Table 8.8 *Number of Staff in Education, Health and University*

	Male	Female	% Female
Education Staff	1,78,230	1,79,836	50.20
Health Staff	21,106	10,916	34.03
University Staff	5,391	1,334	19.80

Source: Ministries of Education, Higher Education and Civil Service of the Kingdom of Saudi Arabia, 2002–2003.

numbers of male and female students are almost equal, this reveals the greater burden that women staff members have to bear.

The total KACST supported projects up to and including 2003 is 1626 (Table 8.6). No woman ever gained the national projects' grants and only two women (2.9 per cent) were among the grand projects' programme winners; 12.7 per cent of the small projects' programme was awarded to projects which had women principal investigators, while 55 per cent of the graduate students' funds were granted to women (Table 8.7).

Within the workforce environment, Saudi women constitute 18 per cent of the total, contrary to the report by the United Nations Development Plan (UNDP) that women comprise 15 per cent of the labour force, which is barely over half of the regional average. Within the Arab Region, women constitute 29 per cent of the total work force (ESCWA, 2003). Table 8.9 shows the pattern of promotion of government employees in the year 2002, which emphasises the restriction of women in achieving key positions such as ranks 12 and 13 in the government positions.

In spite of the women's achievement at educational level, they are still not fully recognised in the government workforce. Table 8.9 demonstrates the gender-rank relationships. Highest ranks are held by men, it requires a man at the highest rank of decision-making, even if he possesses lower educational background than his female counterpart. Women's empowerment and their full participation on the basis of equality in all spheres of society, including participation in the decision-making processes and access to power, are fundamental for the achievement of equality, development, and peace (United Nation, 2001).

Table 8.9 *Pattern of Employment at Various Ranks in Saudi Arabia*

	Male	Female	% Female
2	169	2	1.2
3	968	27	2.7
4	1657	75	4.3
5	2455	121	4.6
6	3393	100	2.3
7	2982	109	3.5
8	2651	59	2.17
9	2043	28	1.3
10	1321	18	1.3
11	549	2	0.3
12	308	0	0
13	142	0	0

Strategies for Women in the Parliament in the Arab World

The historical socio-economic and political conditions of a society are essential aspect of understanding a woman's position in the society. The political participation of women in the Arab region is still low compared to some Asian, sub-Saharan Africa, and Pacific countries. In the Arab Region out of 3010 seats, women hold 176 seats (5.8 per cent) (ESCWA, 2003). In September 2000, the Saudi government signed the agreement to the Convention on the Elimination of All Forms of Discriminations against Women (CEDAW). In Saudi Arabia women are still fighting to have seats in the Shurah Parliament to ensure the full implementation of the human rights of women and the unborn female child as an integral and indivisible part of all human rights and fundamental freedom.

Conclusion

King Faisal International Prize has been given to 45 laureates; out of this 6.7 per cent are women. While out of the 761 total Nobel Prize laureates only 4.1 per cent are women. This emphasises the global nature of the problem faced by women in science.

According to ESCWA, Saudi illiteracy rate is 15.4 per cent for men and 20.2 per cent for women with a gender gap of 13.8 per cent. The average gender gap mean value of all the Arab countries is 22 per cent (ESCWA, 2003). This shows that despites the educational advancement in some Arab countries, the gender gap is still higher than Saudi Arabia. Figures gathered on the Saudi women are more optimistic when compared with the relatively advanced Arab and Western countries. Certainly Saudi women should be praised for the achievements they have managed to accomplish within the limited period of 1960 to 2002. Once out of school, however, the opportunities available to Arab women are diminished. The fields of Science and Technology, particularly engineering, are male territory. In Saudi Arabia, and countries such as Jordan and Kuwait, women make up only about 20 per cent of total scientific researchers. That gender disparity is reflected also in Saudi government grants for research and in promotions within the government workforce.

However, it is clear that the gender gap issue is an international phenomenon and needs to be addressed effectively. It is highly essential to design, implement, and monitor the full participation of women. Effective, efficient, and mutually reinforcing gender-sensitive policies and programmes, including the development policies at all levels that will foster the empowerment and advancement of woman, are thus essential.

References

Central Department of Statistics (2002). *Statistical Year Book.* 38th edition. Central Department of Statistics, Ministry of Economy and Planning, Kingdom of Saudi Arabia.

—— (2003). *Statistical Year Book.* 39th edition. Central Department of Statistics, Ministry of Economy and Planning, Kingdom of Saudi Arabia.

Directorate of Information (2002). King Abdulaziz City for Science Technology. Riyadh, Saudi Arabia.

European Commission (1999). *A Report from the European Technology Assessment Network (ETAN) Expert Working Group on Women and Science,* Expert Working Group on Women and Science, European Communities, Brussels, Luxembourg.

ESCWA (2003).*Women and Men in the Arab Countries.* Economic and Social Commission for Western Asia (ESCWA), United Nations.

Hassan, F. (2003). 'Essay in Science and Society: Islamic Women in Science', *Science,* 2000, 290 (5489): 55–56.

Islam, S. (2001). *Effat Al Thunayan: History and Achievements.*

King Abdulaziz University (2002). *Years Report.*

King Faisal Foundation Public Relation Department (n.d.). Available online at www.elfi.ruhr-uni-bochum.de/begleit/doks/Faisal.htm (retrieved on 28 March 2004).

Lewis, A. and A. Blenkinsopp (1995). 'Pharmacists' Contribution to the Women's Health in the United Kingdom'. *Gender Related Health Issues: An International Perspective.*

Ministry of Civil Service (2002a). *Report: Achievements for the year 1422/1423H.*

—— (2002b). *Report: In numbers for the year 1422/1423H.*

Onsman, Andrys (2011) 'It is Better to Light a Candle than to Ban the Darkness: Government led Academic Development in Saudi Arabian Universities', *Higher Education,* Volume 62, Number 4, pp. 519–532.

Sedgwick, R. (2001). 'Education in Saudi Arabia', *World Education News and Review.* 14(6): 1–8.

United Nation (2001). *Beijing Declaration and Platform for Action with the Beijing + 5 Political Declaration and Outcome.* New York: Department of Public Information, United Nations.

UNDP-POGA (n.d.). *Program on Governance in the Arab Region: Saudi Arabia – Gender.*

UNESCO (1993). *The Higher Education System in the Arab States, UNESCO.*United Nations Educational and Scientific Research Organization.

—— (1999). 'Women in Science - Quality and Equality Conditions for Sustainable Human Development. Women, Science and Technology: Towards a New Development?' World Conference on Science. Budapest, Hungary, July.

CHAPTER NINE

Changing the Facts

Gender Dimensions of the South African Public Science System

Luci Abrahams

> Economies need the talent of their entire population, male and female, to create a leading edge workforce...
>
> *Asia Pacific Economic Cooperation (APEC) 1998: p. 3*

FIFTEEN years after the ushering in of democracy in 1994, South Africa faces critical challenges for building a future economy and society which will sustain and develop its people. The decline of mining and resource-based industries over the past two decades, the challenges of global competitiveness for the manufacturing and services sectors, high unemployment and the need to address the causes of poverty have created increased demand for innovation in the private and public sectors. The increased global demand for science and technology inputs to key industries such as automotive manufacturing and pharmaceuticals, emerging sectors such as biodiversity, advanced materials, and 'green' industries, as well as social sciences and technology inputs to poverty reduction has created enormous and challenging opportunity for the South African research community. While increased levels of productive scientific activity can contribute to growth and development, this economic relationship requires a continuously growing scientific population, capable of creating and sustaining levels of innovation that will contribute to positive GDP per capita growth year after year. Furthermore, public policy must not simply encourage an increase in the number of women and men in the scientific population, but must aim to put the talent for scientific and technological innovation at the service of the poor. These are ambitious tasks in an emerging economy. For these reasons, the issues for South Africa science policy today include stating the actions necessary to expand the national knowledge base in terms of size, capability, gender participation, fields of scientific endeavour, and the value of such innovative capacity to society at large. This chapter explores the gender dimensions of the public science segment within the national system of innovation.

South Africa has witnessed innovations in many fields, including hi-tech heart surgery, strategic fuel industry, energy technology, software engineering, bio-informatics and some other spheres (Addison, 2000). These innovations have entered both the local and global markets, enabling South Africa to develop new services and product lines, to attract new sources of revenue and to create jobs in knowledge intensive economic sectors. South Africa has an estimated 40,084 scientists and researchers (DST, 2009: 7) who make the national system of innovation a reality. Sydney Brenner, Nobel Laureate, 2002, for establishing the nematode worm, *Caenorhabditis elegans*, as a 'novel model organism' (The Nobel Assembly, 2002), and Mark Shuttleworth, a pioneer in encryption technology for Internet-based transactions, are the world-renowned South Africans who have made scientific and technological contributions to the global knowledge. South Africa has a large number of scientists and researchers (both female as well as male), who are though less known for their contributions to global knowledge, but participate extensively in the National System of Innovation. Women are not well represented in the science, engineering, and technology sector, but some of them do have very successful careers and have achieved recognition as leaders in these fields. Professor Tebello Nyokong, who heads the Physical-Inorganic Chemistry Department at Rhodes University, for example, is one of the leading female scientists. Diane Hildebrandt is another famous female professor and is one of fourteen A-rated scientists at the University of the Witwatersrand.[1] Nyokong is known for her research in the photodynamic therapy of cancer and work in nano technology innovation and Hildebrandt has a name for her research in process synthesis for competitive industrial processes (NACI, 2008). However, the participation of South African men and women in science, engineering and technology (SET) is lower than that required in order to effectively advance

[1] The National Research Foundation rates scientists as 'leading international researcher' (A), 'internationally acclaimed researcher' (B) and other categories.

the new technology missions of science for competitiveness and science for development (DST, 2002: 54–56). This chapter discusses selected findings of two studies,[2] which are relevant to the South African science policy with respect to building the 'women in science, engineering and technology' (SET) knowledge base.

The Landscape of Innovation in Science, Engineering and Technology

The National System of Innovation (NSI) comprises groups of inter-locking institutions including scientific research councils and performing agencies, higher educational institutions, and innovation-focused operations in private sector firms and state-owned enterprises. South Africa is classified as a lower middle-income country (World Bank, undated). Its GDP for 2009 was estimated at USD 237.6 billion[3] (Stats SA, 2010: 8) and gross domestic expenditure on research and development as a percentage of GDP (GERD) was estimated at 0.93 per cent or USD 2.48[4] for the period 2007–08, the latest reported period (DST, 2009: 7). This investment in Research and Development is a notable increase from 0.76 per cent of GDP in 2001, and compares well to around 0.9 per cent for Brazil (2004 data), 1.34 per cent for China (2005 data) (OECD, 2007), and 0.94 per cent for India (2004–05 data) (DST India, 2005)[5] – countries which, like South Africa, aim to reshape their economies through, among other measures, increasing their R&D intensity.[6] However, South Africa's aim to achieve a GERD of 1 per cent by 2008 (Mangena, 2005) has not been met. India, on the other hand, has announced that it aims to increase GERD to 2 per cent of GDP in the next five years (CSIR India, 2010). Figure 9.1 shows the 2004 and 2005 GERD data for OECD countries and a selection of developing countries

[2] The data used in this chapter is drawn from a large research study in 2003 and a follow-up study in 2008, both conducted by the Centre for Research on Science and Technology, CREST) and reported in *Facing the Facts: Women's Participation in Science, Engineering and Technology 2004* and *2009*.

[3] ZAR 1782.0bn using a conversion rate of 7.5 ZAR to the USD.

[4] ZAR18.6bn reported in the National Survey of Research and Experimental Development 2007/8 equivalent to USD2.48bn at conversion rate of 7.5.

[5] The OECD (2007) data records 0.69 for India, 0.91 for Brazil and 1.23 for China, but OECD notes that the data is underestimated for India, Brazil, and South Africa.

[6] Also see Westholm G, Tchatchoua B & Tindemans P (2004), as well as Ping, Zhou & Leydesdorff (2006) for a discussion of R&D expenditure in India and China respectively.

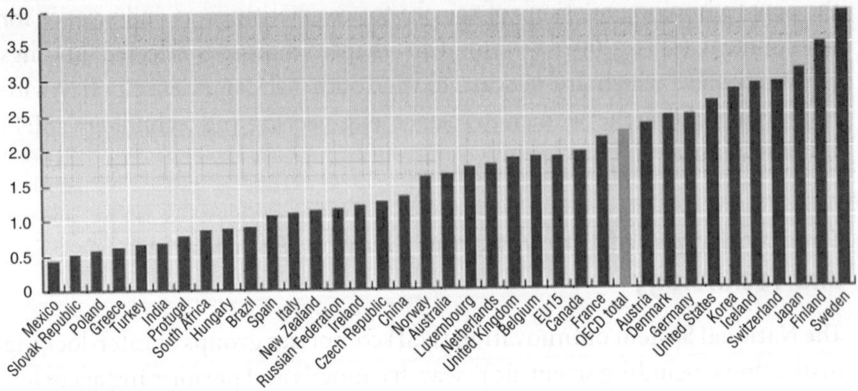

Figure 9.1 Gross domestic expenditure on R&D (GERD) as a percentage of GDP* (35 countries, EU15, OECD total)

Source: OECD, 2007.

**As a percentage of GDP, 2005 or latest available year.*

including the so-called BRICS (Brazil, Russia, India, China, and South Africa) countries.[7]

Given the economic constraints presented by the existence of a small, but highly educated workforce, its history as a largely resource-based economy and other dissimilarities with countries like Brazil, India, and China, South Africa needs suitable policy instruments to ensure a constant rate of increase in its GERD over time. The growth of Research and Development (R&D) workforce in general and the participation of greater numbers of women in R&D will be a key requirement for increasing the capacities of its national system of innovation in order to absorb financial inputs and translate them into scientific, economic, and social value. The 2007–08 national R&D survey[8] reports a small increase in the percentage of women researchers as a proportion of the total number of researchers to 40.3 per cent (DST, 2009: 7).

On the policy side, the White Paper on Science and Technology (DACST, 1996) has established new technology missions – science for competitiveness and science for development. The National Biotechnology Strategy (DACST, 2001) laid the foundations for the creation of a biotechnology industry through public funding of biotechnology regional innovation centres and other measures. Advanced Manufacturing Technology Strategy aimed at introducing applications in advanced manufacturing to secure competitive

[7] Brazil, Russian Federation, India, China, and South Africa.

[8] Conducted according to the Frascati Manual.

advantage and an increase in the Competitive Industrial Performance Index (CIPI),[9] CIPI from 0.10 in 2002 to 0.20 in 2014 (DST, 2003a: 30). The 2018 Ten-year Innovation Plan (DST, 2008) sets its vision on participation in the global pharmaceutical industry, satellite-based services, a sustainable energy sector, becoming a leading participant in climate science and making a contribution to the goal of halving poverty. There has been a concomitant shift in focus in publicly funded science councils and universities with increasing activity in these focus areas, but apparently on a small scale.

The strategies outlined above, active mega-projects such as the South African Large telescope (SALT), the largest optical telescope in the southern hemisphere and the introduction of a rapid-rail system, the Gautrain, reflect South Africa's engagement with the challenge of becoming an 'emerging knowledge economy'. However, the key productive sectors require a greater scientific human resource pool to draw on. We can take example of the automotive manufacturing and cultural industries sectors to illustrate the need to substantially increase the available pool of R&D workers in order to change the country's positioning in the global production system. While South Africa contributes less than a million units to global vehicle production, the automotive sector contributes 7.5 per cent to GDP and has shown strong export growth since 1997 (South Africa.info, undated). The automotive manufacturing and vehicle sales sector employs approximately 118,000 people (Galbraith, 2007: 7). The export value of vehicles and components rose ten-fold in the ten year period 1995–2005 to USD 6.04 billion[10] (ibid.). Yet, in 2009, the following still applies:

> *It remains, however, essentially a manufacturing and assembly industry, with much or all of the research, development and design, and to a lesser extent, testing work, being done abroad (Europe, Japan and the USA) (DST, 2003b: 6).*

Long-term human resources challenges for the automotive sector are identified as there is a need for people highly skilled in design and engineering who can conduct the R&D necessary for producing advanced materials, especially electronic materials.

> *Competence in and knowledge of the fundamental sciences of these materials and their processing will be a prerequisite for participation in this field (ibid.: 14).*

[9] A competitiveness ranking designed by the United Nations Industrial Development Organisation.
[10] ZAR 45.3 billion.

In the cultural industries sector, employing approximately 200,000 people, many from rural communities (ibid. 22), R&D and technological innovation has been very limited. The National Craft Development Agency aims to build R&D capacity for cultural industries in, *inter alia*, advanced materials research and in cleaner production (ibid: 34–35). These and other areas are the ones in which South Africa needs to grow its knowledge base in the future years and in which women's participation needs to be sought.

These human resource requirements for research, design, and new product development are unlikely to be met by male engineering, science and technology graduates alone. Yet, a review of the statements on human capital development in selected public policies and strategies, such as the White Paper (1996), Biotechnology Strategy (2001), R&D Strategy (2002), Advanced Manufacturing Strategy (2003) and Innovation Plan (2008), shows a very limited treatment of the public and private sector initiatives required to expand the national R&D knowledge base in terms of size, capability, fields of scientific endeavour, and gender. In the 2018 innovation plan, there is only a single reference to women scientists and there is no reference to gender participation in the innovation system. The approach taken on human capital is that it is a foundation requirement and the number of PhD graduates must be increased five-fold in twenty years (DST, 2008: 4–8 & 25–27). This is a rather restrictive view, and an apparent misalignment with the wide range of SET human resources needs, in the formal and informal sectors, from technology-savvy users and professionals to R&D workers. Despite these policy limitations, there has been a notable increase in the participation of women in the public science system over the period of 1991–2005.

Facing the Facts: Deconstructing Talent

Women's Participation in the Public Sector Science Engineering and Technology (SET) Workforce

This section reviews the state of women's participation against the backdrop of SET human resource needs. The 'Facing the Facts' reports (NACI, 2004; NACI, 2009) provides a profile of gender distribution in the public sector science, engineering, and technology workforce. The studies[11] conducted

[11] The results are reported in full in Bailey T. (ed.) (2004). Women in science, engineering and technology in South Africa, Final Report to the South African Reference Group on Women in Science and Technology, Main Report Volume 1 and in Ritter, Boshoff & Mouton (2008) Baseline report on women in science, engineering and technology.

in 1992–2001 and 2001–2005, on which these reports are based, consider participation in the public science system from the point of view of (a) the current and future employment of men and women as instructors and researchers in the higher education sector, and (b) as research and development personnel in government science, engineering, and technology institutions. The findings, which are of interest here, relate to (a) the gender distribution of enrollments and graduations, which reflects SET human resource potential and (b) the gender distribution of permanent instructional and research staff in universities and R&D staff in science councils, reflecting the characteristics of the current SET workforce. The main data sources for these segments of the study were data supplied by the higher education institutions, science councils, and the South Africa Knowledgebase and included (a) the South African Post-Secondary database (SAPSE); (b) Higher Education Management Information System (HEMIS), both held by the Department of Higher Education and Training,[12] and the data at Centre for Research on Science and Technology (CREST[13]).

Science Engineering and Technology (SET) Potential

The SET workforce, current and future, is drawn from the stock of higher education students who were enrolled and graduated during the period 1992–2005, particularly the post-graduate cohorts. In this period, there was a marked shift in the representation of women, in both enrollments and graduations in the higher education sector. In 1992, women constituted 43 per cent of all student enrollments, but in 2001, this percentage had increased to 53 per cent. Similarly, in 1992, 48 per cent of all graduates were women and in 2001, this percentage had increased to 58 per cent. Female students, then, were in the majority in 2001 as far as enrollments and graduations are concerned.

The general trend of increase in women's participation in terms of enrollments and graduations continued through 2000 to 2005. Women's participation was greater than their share in the population, i.e, above 52 per cent at the undergraduate and lower postgraduate levels (honours). Although, as in the decade of 1992–2001, women's participation at the upper postgraduate levels (masters and doctoral degrees) was lower than

[12] Previously the Department of Education.

[13] Centre for Research on Science and Technology, http://academic.sun.ac.za/crest/

Table 9.1 *Gender specific indicators for SET potential (2000 and 2005)*

Indicator	2000	2005
Share of female students enrolled for a first bachelors degree	58.0%*	54.8%
Share of female students graduating with a first bachelors degree	63.8%*	57.9%
Share of female students enrolled for a post graduate diploma/certificate	54.9%	62.6%
Share of female students graduating with a post graduate diploma/certificate	56.2%	65.0%
Share of female students enrolled for an honours degree or equivalent	58.7%	60.3%
Share of female students graduating with an honours degree or equivalent	56.5%	59.3%
Share of female students enrolled for a masters degree or equivalent	41.5%	45.4%
Share of female students graduating with a masters degree or equivalent	41.2%	44.8%
Share of female students enrolled for a doctoral degree or equivalent	38.0%	41.4%
Share of female students graduating with a doctoral degree or equivalent	40.9%	43.9%

Source: Ritter, et al. 2008: 51
*2001 data

their demographic share, there was a steady increase towards the 52 per cent benchmark[14] (Table 9.1).

Though there are positive trends in women's participation in higher education, these figures mask some very significant disparities in particular facets of the higher educational experience. These are apparent, for example, when we consider university enrollments and graduations by (a) gender and race and (b) gender and the level of study. There is a distinct drop-off in women's participation from the Masters through Doctoral levels of study, the levels at which training for the future R&D workforce occurs (Table 9.2). At the doctoral level, as regards race, participation is dominated by the white male and female. In terms of gender, the participation is dominated by the white and black male. Ideally, black women should participate at the benchmark of

[14] The '52 per cent benchmark' relates to the demographic weighting of women in the South African population and is adopted here as a broad target for women's participation in SET. Of course, women and men may participate at varying levels in SET either above or below their respective gender representation in the population, for good or bad reasons.

Table 9.2 *Doctoral (or equivalent) Enrollments by Gender and by Race (2005)*

Race	Women			Men			
	Number of women	Percentage race distribution women (%)	Percentage women within race group (%)	Race	Number of men	Percentage race distribution men (%)	Percentage men within race group (%)
African	1,081	27.7	33.0	African	2,194	39.7	67.0
Coloured	242	6.2	42.3	Coloured	330	6.0	57.7
Indian	344	8.8	45.6	Indian	410	7.4	54.4
White	2,229	57.1	46.3	White	2,582	46.7	53.7
Unspecified	9	0.2	40.9	Unspecified	13	0.2	59.1
Total	3,905	100.0	41.4	Total	5,529	100.0	58.6

Source: Ritter, M., Boshoff, N & Mouton, J, 2008: 62

Table 9.3 *Doctoral (or equivalent) Enrollments by Gender and the Field of Study (2005)*

Field of study	Male	Female	Total	F%	M%
Natural and agricultural sciences	1,331	838	2,169	38.6	61.4
Engineering sciences and applied technologies	619	108	727	14.9	85.1
Health sciences	434	668	1,102	60.6	39.4
Humanities	1,278	630	1,908	33.0	67.0
Social sciences	1,870	1,666	3,536	47.1	52.9

Source: Adapted from Ritter, et al. 2008: 62

79.5 per cent of the total female student population as per their share of the female population (StatsSA 2009: 4), as against the 27.7 per cent recorded.

Regarding gender and the field of study, it is important to note that the engineering sciences and applied technologies have a significantly lower share of graduations as compared to other disciplines of study. The health sciences share of graduations is also relatively low for a country that has many health challenges including high rates of HIV-AIDS and tuberculosis and high child mortality rates (AFDB, 2009: 2–3). Women dominate in the health sciences and have a relatively high share of graduations in the natural and agricultural sciences and the humanities. The really stark figure is women's participation in the engineering sciences and applied technologies at 14.9 per cent (Table 9.3). This category ranges from participation levels at 7.6 per cent for mechanical engineering and 9.7 per cent for electrical and electronic engineering as against 21.4 per cent for mining engineering and 20.5 per cent for other engineering and applied technologies (Ritter, Boshoff & Mouton, 2008: 62).

These figures provide the foundation for the assertion that a number of shifts are required in women's participation in order to build SET potential:

(a) A much larger proportion of the general population of men and women is needed to participate in higher education to increase SET potential and the South African knowledge base. The intention stated in the 2018 Innovation Plan to increase the number of PhDs five-fold in a twenty-year period implies a concomitant increase at all levels of higher education study.

(b) The participation of African women in higher education is significantly lower than their population share from first bachelor's degree through to doctoral level (Ritter, et al., 2008: 55–63). Interventions in the schooling system are required to attract young black women to post-school and university level study, with particular measures directed towards attracting young black women to SET careers. African women are a major demographic grouping with untapped SET potential.

(c) The conversion rate from completion of a first Bachelor's degree to completion of higher degrees requires attention for both men and women. A much greater proportion of the student population is needed to participate at the upper postgraduate levels. A major barrier here is the transition from diploma/certificate/honours to Masters level with only 10 per cent of the student population graduating with Masters degrees in 2005 (Table 9.4). This is the point at which women and men would commence a career track to form future generations of the R&D and academic workforce.

Table 9.4 *Proportion of Graduates across Degree Levels (2005)*

Degree level	Total	Male	Female	Proportion male of total	Proportion female of total
First Bachelor's degree	46,557	19,624	26,933	25.2%	34.6%
Diploma/Certificate	6,783	2,376	4,407	3.0%	5.7%
Honours	15,440	6,290	9,150	8.0%	11.8%
Masters	7,894	4,356	3,538	5.6%	4.5%
Doctoral	1,176	660	516	0.8%	0.7%
Total	77,850	33,306	44,544	42.6%	57.3%

Source: Adapted from Ritter, et al. 2008: Tables on 55–63.

Science Engineering and Technology (SET) workforce

The SET workforce discussed here includes instructional/research staff in the higher education sector and R&D personnel in the science council sector. Despite steady increases in numbers between 1992 and 2001, women still constituted less than 50 per cent of permanently employed instructional/ research staff in higher education (Table 9.5) and less than 50 per cent of R&D personnel in the science councils in 2001 (Bailey, 2004: 96).

When we consider the distribution of women across ranks at universities, Table 9.6 shows that, while the total number of permanent instructional/ research staff remained relatively stable at between nine and eleven thousand people in the decade of 1992–2001, the percentages of female instructional and research staff at universities increased across the various ranks. The exception was that of junior lecturer, where a decrease of 4 per cent was recorded. Women were adequately represented (more than 50 per cent) in the three lower ranks (lecturer, junior lecturer, below junior lecturer) but were still significantly under-represented in the three most senior ranks of professor, associate professor and senior lecturer, although the situation improved between 1992 and 2001. Women's participation increased slowly at senior lecturer levels and above, the levels at which academics are engaged in large research projects and in upper post-graduate supervision (Table 9.6), indicating the untapped potential of women's contribution to the higher education and broader science systems.

By 2005, the total number of instructional/research staff in the higher education sector had increased by nearly a third from 9,792 in 2001 to 14,288 in 2005. However, the participation of women at the professorial level had increased only marginally, by 4 per cent. Similarly the participation of

Table 9.5 *Gender Distribution of Permanent Instructional/Research Staff in Higher Education (1992, 1995, 1998 and 2001)*

Gender	Year				% Growth (1992–2001)
	1992	1995	1998	2001	
Women	3,770 (30.0%)	4,480 (32.8%)	4,961 (36.1%)	5,303 (39.7%)	40.7%
Men	8,786 (70.0%)	9,170 (67.2%)	8,774 (63.9%)	8,057 (60.3%)	-8.3%
Total	12,556	13,650	13,735	13,360	6.4%

Source: Adapted by Boshoff (2006) from Bailey T (2004)

Table 9.6 *Women as a Percentage of Permanent Instructional/Research Staff in Universities by Rank (1992, 1995, 1998 and 2001)*

Rank	1992		1995		1998		2001	
	Total	*W%T*	Total	*W%T*	Total	*W%T*	Total	*W%T*
				Universities				
Professor	2,098	6.6%	2,132	7.9%	2,014	10.0%	1,795	14.5%
Associate professor	974	14.8%	1,094	18.2%	1,155	22.1%	1,056	25.7%
Senior lecturer	2,951	26.0%	3,039	28.4%	2,951	34.2%	2,552	37.2%
Lecturer	3,376	45.7%	3,728	48.2%	3,475	50.3%	3,349	54.4%
Junior lecturer	597	60.0%	661	56.3%	740	55.1%	704	56.4%
Below junior lecturer	43	51.2%	69	53.6%	71	62.0%	46	69.6%
Undesignated	172	41.9%	182	47.8%	254	56.7%	290	56.2%
Total of universities	10,211	29.8%	10,905	32.3%	10,660	35.7%	9,792	39.8%

W%T is Women as a percentage of the total number of instructional/research staff
Source: Bailey T (2004), adapted from table at 83 by Galant, J

Table 9.7 *Employment Ranking for the Higher Education Sector by Gender (2005)*

Rank	Male	Female	Total	M%T*	F%T**
Professor	1,630	370	2,000	81.5	18.5
Associate professor	997	424	1,421	70.2	29.8
Senior lecturer	2,451	1,597	4,048	60.5	39.5
Lecturer	3,018	3,139	6,157	49.0	51.0
Junior lecturer	289	373	662	43.7	56.3
Total	8,385	5,903	14,288	–	–

Source: Adapted from Ritter, et al 2008: 82 Table D2
*M%T = male as a percentage of the total
**F%T = female as percentage of the total

women at the entry level to the professoriate (senior lecturer) increased by 2 per cent only (Tables 9.6 and 9.7). Thus, fewer women than men are entering into academic careers, whether through low numbers of applicants, decisions made in the selection process or other factors.

With respect to the scientific fields, female instructional/research staffs in the higher education sector are best represented in the health sciences, social sciences, and the humanities. Significantly, in 2001, less than 30 per cent

Table 9.8 *Women as a Percentage of Instructional/Research Staff in Higher Education by Scientific Field (2001)*

Scientific field	Total	Women as % of total
Higher education		
Social sciences & humanities	7,339.4	45.9%
Natural sciences & engineering	4,174.1	27.7%
Health sciences	1,938.2	51.1%
Total of higher education	13,494.9	41.0%

Source: Boshoff (2006) adapted from Bailey T (2004).

Table 9.9 *Women as a Percentage of Higher Education Instructional/Research staff in the Natural Sciences and Engineering (2001)*

CESM category of natural sciences and engineering	Total	Women as % of total
Agriculture and renewable natural resources	330.9	23.9%
Architecture and environmental design	244.3	28.1%
Computer science and data processing	613.1	45.4%
Engineering and engineering technology	1,061.5	10.2%
Life sciences and physical sciences	1,356.8	31.4%
Mathematical sciences	567.5	34.1%
Total	4,174.1	27.7%

Source: Boshoff (2006) adapted from Bailey T (2004).

of instructional/research staff in the natural sciences and engineering were women (Table 9. 8).

Looking within the sub-fields of the natural sciences and engineering, we find that female instructional and research staffs are best represented in the sub-field of computer science and data processing, constituting 45 per cent in 2001. Their representation in the sub-field of engineering and engineering technology is extremely limited – only about 10 per cent of all staff in this sub-field are women (Table 9.9).

By 2005, only 2.7 per cent of the professors and only 6.7 per cent of the associate professors in the engineering sciences and applied technologies were women[15] (Ritter, et al., 2008: 83).

[15] The data and analytical categories used were different for the 2003 and 2008 studies, and this has been factored into the analysis made above.

Table 9.10 *Gender Specific Indicators for R&D Workforce (2000 & 2005)*

Indicator	2000*	2005
Share of female researchers	35.0%	39.2%
Share of female technicians/technologists	31.0%	34.0%
Share of female R&D support staff	43.8%	48.4%

Source: Ritter, et al. 2008: 64, Table B1
*2001 data

Research and Development (R&D) Workforce in Science Councils

The participation of women in the R&D workforce in science councils increased from 35 per cent in 1992 to 42 per cent in 2001 (Bailey, 2004: 96). By 2005, women's contribution was approaching 52 per cent at the level of R&D support staff, but still far off the 52 per cent mark for researchers and technicians/technologists (Table 9.10).

In summary, the studies reflect that women's participation in public sector SET requires the attention of policy-makers and institutional decision-makers on a wider scale. It shows that the overall number of women students and staff in these sectors has increased steadily over the period 1992–2005. However, the data reveals some major disparities within the sector at various levels of analysis such as fields of study, levels of study, and rank distribution, with women falling out of the potential SET workforce at the lower postgraduate exit (diploma/certificate/honours) level.

Women in Science: The situation in Asia-Pacific Economic Cooperation (APEC) and European Commission (EC)

In the twentieth century, the natural sciences and engineering were key fields for innovation across the globe and they continue to be so in the twenty-first century, producing inventions, technologies and services that touch the lives of billions of people and contribute to economic change. However, these innovations have seldom benefited from the insights, knowledge, and perspectives of women as engineers and scientists. While on the one hand, the demand for a larger global scientific workforce requires the greater participation of both men and women, on the other hand, women can also contribute engineering intelligence to the world, not by virtue of perspectives based merely on their biology, but by virtue of perspectives based on their particular life experiences. In a discussion on universities in the knowledge economy, Sheehan argues 'when knowledge is recognised as a primary economic good,... What knowledge is generated, who controls it and who is trained to use it, become matters of central importance throughout the broader community' (Sheehan, 2002: 29). The approaches of the Asia-Pacific Economic Cooperation (APEC)[16] and the European Commission (EC)[17] offer interesting perspectives for reference, as these represent relatively advanced, yet varying approaches to increasing women's participation in science and technology.

The APEC First Ministerial Meeting on Women, held in 1998, discussing the theme of women in industrial science and technology agreed that, 'Economies need the talent of their entire population, male and female, to create a leading edge workforce...' (APEC, 1998: 3). It was noted that women's participation in formal science and technology education was lower than men's participation. Recommendations from the meeting emphasised the need for the collection of gender-disaggregated data in order to highlight the contributions of respective genders in the broadest range of areas including industrial science and technology and to provide an information base for policy-making. Member countries were advised to use the 'Framework for the Integration of Women in APEC' (APEC, 1999) and APEC guides on gender analysis and data collection to guide their gender focus at the country level. In 2002, APEC established the Gender Focal Point Network to promote the participation of women in the economic and social life of the region (APEC, 2002: 15). Through the period 2004 to 2009, continuous reference is made to increasing the participation of 'women and ethnic communities' in R&D, in the SET workforce, in micro-enterprise development. APEC plans and reports include a range of statements and requirements for women's participation, including gathering and utilising gender-disaggregated data to encourage greater understanding of the issues, and training women for enhanced participation in entrepreneurship, e-business, and the digital economy (APEC 2008; APEC 2009). Despite this encouragement, no studies of women's participation are readily available for public download.

The European Commission (EC) adopted 'The Communication on Women and Science: Mobilising Women' to enrich European Research in 1999,

[16] 21 member countries on four continents including Australia, China, Indonesia, Japan, the Republic of Korea, Malaysia, Mexico, Russia, Thailand, and the United States.

[17] 25 member countries.

encapsulating broad directives to nurture women's participation. This was followed by the ETAN report on Women and Science in 2000 – highlighting the poor participation of women in the public science system (EC, 2000). The Women in Industrial Research (WIR) study conducted in 2003 drew attention to the fact that, *inter alia*, women in EU member states were obtaining nearly 40 per cent of all new PhDs (2000 figures), but only 20 per cent of new PhD's in engineering and 19 per cent in computing (EC, 2003b). The EC 'She Figures 2003' presents a statistical overview across five themes – numbers of women participating in the science systems across the EU countries, horizontal segregation (gender differences across scientific fields), vertical segregation (seniority at various levels of the hierarchy), pay gap and fairness, and success rate.[28] Eurostat collects and publishes figures on an annual basis. Another valuable source of data for policy-making is UNESCO's statistics on the gender gap in science and technology (UNESCO, 2007).

Common themes across APEC and the EC are gender mainstreaming in science, technology, and all the domains of economic as well as the social activity, and the value of collecting, analysing, and utilising gender-disaggregated data to design appropriate public and institutional policies. There appear, however, subtle differences of approach. The EC pushes statistics at member countries in an assertive manner (hard sell), while APEC provides frameworks and guides to member countries, which they may use, apparently at their own discretion (soft sell). Furthermore, while these two regional economic groupings promote both quantitative and qualitative analyses on the positioning of women in science, they emphasise the value of quantitative measures in guiding public and institutional policy, perhaps reflecting Sheehan's (2002) assertion that the broader community (here as represented by governments actions or public policy) have an interest in charting *what knowledge is generated, who controls it, and who is trained to use it.*

Conclusion: Changing the Facts!

Measures and Approaches to Improve Women's Participation in Science Engineering and Technology

In addressing the limited participation of women in SET and innovation, it is necessary to be cautious about making unfounded claims regarding the value (and pricing) of knowledge in a 'knowledge economy'. Brown and Hesketh (2004) argue that there is fierce competition amongst university graduates for jobs at the top end of the management and professional ladder precisely because the market for such skills is growing more slowly than the numbers of university graduates. Thus, public policy aimed at further increasing the numbers of highly educated graduates may be ill conceived. However, they do not address their analysis to SET and innovation, where greater numbers of highly educated people are in demand. Furthermore, the caution that the knowledge economy is not synonymous with hi-technology industries and R&D expenditures (Smith, 2002) must be noted. It is nevertheless the case that SET innovation, whether low, medium or high-tech, requires people and their tacit knowledge, mental skill and expertise, experience, insight, and capacity to utilise existing and generate new knowledge – what Melody calls the 'foundation resource for building network knowledge economies' (Melody, 2002). Research and development for advanced materials and cleaner production, whether for application in the automotive manufacturing sector or cultural industries, requires new intellectual labour and application of new knowledge and varied talents to produce these innovations.

If the contribution of innovation to the economic and social development is dependent on the rate of introduction of new knowledge in the economy and the levels of knowledge intensity vested in products and services (Stiglitz, 1999: 21–23), then in South Africa, the further evolution of R&D and non-R&D innovation as a path to international competitiveness and local development will be partly dependent on the participation of its female population, since women constitute 52 per cent of the population of 49.3 million[18] people (StatsSA, 2009: 3 and 7). African women constitute 79.5 per cent of the total female population (ibid: 4) and therefore present a major source of untapped potential for SET capacity. Indeed, while in demographic terms, African women should dominate the scientific workforce, the existing sociological paradigm in South Africa negatively affects their participation with respect to sex and field of study.

Between 1996 and 2008, South Africa introduced a range of science policy aimed at building new knowledge sectors such as biotechnology, while rejuvenating old sectors such as automotive manufacturing and craft industry. It acknowledged the requirement for SET human capital, but has given little attention to the specifics as regards particular measures and interventions to upscale capacity or to address the relevant gender and race dimensions. Against this backdrop of the need for highly educated human resources, the

[18] Medium variant of population estimate.

'Facing the Facts' studies suggest that despite a general improvement over the period 1992 to 2005, there are systemic inhibitors to women's participation in SET, R&D and innovation (a) in the sphere of building future generations of SET human capital; and (b) in the sphere of effective participation of the women in the SET workforce. Addressing the existing inhibitors will require major shifts in policy thinking about what knowledge is generated by women, how women are engaged in controlling it and which women are trained to use it. Should South Africa import engineers for the energy sector or train greater numbers of men and women in engineering?

The actions required to 'change the facts'!

Firstly, South African SET strategies and interventions must take specific measures to increase participation of women in the upper postgraduate sphere, to attract women into executive leadership and decision-making positions in the science system, and to lead and participate in large-scale projects. The precipitous exit of young women from senior postgraduate studies must be stopped if SET potential is to increase to the levels required for future economic growth and social development. Such measures will require attention to the attractiveness of entry and progression in SET careers, to throughput, and to attractiveness of those fields of study where women are currently in the vast minority. Fields such as advanced manufacturing in the automotive sector or technology focus areas such as cleaner production technologies are good candidates for attracting women into graduate research.

It is also not too late to pay attention to the needs of women in the current SET workforce as regards the attractiveness of remaining in and fostering SET careers from mid-career to retirement. Mid-career women in SET in higher education, in science councils, scientific performing agencies and in industry, can engage in new opportunities, projects and create value across a wide range of fields of study, whether this is related to their historical career trajectory or a completely new endeavour. Switching SET career tracks through re-entering post-graduate studies has become more possible with the greater interest in and growth of multi-disciplinary research.

Secondly, building the interface between science and technology policy on the one hand, and economic and social development policy on the other hand, can engage larger numbers of men and women in productive SET, not only in research centres, but also in productive centres like factories, mines, and services companies. This approach would enhance the health and performance of locally and globally competitive sectors. Engaging larger numbers of women in SET research in those sectors, which are already dominated by women, such as the public health sector, should be a simultaneous endeavour.

Thirdly, understanding the dimensions of the challenge is a key to formulating responsive policies and practice across all institutions. The collection, analysis, publishing, and utilisation of gender-disaggregated data to inform public and institutional policy can enable South Africa's firms and public institutions to manage their talent – both male and female – effectively, towards a constant rate of increase in R&D intensity, working to achieve selected development goals. The EC 'She Figures 2003', possibly the most comprehensive set of statistics on women in science for its time, opens with the quotation from Dr Hilary Rose, University of Bradford, United Kingdom:

> *Statistics are central, the word comes from numbers for the state, if you wish to have policy, you have to have competent statistics. No statistics, no problem, no policy. You just get gestures. Statistics help identify problems and can monitor the effectiveness of remedies* (EC, 2003a: 15).

South Africa currently collects limited data on women in SET in the HEMIS surveys, in the National Survey on Research and Experimental Development and in the 'facing the facts' studies reported on here. Collection and analysis of more detailed data sets can provide the basis for interventions at the level of sub-disciplines and newly emerging multi-disciplinary areas such as cleaner production technologies for automotive manufacturing.

The 'Facing the Facts' study illustrates that a generic approach to building women's participation in the National System of Innovation (NSI) will not necessarily address the complex array of systemic barriers to women's advancement in, for example, traditionally male disciplines in the natural sciences and engineering. An evolving grid of quantitative and qualitative studies is necessary to build gender analysis of the national system of innovation (NSI). It is essential to track the gender-wise participation and performance over the time, in order to form medium and long-term policy at the institutional and the national levels. The first study on women in industrial SET has been conducted and another study on the role of women in SET sectors, commencing with the agricultural and agro-processing sectors has been initiated. While the data analysed here presents a view on the state of the public science system, the analysis applies equally across the national system of innovation in its entirety, as these systems co-exist in the same socio-cultural milieu.

South Africa may, in the future, shift its policy, institutional and cultural practices in the twenty-first century towards greater participation of women in science for societal benefit. In so doing, it will need to address itself to three focus points: (a) the demographics versus the sociology of women's participation, meaning the extent to which women are actively encouraged to participate in the national system of innovation across all fields of study and research production; (b) the consistent application of Weberian meritocratic principles with respect to the administrative processes pertinent to equal treatment; and (c) addressing the weaknesses in cultural and attitudinal paradigms that may require affirmative treatment of either men or women.

Acknowledgement

All tables for this chapter are extracted or adapted from the CREST studies. I am extremely grateful to Jaamiah Galant for some of the tables and her inputs to the first draft of this chapter.

References

Addison, G (2000) *The Hidden EDGE: South Africa's Quest for Innovation*. The Engineering Association, Yeoville.

AFDB (2009) *Gender, Energy and Development: Gender, Poverty and Environmental Indicators on African Countries*, Economic and social statistics division of the Statistics Department, African Development Bank (AFDB), Tunis, www.afdb.org/en/knowledge/publications/gender-poverty-and-environmental-indicators-on-african-countries-2009/, retrieved on 15 December 2009.

APEC (1998) Joint Ministerial Statement, Asia Pacific Economic Co-operation (APEC) Ministerial Meeting on Women, Makati, Philippines. http://www.apec.org/apec/ministerial_statements/sectoral_ministerial/women/1998_women.html retrieved on 13 February 2005.

APEC (1999) Framework for the integration of women in APEC, Asia Pacific Economic Co-operation (APEC) Gender Focal Point Network, http://www.apec.org/apec/apec_groups/som_committee_on_economic/som_special_task_groups/gender_focal_point_network.html, retrieved on 6 March 2010.

APEC (2002) APEC Outcomes and Outlook brochure. Asia-Pacific Economic Cooperation, APEC Secretariat, Singapore.

APEC (2008). APEC Industrial science and technology working group: Terms of reference and workplan 2008, presented at the 34th Industrial science and technology working group meeting, Hong Kong, 27–28 March 2008, retrieved 6

March 2010 from http://www.apec.org/apec/apec_groups/som_committee_on_economic/working_groups/industrial_science_and_technology.html

APEC (2009) APEC senior officials report on economic and technical co-operation, November 2009, APEC secretariat, Singapore, http://publications.apec.org/publication-detail.php?pub_id=943, retrieved on 6 March 2010.

Bailey T (Ed) (2004) *Women in Science, Engineering and Technology in South Africa*, Final CREST report to the South African Reference Group on Women in Science and Technology, Volume 1, Centre for Research on Science and Technology (CREST), Stellenbosch.

Boshoff, N (2006) The representation of women academics in higher education in South Africa: progress in the pipeline? *South African Journal of Higher Education*, 19(2): pp. 359–377.

Brown, P and Hesketh, A (2004) *The Mismanagement of Talent: Employability and Jobs in the Knowledge Economy*. New York: Oxford University Press.

CSIR India (2010) India to step up R&D expenditure, CSIR on the web, http://csirindia.wordpress.com/2010/02/05/india-to-step-up-rd-expenditure/, retrieved on February 2010.

DACST (1996). White Paper on science and technology: Preparing for the 21st century, Department of Arts, Culture, Science and Technology (DACST), Pretoria, retrieved on 15 December 2009 from www.naci.org.za/pdfs/whitepaper_st.pdf

DACST (2001). A national biotechnology strategy for South Africa, Department of Arts, Culture, Science and Technology (DACST), Pretoria, June 2001, retrieved 15 December 2009 from http://naci.org.za/pdfs/workshops/julNBAC/BiotechnologyStrategy2001.pdf

DST (2002) South Africa's national research and development strategy, Department of Science and Technology, Pretoria, www.space.gov.za/ws2007/rd_strat.pdf, retrieved on 6 March 2010.

DST (2003a) A National Advanced Manufacturing Technology Strategy, Volume 1, Department of Science and Technology, Pretoria, http://www.amts.co.za/amts_vol1_2005_03_31.pdf, retrieved on 11 February 2005.

DST (2003b) A National Advanced Manufacturing Technology Strategy, Volume 2, Department of Science and Technology, Pretoria, www.amts.co.za, retrieved on 11 February 2005.

DST (2008) Ten year innovation plan, Department of Science and Technology (DST), Pretoria, www.dst.gov.za/publications-policies/strategies-reports/The%20-Ten-Year%20Plan%20for%20Science%20and%20Technology, retrieved on 15 December 2009.

DST (2009). *National survey of research and experimental development 2007/08: High level key results,* Report prepared by the Centre for Science, Technology and Innovation Indicators of the Human Sciences Research Council, Department of Science and Technology (DST), Pretoria, retrieved on 15 December 2009 from http://www.hsrc.ac.za/CCUP-RnD-7.phtml.

DST India (2005) Major highlights, Department of Science and Technology India, http://dst.gov.in/majorhighlights.pdf, retrieved on February 2010.

EC (2000). Science policies in the EU: Promoting excellence through mainstreaming gender equality, European Technology Assessment Network (ETAN) report on women and science, The European Commission, Brussels, retrieved on 6 March 2010 from ftp://ftp.cordis.europa.eu/pub/improving/docs/g_wo_etan_en200101.pdf

EC (2003a). She Figures 2003, The European Commission, Brussels, retrieved 15 February 2005 from http://europa.eu.int/comm/research/science-society/women/wssi/publications_en.html

EC (2003b). Women in Industrial Research – A wake up call for European Industry, The European Commission, Brussels, retrieved on 15 February 2005 from http://europa.eu.int/comm/research/science-society/women/wir/report_en.html

Galbraith, L (ed) (2007) *South African automotive yearbook 2007*, 10th edition, Balgair Publications, Cowies Hill.

Mangena, M (2005) Address by the Minister of Science and Technology at the media launch of the results of the 2003/04 South African survey of research and experimental development, National Parliament, Cape Town, 13 April 2005, http://www.info.gov.za/speeches/2005/05041415151001.htm, retrieved on 21 June 2005.

Melody, W H (2002) *The triumph and tragedy of human capital: Foundation resource for building network knowledge economies*, LINK Centre, University of the Witwatersrand, Johannesburg, http://link.wits.ac.za/papers/wm20020918.htm, retrieved on 13 February 2005.

NACI (2004) Facing the facts: Women's participation in science, engineering and technology, Report of the South African Reference Group on Women in Science and Technology, National Advisory Council on Innovation (NACI), Pretoria.

NACI (2008) Changing perceptions of women in science, engineering and technology, Report of the Science, Engineering and Technology for Women Committee, National Advisory Council on Innovation (NACI), Pretoria.

NACI (2009) Facing the facts: Women's participation in science, engineering and technology 2009, Report of the Science, Engineering and Technology for Women committee, National Advisory Council on Innovation (NACI), Pretoria.

OECD (2007). *OECD factbook 2007 – Economic, environmental and social statistics*, OECD Science and Technology, Organisation for Economic Co-operation and Development (OECD), Paris, retrieved on 28 February 2010 from http://www.thepresidency.gov.za/learning/reference/factbook/07-01-01.htm

The Nobel Assembly (2002) Press release: The 2002 Nobel Prize in Physiology or Medicine, 7 October 2002, http://nobelprize.org/medicine/laureates/2002/press.html, retrieved on 13 February 2005.

Ritter, M Boshoff, N & Mouton, J (2008) Baseline report on women in science, engineering and technology: Populating a monitoring and evaluation framework to benchmark the performance of women in the national science system, Centre for Research on Science and Technology (CREST), Stellenbosch.

Sheehan, P (2002). Universities in the knowledge economy, in Mansell, R, Samarajiva, R and Mahan, A (eds) *Networking knowledge for information societies: Institutions and intervention*, 28–33, DUP Science, Delft University Press, The Netherlands.

Smith, K (2002) *What is the 'Knowledge Economy'? Knowledge Intensity and Distributed Knowledge Bases*, The United Nations University, Institute for New Technologies, Discussion Paper Series, June 2002, http://www.intech.unu.edu/publications/discussion-papers/2002-6.htm, retrieved on 10 February 2003.

South Africa Information (undated) South Africa's automotive industry, http://www.southafrica.info/about/416221.htm#auto, retrieved on 6 March 2010.

Statistics South Africa (2010) *Gross Domestic Product*. Statistical Release P0441, 23 February 2010, Statistics

StatsSA (2009). *Mid-year Population Estimates*. Statistical Release P0302, Statistics South Africa (StatsSA) retrieved on 30 September 2009 from http://www.statssa.gov.za/publications/P0302/P03023009.pdf

Stats SA (2010). *Gross Domestic Product*. Statistical Release P0441, 23 February 2010, Statistics South Africa (StatsSA), retrieved on 2 December 2009 from http://www.statssa.gov.za/publications/P0441/P04413rdQuarter2009.pdf

Stiglitz, J (1999) *Public Policy for a Knowledge Economy*, Department for Trade and Industry and Center for Economic Policy Research, London, 27 January 1999, www.avj.ro/node/181669/docs/knowledge-economy.pdf, retrieved on 15 December 2009.

UNESCO (2007). A global perspective on research and development, Fact Sheet, UNESCO Institute for Statistics, October 2007, retrieved on 28 February 2010 from www.unesco.org/science/psd/wsd07/global_perspective.pdf

Westholm G, Tchatchoua B & Tindemans P (2004). The great global R&D divide in *Multinational Monitor*, July/August 2004, Volume 25, No. 7, The political economy of R&D (online version), retrieved on 15 December 2009 from http://multinationalmonitor.org/mm2004/072004/westholm.html

World Bank (no date) Country classification, http://www.worldbank.com/data/countryclass/classgroups.htm#Lower_middle_income, retrieved on 11 February 2005.

Zhou, P & Leydesdorff, L (2006) The emergence of China as a leading nation in science, *Research Policy 35(1), 2006*, 83–104 http://users.fmg.uva.nl/lleydesdorff/ChinaScience/, retrieved on 8 July 2005.

CHAPTER TEN

Demographic Inertia and the Glass Ceiling in American Science (1970–2000)

J. Scott Long

This chapter intends to document some of the fundamental changes in American women's full participation into science, changes that will have major impacts on the 'face of science and engineering' for decades to come. 'Demographic inertia' in this chapter refers to the long-term impact of combined demographic forces on women's representation in science. This chapter highlights the many dimensions of the changing representation of women in US science and engineering (S&E). Using data from two National Science Foundation (NSF) databases – the Survey of Earned Doctorates (SED) for new PhDs and the Survey of Doctoral Recipients (SDR) for the S&E doctoral workforce – it brings together data on the educational background and demographic characteristics of three decades of new PhDs and then examines their careers as described by the data. After considering changes in the percentage of PhDs to women, changes in their labour force participation, and reasons for the greater proportion of female scientists and engineers with less than full employment, I show how these changes have important implications for major career outcomes within academia. The importance of examining achievement within the context of the entire career and to understand the effects of demographic changes on movement into more advanced positions has been considered in a variety of recent studies (Hargens and Long, 2002; Long 2001; Morgan, 1998). The current chapter applies these ideas to rank advancement for women in American science. The results show that there has been a steady convergence, making the careers of men and women increasingly more similar. But, there remain consistent differences that leave women with less achievement, salary, and position.

The history of science in the twentieth century has shown that female scientists encountered many barriers to upward progress that prevented them from attaining those positions for which they are fully qualified. The chilly, hostile or simply patronising attitudes that women faced during much of this period is documented in many sources, such as Rossiter's (1982, 1995) excellent history of women in science and in reports such as the Massachusetts Institute of Technology's (MIT) study of women within that institution (Committee on Women Faculty in the School of Science, 1999). Within the United States, however, there were important changes beginning in the late 1960s that affected the opportunities for women in science. Rossiter (1995, p.382) writes:

> ...starting in 1968 and essentially complete by 1972, there was a legal revolution in women's education and employment rights. It promised, even seemed to guarantee, broad ramifications for women's careers in science and engineering, but its full implementation would require many battles in the years ahead. One era had ended and a new, more equitable one was beginning.

The legal revolution highlighted by Rossiter grew out of the resurgence of the women's movement in the late 1960s and dramatic changes in our society's view of the role of women at home and in the workplace. The effects of these changes are reflected in the rapid and remarkable increases in the presence and participation of women in Science and Engineering (S&E).

From 'Scarcity to Visibility'[1]

Women were barely represented among doctoral scientists and engineers in US till the 1960s. There has been remarkable changes in the representation of women in S&E in US since the early 1970s. Results are presented for five broad fields of science and engineering – mathematical sciences, physical sciences, engineering, life sciences, and the social/behavioural sciences. While the analyses include both science and engineering, I often use the shorter term 'science' rather than 'science and engineering' to refer to the fields combined. Similarly, the term 'scientist' is sometimes used as shorthand for 'scientist or engineer'. The primary data are from the 1973, 1979, 1989, and 1995 panels of the SDR, a biennial survey

[1] This sub-title is being used from Long (2001).

of a sample of doctoral scientists, engineers, and humanists who completed the SED. Additional data from the SED were also used. For technical details, see NSF (1920–95, 1973–95, 1997). Two types of results are presented. First, I present simple descriptive statistics. For example, I might show the percentage of women who are working full time in each of the four panel years. This type of information does *not* include controls except for those explicitly mentioned. Second, I use regression models (see Long 1997 for full details) to compare the predicted career outcomes of men and women after controlling statistically for differences in background characteristics. This statistical adjustment is critical for understanding gender differences in tenure and academic rank since these outcomes are so dependent on the age of the scientist.

The PhD and Labour Force Participation

Information on the percentage of women with PhDs by field and year is summarised in Figure 10.1.[2] There are several important differences across the fields. Engineering is the most male-dominated field, while mathematics

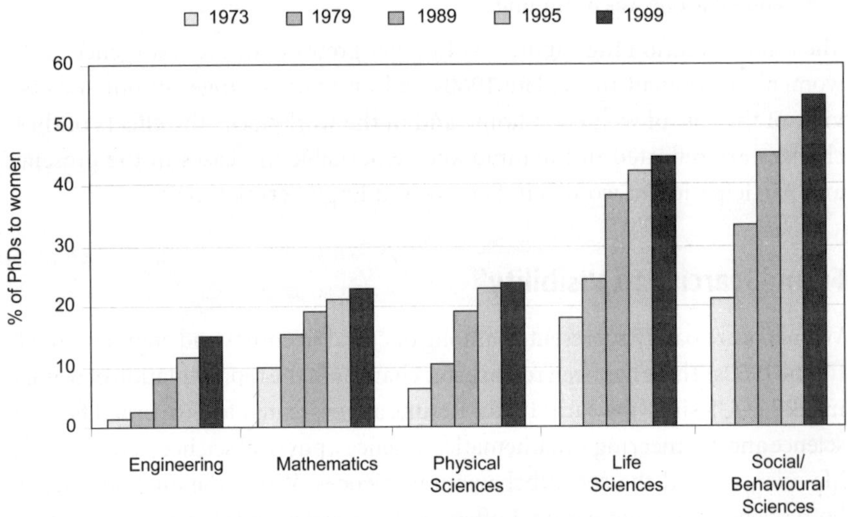

Figure 10.1 Per cent of PhDs Awarded to Women, by Field and Year
Source: Long (2001).

[2] In assessing these graphs and those that follow, keep in mind that the survey years are not evenly spaced. Everything being equal, we would expect larger changes from 1979 to 1989 than from 1973 to 1979 or from 1989 to 1995.

has the slowest growth in the participation of women. Since 1973, the physical sciences saw a tripling in the number of PhDs earned by women, although women still make up less than 25 per cent of the total degrees. In 1973, 17 per cent of the degrees in the life sciences were awarded to women, doubling to 45 per cent by 1999. The representation of women is greatest in the social/behavioural sciences, with 20 per cent of the degrees in 1973, and by 1999 the proportion of women was just over 55 per cent. Overall, from 1970 to 1999, there were significant advances in the entry of women into S&E. Combining our five fields, there were 350 per cent more women among new PhDs in 1995 than in 1973. The progress towards gender equity in the receipt of S&E doctorates seems to have resulted from general social trends in women's advancement in higher education; the enforcement of anti-discrimination laws (including equalisation of financial aid); falling interest in S&E among men; and a more rapid increase for women in degrees in scientific compared to non-scientific fields. Still, as the proportion of doctoral degrees to women in the social/behavioural and life sciences approaches parity, women remain but a small fraction of doctorates in engineering and mathematics.

Increases in the number of women among new PhDs do not translate directly into increases in the proportion of women in the S&E labour force. One reason for this is that each new cohort of PhDs is only a small fraction of the total number of scientists in the labour force. This is shown in Figure 10.2,

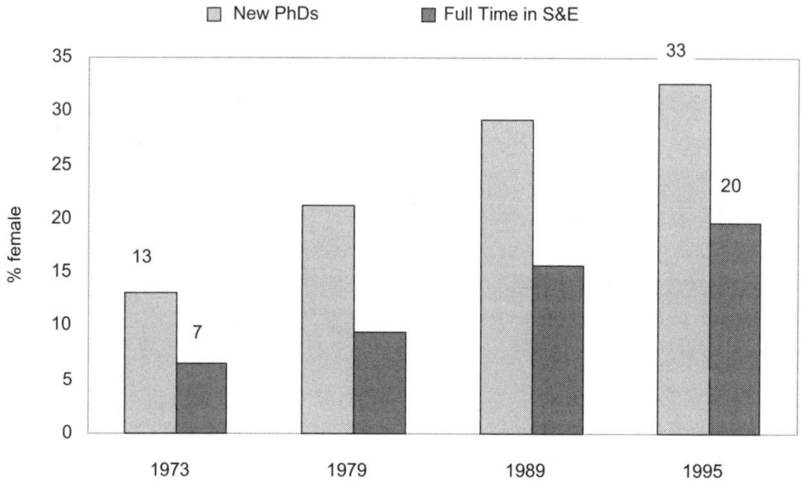

Figure 10.2 The Per cent of Women among New PhDs and the Per cent of Women in the Full-time S&E Labour Force, for Combined Fields, by Year of Survey

Source: Long (2001).

which combines all fields. The proportion of women among new PhDs increased by 20 percentage points from 13 per cent in 1973 to 33 per cent in 1995, while the proportion of women working full time in S&E increased by 13 points from 7 per cent in 1973 to 20 per cent in 1995. While there has been substantial movement towards parity, the full-time participation of women remains far below 50 per cent in all fields, with 32 per cent in the social and behavioural sciences, 27 per cent in the life sciences, 10 per cent in the physical sciences, 12 per cent in mathematics, and 5 per cent in engineering.

A second reason for the lesser participation of women in the full-time S&E labour force is that female scientists are much more likely than men to be less than fully employed, as shown by Figure 10.3. The total height of each bar shows the per cent of scientists who are not working full time, with the divisions within each bar indicating the specific labour force status. In 1973 women were 20 percentage points more likely to be less than fully employed, decreasing to an 11-point difference in 1995. Still, 17 per cent of the female doctorates do not have full-time employment, compared to only 6 per cent of the male doctorates. Part-time employment for women decreased between

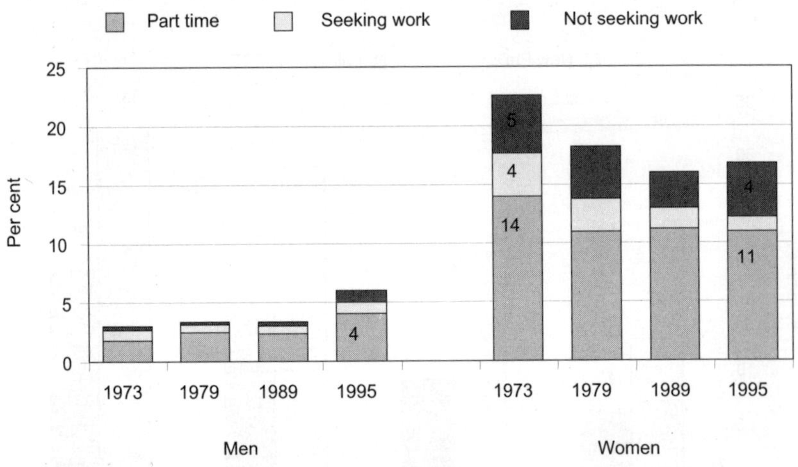

Figure 10.3 Employment Status of Those not Working Full Time for Combined Fields, by Sex and Year of Survey

Source: Long (2001).

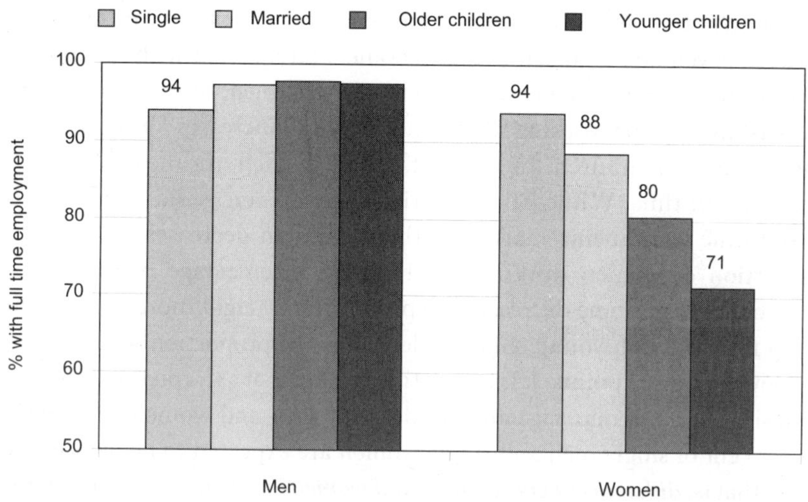

Figure 10.4 Predicted Per cent with Full-time Employment in 1995, by Sex and Familial Status

Source: Long (2001).

1973 and 1979, leveling off around 12 per cent. Since 1973, there was a steady decrease in the per cent of women who were seeking work, from 4 per cent in 1973 to just over 1 per cent in 1995. By 1995, the difference between male and female PhDs in the per cent seeking work was reduced to less than half a point from a difference of three points in 1973. There are much larger differences between male and female doctorates in the per cent who are out of the labour force (that is, not employed and not seeking work), shown by the top segment of each bar. These individuals are fully trained scientists and engineers who have not retired but who are no longer pursuing jobs in their field of training.

For women, less than full employment is strongly associated with familial obligations, while for men marriage and family slightly increase the chances of working full time. Figure 10.4 plots the per cent of male and female scientists and engineers in 1995 who are predicted to work full time either in or out of S&E according to familial status.[3] Four familial statuses are considered – single

[3] Predictions are for an individual who is average on all of the variables in a logit model for labour force participation. See Long (2001) for details.

without children, married without children, married with one or more child between seven and eighteen living at home, and married with one or more child younger than seven living at home. Among men, those who are single are least likely to be working full time, with small increases for married men and those with children. In contrast, single women are most likely to be working full time. While being married slightly increases a man's chances of full-time work, being married without children decreases the predicted proportion of women working full time by 5 percentage points. Having older children at home decreases the proportion by eight more points, while being married with young children decreases the proportion with full-time employment by 22 points. It is interesting to note that as a consequence of the opposite effects of marriage and children for men and women, an identical 94 per cent of single men and single women are expected to be working full time. That is, *differences between men and women in labour force participation are eliminated if we compare single men to single women.*

In 1995, the SDR asked scientists and engineers why they were working part time. Figure 10.5 plots the per cent of men and women who cite family

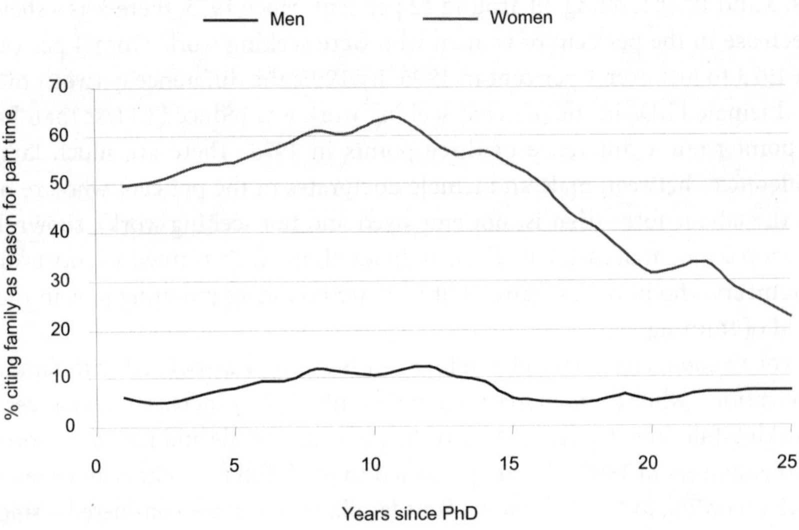

Figure 10.5 Percentage who Cite Family Reasons for Working Part Time, by Sex and Year of Survey
Note: Values are moving averages across five-year periods.
Source: Long (2001)

reasons for part time employment against how many years it is since the PhD was obtained. If we assume that the PhD was received at age thirty, the horizontal axis corresponds to biological ages from 30 to 56. Immediately after the PhD, women cite family obligations 50 per cent of the time. This rate increases till it peaks in year 11 at around 65 per cent. The rate decreases steadily from this point on. For men, the percentages generally stay below 10 per cent, with a slight increase occurring from years five to fifteen.

Overall, differences between men and women in full-time labour force participation add up to less accumulated work experience and less valuable experience for women over the course of their careers, a factor that is important for understanding the gender differences in career outcomes that are described in the section on The Academic Outcomes section below. While there has been improvement since 1973, *female PhDs continue to be substantially less likely than men to be fully employed in scientific and engineering occupations, with roughly 10 per cent of the potential professional work of female doctorates being lost.*

In the following analyses, it is essential to remember that the results apply only to those scientists and engineers who are fully employed. Thus, even if we found no gender differences in the career outcomes of scientists who are fully employed (which we do *not* find), there would still be important differences between male and female scientists and engineers in their success in moving into full-time employment. To the extent that the glass ceiling operates by discouraging women from entering academia or by increasing their chances to leave academia, the following results will underestimate the disadvantage for women.

The Academic Outcomes

Women are proportionally more likely than men to be in academic jobs in all but one field. Even though the largest proportion of female PhDs is found in the social and behavioural sciences, this is the only field with a greater proportion of men than women in academia. Women with PhDs in engineering and life sciences are the most likely to be academic, with little change over time. In mathematics and the physical sciences, gender differences in full-time academic employment have nearly disappeared by 1995. While there is an increasing representation of women in each field, substantial variation exists across fields in the proportions, numbers, and rates of increase of female academics. In the life and social/behavioural sciences, the per cent of full-time academics who are women increased by nearly 20 percentage points from 1973 to 1995. As a result of the greater overall increase in the number of life scientists during this period, by 1995 there were more women in the life sciences than in the social and behavioural sciences. In other fields, the increase in the per cent of women was only between six and seven points. Even by 1995, women were only 6 per cent of the full-time academic work force in engineering, with less than 2,000 full-time female engineers. In mathematics and the physical sciences, women's representation exceeded 10 per cent by 1995, but in mathematics the number reached only 2,000 and in the physical sciences just over 4,000. In the life sciences, the proportion of women approached 30 per cent by 1995.

While the number and proportion of women has increased steadily in all fields and types of institutions, *it is also critical that women hold positions of similar status to those of men within these institutions.* In this section, I examine gender differences at each rung of the academic ladder, beginning with differences in having tenure track positions (that is, faculty positions where tenure can be attained) compared to less prestigious and less secure off-track positions. For those who are faculty, we consider who has tenure and who does not. Finally, we consider advancement to the highest rank, that of full professor. Figure 10.6

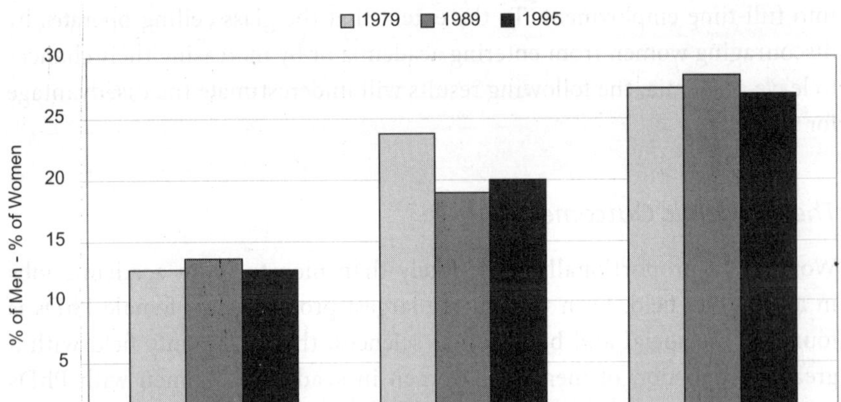

Figure 10.6 Differences in the Per cent of Men with a Given Outcome and Per cent of Women, by Year

Note: Positive values indicate that men are more likely to have the specific outcome.
Source: Long (2001).

plots differences in the per cent of men with a given outcome and per cent of women, by year.[4] Women are proportionally less likely to have each of these desirable outcomes. Overall, these results suggest very little progress in the advancement of women to the higher level of academia and are consistent with the idea of a glass ceiling whereby women are proportionately less likely to advance in rank.

The smallest difference, roughly 13 percentage points, is in the per cent of men and per cent of women with on-track positions. There is some improvement in the differences for being tenured, but by 1995, men are 20 points more likely to be tenured. The consistently large differences for promotion to full professor, over 25 percentage points, might suggest the presence of a 'glass ceiling' in academia. However, much of the lack of progress in the representation of women in each of these outcomes is due to the demographic effects of the rapidly increasing numbers of women.

The rapid change in the per cent of academic positions held by women is largely the result of increases in the proportion of new PhDs who are women, which has important implications for the age structure in academia. The average academic woman received her degree more recently than the average academic man, and the difference between the average career age (that is, years since the PhD) for men and women is increasing. In 1973, the mean career age for women was 9.5 years and 11.1 years for men; in 1979, 8.9 for women and 12.7 for men; in 1989, 10.6 for women and 15.9 for men; and in 1995, 11.2 and 17.0. The effects of changes in the growth of academia and the increased entry of women can be seen with a population pyramid (see Shyrock et al., 1973: 236–45 for details), which is a pair of horizontal histograms, one for men and one for women. Each bar represents the percent in an age-sex group (for example, women between one and three years from the PhD) relative to the size of the total population. The shape of a pyramid reflects the number of each sex entering the population (for example, new PhDs) and the number leaving through death or retirement. For example, if the same number of new PhDs was hired each year and there was no attrition until the age of retirement, the pyramid would be a rectangle. Or, if the size of new cohorts is increasing with increasing attrition among older members of the population, the pyramid would be triangular.

[4] Data on these outcomes was unavailable in the 1973 survey.

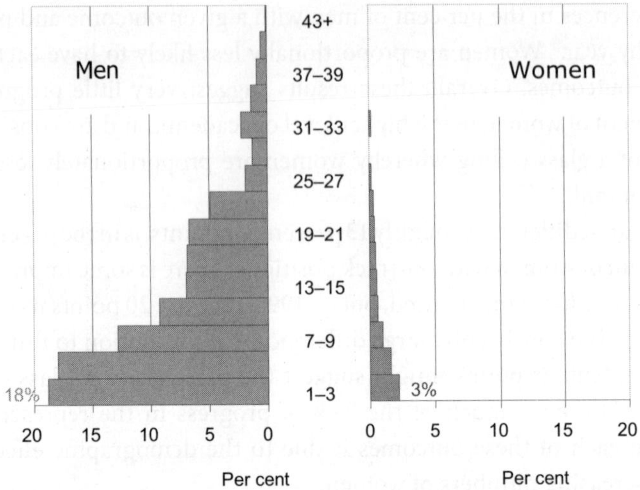

Figure 10.7 Distribution of Career Ages of Scientists in the Full-time Academic Labour Force in 1973

Note: Bars show the per cent of the total population in a given category. For example, in 1973 18 per cent of all scientists were men 1–3 years from the PhD; 2.5 per cent of all scientists were women 1–3 years from the PhD.
Source: Long (2001).

Comparing Figure 10.7 for 1973 and Figure 10.8 for 1995 shows the dramatic changes in the age structure of academia. In 1973, men represented a much larger proportion of all academics, as indicated by the greater area of the left hand side of the pyramid. The longer bars for young scientists, those with career ages of 1–3 and 4–6, show that incoming cohorts were larger than prior cohorts. By 1995, the proportion of men overall is reduced, as indicated by the lesser difference in the areas of the two halves of the pyramid. The size of incoming cohorts of men has stabilised, as reflected by the similar sizes of the bars for men between age categories 1–3 and 25–27. For women, in contrast, each younger cohort is larger than older cohorts. For example, in 1973 the youngest group of men represented nearly 20 per cent of all academics, while the youngest group of women represented less than 3 per cent. By 1995, new female PhDs grew to 4 per cent of academics, while new men dropped to less than 8 per cent.

Types of positions in academia are closely tied to professional age (see, for example, Long et al., 1993), and accordingly the changing age structure has important implications for gender differences in career outcomes. Early in the

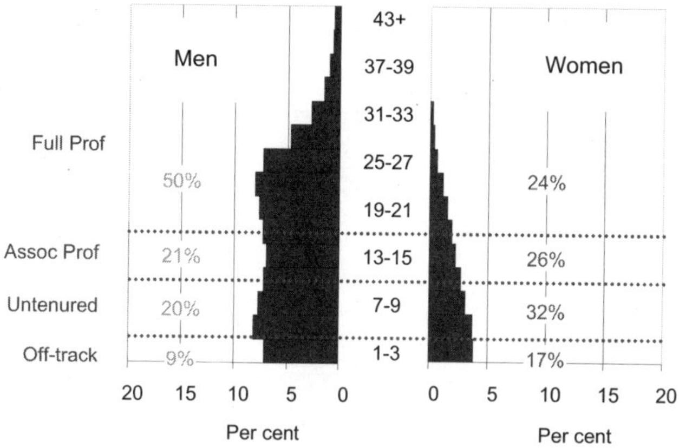

Figure 10.8 Distribution of Career Ages of Scientists in the Full-time Academic Labour Force in 1993

Note: Bars show the per cent of the total population in a given category. For example, 9 per cent of all scientists were men 1–3 years from the PhD; 17 per cent of all scientists were women 1–3 years from the PhD.
Source: Long (2001).

career, many scientists are off-track while taking a postdoctoral fellowship. Let's assume that such positions are most likely to occur during the first three years after the PhD. The lower portion of Figure 10.8, marked 'Off-track', shows that 9 per cent of academic men are within three years of the PhD, compared to 17 per cent of female academics. Accordingly, all else being equal, we would expect more women to be off-track. Similarly, 20 per cent of the men and 32 per cent of the women are within four to nine years of the PhD. This is the academic age in which scientists would be on-track, but too young to have earned tenure. So, again, all else being equal, we would expect more women to be untenured, or equivalently, more men to be tenured. Finally, 50 per cent of the men and only 24 per cent of the women have reached the age where we might expect promotion to full professor.

Given that tenure and rank are time dependent, we would expect women to be found less frequently in advanced ranks. To see the effects of age on gender differences in career outcomes, logit models were used to predict the per cent of men and per cent of women at each career outcome, controlling for years since the PhD.[5] The results of these analyses are shown in Figures 10.9, 10.10, and 10.11.

[5] These models also controlled for type of academic institution and field, but by far the most important factor was age. See Long (2001) for details.

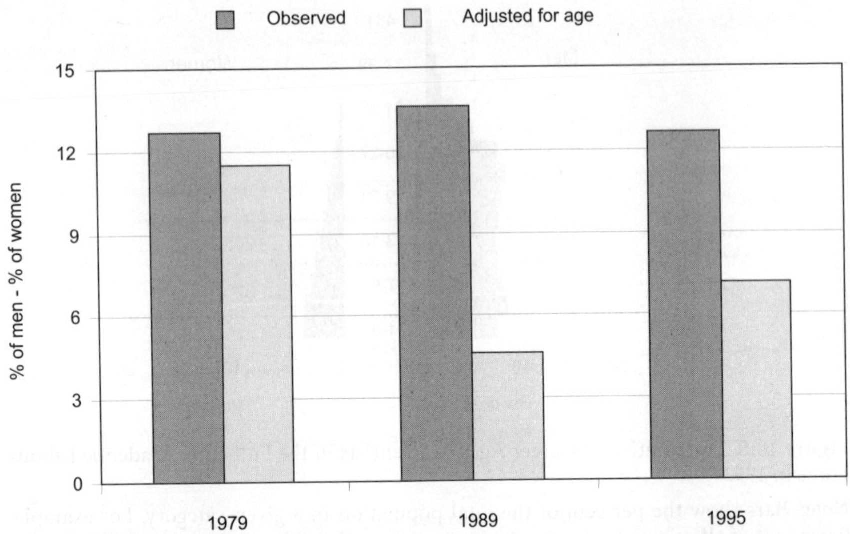

Figure 10.9 Difference between Men and Women in the Observed Proportions with Faculty Positions and the Adjusted Predictions after Controlling Only for Years since the PhD

Note: Predictions are for the fifeteenth year of the career. Data were not available in 1973.
Source: Long (2001).

The most fundamental distinction among academic positions is between tenure track positions and off-track positions. Scientists with tenure track positions have the possibility of advancing through the faculty ranks and attaining the job security provided by tenure. In comparison, off-track positions have lower pay, fewer resources, and less security. They include temporary teaching positions, research positions funded by soft money, visiting scholars, adjunct faculty without tenure track appointments elsewhere, postdoctoral fellows, and lower level administrative positions. The greater likelihood of women being in off-track positions is well known (Ahern and Scott, 1981; Haley-Oliphant, 1985; Reskin, 1978: 1239; Zuckerman, 1987: 133) and is confirmed by our data.

From 1979 to 1995, the per cent of all full-time academic jobs that were on-track decreased from 84 per cent to 79 per cent.[6] As shown by the 'observed' bars in Figure 10.10, throughout this period men had a 14 percentage point advantage over women in obtaining tenure track positions. While this

[6] The 1973 SDR did not collect information on whether a position was on a tenure track. Accordingly, analysis of tenure track status, tenure, and rank does not include data from 1973.

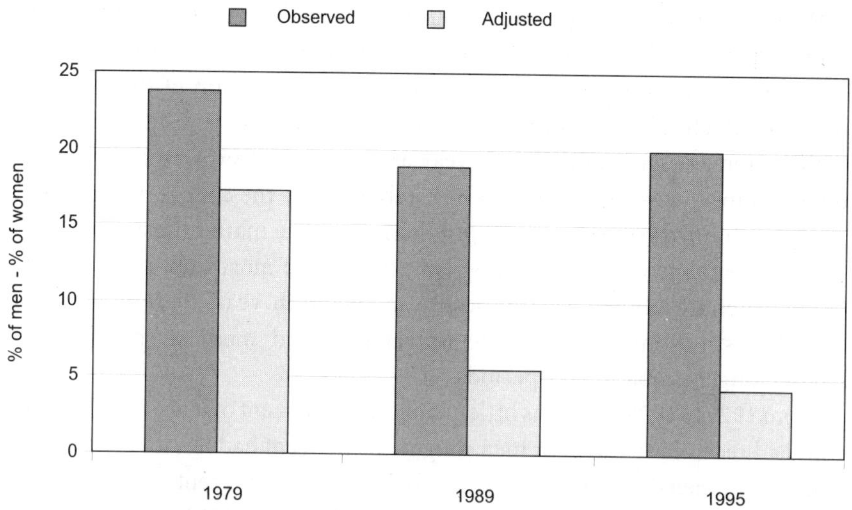

Figure 10.10 Difference between Men and Women in the Observed Proportion with Tenure and the Adjusted Prediction after Controlling for Field, Career Age, and Carnegie Type of Employer, by Year of Survey

Note: Predictions are for the tenth year after the PhD. Data were not available in 1973.
Source: Long (2001).

suggests that there has been little progress for women in becoming members of the faculty, these overall figures mask broad differences in the availability of faculty positions by field, type of institution, and at different stages of the career. To adjust for age differences, we estimated logit models that predict being in a faculty position after statistically controlling for career age. Differences in the adjusted proportions of men and women in their fifeteenth career year are shown by the 'adjusted' bars. In 1979, there was only a small decrease from the observed to the adjusted difference. That is, the observed gender difference in the per cent with tenure track positions *cannot* be explained by the younger age of female academics. By 1989, however, the observed difference was substantially reduced by adjusting for age, with a somewhat smaller reduction in 1995. Our results suggest that much, but not all, of the difference between men and women in their success in becoming faculty is due to differences in the stage of the career. The recent entry of women into S&E has contributed to the smaller per cent of women who are faculty. If current trends in PhD production and the job market for faculty continue, we expect that there will be increases in the per cent of women with faculty positions in the next decade.

While the process of granting tenure has varied over time and there are some differences across the types of institutions, the process normally occurs as follows. A person enters the academia upon completion of the doctorate or a postdoctoral fellowship with the initial rank of assistant professor without tenure. During the sixth year as an assistant professor, faculty are reviewed for tenure based on criteria established by the college or university. In research universities, research productivity is the main criterion, while in baccalaureate institutions teaching is normally the most important activity. While a tenure review usually occurs in the sixth year, there is variation among institutions, with private universities and medical schools often having longer probationary periods.

From 1979 to 1995 there was little change in the percent of tenure track faculty who had received tenure. For men roughly 80 per cent had tenure in each year, while for women the number increased slightly from 56 per cent in 1979 to 62 per cent in 1989, dropping to 60 per cent in 1995. Throughout this period, a nearly constant 20 percentage points more men than women were tenured, as shown by the 'observed' bars in Figure 10.11. This appearance of a lack of progress for

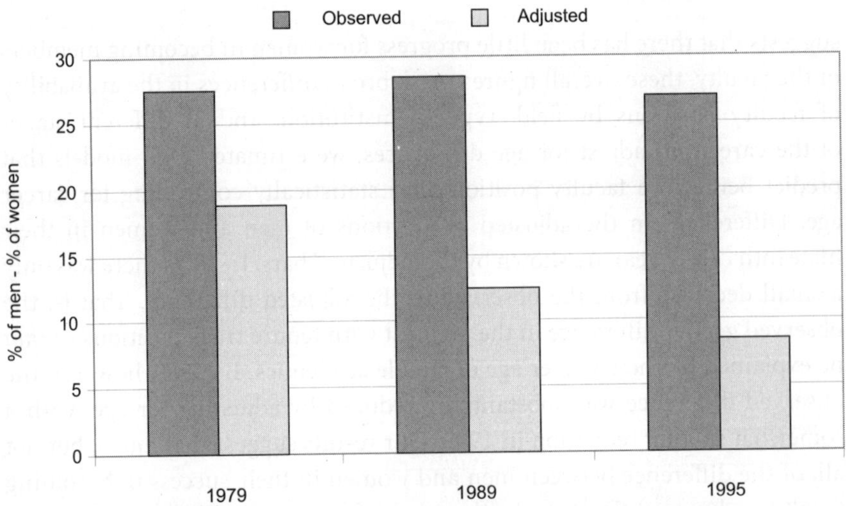

Figure 10.11 Difference between Men and Women in the Observed Proportion of Full Professors and the Adjusted Proportions Controlling for Field, Career Age, and Carnegie Type of Institution, by Year of Survey

Note: Adjusted proportions are for 20 years after the PhD. Data were not available in 1973.
Source: Long (2001).

women is due largely to the shifting age structure for women. The increasing entry of women into academia means that they are younger compared to men, which makes the average female faculty member less likely to be tenured compared to the average male faculty member. Further, there are differences in tenure rates across types of institutions, with tenure being least likely in the medical institutions, followed by doctoral, baccalaureate, research I, master's, and finally research II institutions. Since men and women have different proportional representation by Carnegie type of institution, it is important to take this into account when examining gender differences in receiving tenure. Adding statistical controls for professional age, type of institution, field, and other variables substantially reduces gender differences in the predicted proportion of scientists and engineers with tenure, as shown by the light bars in Figure 10.11. In 1979 the observed difference was reduced from 24 points to an adjusted difference of 17 points; in 1989 the observed difference of 19 points was reduced to six points, dropping to four points in 1995. Overall, by 1995 gender differences in being tenured are largely the result of differences in career age and to a lesser extent to differences in types of employing institutions. *However, even after these controls, men continue to be more likely to be tenured.*

Progress in the academic career is marked by advancement in rank, culminating in promotion to full professor, and it is in this outcome that past research has provided the strongest evidence for the unequal treatment of women in academia (Ahern and Scott, 1981; Astin and Bayer, 1979; Cole, 1979; Hurlbert and Rosenfeld, 1992; Long et al., 1993; Perrucci et al., 1983; Rosenfeld and Jones, 1986, 1987; Sonnert and Holton, 1995 ; Szafran, 1984). This is a central problem since with rank advancement comes the prestige, resources, and authority that are critical for a successful career in science.

The 'observed' bars in Figure 10.11 show a nearly constant difference of 27 points in the per cent of men and per cent of women who are full professors. Before concluding that there has been no progress in promotion to full for women during the past eighteen years, it is essential to keep in mind that academic rank, like tenure, is highly dependent upon career age. The adjusted bars show the effects of statistically controlling for differences in the age structures for male and female faculty, as well as differences in field and Carnegie type. The age-adjusted differences show a decrease in the over-representation of men from 20 points in 1979 to less than 10 points in 1995. Still, even after controlling for gender differences in career age, field of employment, and type of institution, *men continue to have an almost 10 percentage point advantage in being full professors.*

Overall, while there has been little change in the observed differences (that is, without controls) in the per cent of men and the per cent of women with a given career outcome, this reflects in large part the rapid increase in the entry of women into S&E. Controlling for age, the gender differences in outcomes are greatly reduced by 1995. Still, it is important to keep in mind that non-trivial differences remain that show the greater advancement of men in academia.

Conclusion

This chapter has documented some of the truly remarkable changes that have occurred in the representation of American women in S&E. In all aspects of the career – from the receipt of the PhD to entry into the labour force to attaining the rank of full professor – women are an increasing presence, both in absolute number and as a proportion of all scientists and engineers. As positive and encouraging as these changes are, it is equally clear that substantial differences remain. Women as a group remain less well represented and less successful than men in every dimension of the career that we have examined. It is fitting to close with the cautiously optimistic conclusions of the report from the National Academy of Sciences, *From Scarcity to Visibility: A Study of Gender Differences in the Careers of Doctoral Scientists and Engineers* (Long, 2001:10): 'But while women have clearly made enormous gains in their participation in science and engineering, it is also clear that these advances represent neither unconditional success in overcoming gender inequalities nor assurance of continuing progress in the future.'

References

Ahern, Nancy C. and Elizabeth L. Scott (1981). *Career Outcomes in a Matched Sample of Men and Women Doctorates*. Washington, D.C.: National Academy Press.

Astin, Helen S. and Alan E. Bayer (1979). 'Pervasive Gender Differences in the Academic Reward System: Scholarship, Marriage and What Else?' in D.R. Lewis and W.E. Becker (eds), *Academic Rewards in Higher Education*, pp. 221–29. Cambridge, MA: Ballinger Publishing Co.

Cole, Jonathan R. (1979). *Fair Science*. New York: Free Press.

Committee on Women Faculty in the School of Science (1999). *A Study on the Status of Women Faculty in Science at MIT*. Boston: Massachusetts Institute of Technology (MIT).

Haley-Oliphant, Ann. E. (1985). International perspectives on the status and role of women in science. In J. Butler-Kahle (ed.), *Women in Science: A Report from the Field*, pp. 169–182. Philadelphia: Falmer Press.

Hargens, L. and J. Scott Long (2002). 'Demographic Inertia and the Representation of Women and Minorities on Higher-education Faculties', *Journal of Higher Education*, 73(4): 494–517.

Hurlbert, Jeanne and Rachel A. Rosenfeld (1992). 'Getting a Good Job: Rank and Institutional Prestige in Academic Psychologists' Careers', *Sociology of Education*, 65(3): 188–207.

Long, J. Scott (1997). *Regression Models for Categorical and Limited Dependent Variables*. Thousand Oaks, CA: Sage Publications.

—— (ed.) (2001). *From Scarcity to Visibility: A Study of Gender Differences in the Careers of Doctoral Scientists and Engineers*. Report of the Panel for the Study of Gender Differences in the Career Outcomes of Science and Engineering PhDs. National Academy of Sciences. Washington, DC.

Long, J. Scott, Paul D. Allison, and Robert McGinnis (1993). 'Rank Advancement in Academic Careers: Gender differences and the Effects of Productivity', *American Sociological Review*, 58(8): 703–22.

Morgan, Laurie (1998). 'Glass Ceiling Effect or Cohort Effect? A Longitudinal Study of the Gender Salary Gap for Engineers, 1982–1989', *American Sociological Review*, 63(4): 479–93.

NSF (National Science Foundation) (1920–95). *Survey of Earned Doctorates*. Washington, DC.: National Science Foundation (NSF).

—— (1973–95). *Survey of Doctorate Recipients*. Washington, DC.: National Science Foundation (NSF).

—— (1997). *Characteristics of Doctoral Scientists and Engineers in the United States: 1995*. Washington, DC.: National Science Foundation (NSF), Division of Science Resources Studies.

Perrucci, Robert, Kathleen O'Flaherty, and Harvey Marshall (1983). 'Market Conditions, Productivity, and Promotion among University Faculty', *Research in Higher Education*, 19(4): 431–49.

Reskin, Barbara F. (1978). 'Scientific Productivity, Sex, and Location in the Institution of Science', *American Journal of Sociology*, 83(5): 1235–43.

Rosenfeld, Rachel A., and Jo Ann Jones (1986) Institutional mobility among academics. *Sociology of Education*. 59(4): 212–226.

Rosenfeld, Rachel A. and Jo Ann Jones (1987). 'Patterns and Effects of Geographic Mobility for Academic Women and Men', *Journal of Higher Education*, 58(5): 493–515.

Rossiter, Margaret W. (1982). *Women Scientists in America: Struggles and Strategies to 1940*. Baltimore, Md.: Johns Hopkins Press.

—— (1995). *Women Scientists in America: Before Affirmative Action 1940–1972*. Baltimore: Johns Hopkins Press.

Shyrock, Henry S., Jacob S. Siegel, and Associates (1973). *The Methods and Materials of Demography*, vol. 1. Washington, DC.: Department of Commerce.

Sonnert, Gerhard and Gerald Holton. 1995. *Who Succeeds in Science? The Gender Dimension*. New Brunswick, NJ: Rutgers University Press.

Szafran, Robert F. (1984). *Universities and Women Faculty: Why Some Organizations Discriminate More Than Others*. New York: Praeger.

U.S. Department of Labor (1991). *A Report on the Glass Ceiling Initiative*. Washington, DC.: U.S. Government Printing Office.

Zuckerman, Harriet (1987). 'Persistence and Change in the Careers of Men and Women Scientists and Engineers', in L.S. Dix (ed.), *Women: Their Underrepresentation and Career Differentials in Science and Engineering*, pp. 127–56. Washington, DC.: National Academy Press.

CHAPTER ELEVEN

Women in Science in France

Claudine Hermann, Françoise Cyrot-Lackmann, Jeanne Peiffer and Hélène Rouch

This chapter focuses on women in academic science and engineering and the education paths that lead to such careers in France. It situates the milestones in women's education within the context of historic-political struggles surrounding sexual equality in France. It also analyses women's place in scientific public employment using the recent data relating to universities and research institutes. The chapter examines the unique French dual university-*Grande Ecole* (elite higher education school distinct from university) system and explains it with a focus on the failure of co-education in the *Ecoles Normales Supérieures* (elite higher education schools for training teachers and researchers). The final section describes recent government actions to improve the situation for French women in science.

Historical background

Public education for women in France developed in the nineteenth and twentieth centuries (Montreynaud 1992–2000, Préfecture d'Ile-de-France 1995). The *lycées* (secondary public schools) created by Napoléon in 1808 were closed to women. Primary schools for girls were first established in 1836, and the first women's *école normale* (school to educate primary school teachers) was opened in 1838. Girls' education was not compulsory at that time and their curriculum was not comparable to that of boys. The first woman to complete her *baccalauréat* (examination ending secondary school and giving access to university) was Julie Daubié in 1861. She prepared herself for this exam and was allowed to sit for it only after Empress Eugenia intervened.

Milestones in Women's Education

The great strides forward in girls' education came after the establishment of the Third Republic, which followed the defeat of France by Bismarck's Prussia in 1870. In 1882, the Minister of Education, Jules Ferry passed a law requiring public primary education for all children starting at age six in efforts to undermine the influence of the Church in education. The curriculum was designed to educate girls to be 'good' mothers and housewives. Girls' public *lycées* (secondary schools) were opened, and in 1881 the *Ecole Normale Supérieure de Sèvres* was created to educate female teachers for girls' *lycées. Ecole Normale Supérieure de la rue d'Ulm*, which trained male teachers for employment in boys' *lycées*, had been established in 1794. During this time, the curriculum of girls' *lycées* did not give access to university (Hulin 2002). At the end of the nineteenth century, however, a few female students, mostly foreigners, began taking universities courses. In 1881, Blanche Edwards passed the *externat de médecine* (competitive medicine exam to become a medical student in a hospital). Her male classmates were so outraged that they burned her in effigy in the street.

The story of Marie Slodowska-Curie, who emigrated from Poland to France to prepare for a PhD in physics and chemistry, is well known. In 1903, Henri Becquerel, Pierre and Marie Curie were jointly awarded the Nobel Prize for physics for their discovery of radioactivity. After her husband died in an accident, Marie Curie took his place, becoming the first woman to be appointed as professor at the Sorbonne in 1906. Despite the fact that she was the first person ever to receive a second Nobel Prize (1911), awarded alone this time, she was not admitted to the French Academy of Sciences! Being a woman and a foreigner prevented her acceptance into this very select assembly. In fact it was only in 1979 that a first woman, the mathematician Yvonne Choquet-Bruhat, became a member of the Academy of Sciences. Out of a total of 247 current members, only eighteen are females in this prestigious institution.

The diploma for girls' secondary schools was upgraded to a *baccalauréat* in 1919 and *baccalauréats* for girls' and boys' were put on an equal footing in 1924, finally allowing women equal access to universities (at least in theory). Public primary and secondary schools in France gradually became co-educational in the 1960s and early 1970s. Women's and men's *Ecoles Normales Supérieures* were merged in the 1980s in such a way that the percentages of women admitted in physics and particularly in mathematics dropped dramatically (see the section below on Education and the sub-section Top-ranked Grandes Ecoles: a co-education failure story). Women were gradually admitted to French engineering schools; the last engineering school to accept women was the military School for Aviation (*Ecole de l'Air)* in 1978. Later girls started even outnumbering the boys until the third year of university studies, and there were many female teachers and professors, though there were still fewer at the education decision-making positions (Feldman and Morelle 1994; Fave-Bonnet 1996).

The Role of Politics

In order to understand the history of women in French science, one must look at the situation of women in French politics more generally. French women obtained the right to vote in 1944, at the end of World War II. The Parliament had, in fact, voted for this right in the 1930s, but the law was not passed due to the lack of Senatorial support. The left-union (socialists, radicals, communists) *Front Populaire* (Popular Front) government of 1936 had three women deputy-state secretaries: one of them was Irène Joliot-Curie, the daughter of Pierre and Marie Curie, who had been awarded the Nobel Prize for physics jointly with her husband Frédéric Joliot-Curie in 1935 for the discovery of artificial radioactivity. Irène Joliot-Curie served as deputy-state secretary for research. Throughout the second half of the twentieth century, the percentage of women in the French Parliament remained low. Up to 2002 women represented approximately 10 per cent of Parliamentary members, ranking France second-to-last among the European Union Member States. As women became dissatisfied with this unfair situation in the 1990s, a network called *Demain la Parité*[1] (Parity Tomorrow) was created. Its members realised that the problem of the under-representation of women was not unique only to French politics: it occurred more generally in all channels leading to power. One road to decision-making positions in France is to be alumnus (alumna) of a *Grande Ecole.* As is well known, French higher education has two distinct tracks: the universities and the elite *Grandes Ecoles* (literally, 'Great Schools,' the French equivalent of the Ivy

[1] Website of this network : http://www.int-evry.fr/demain-la-parite.

League in the USA). Of the *Grandes Ecoles* (in the areas of engineering, agriculture, business, and administration), the scientific ones are particularly prestigious – former president Giscard d'Estaing and numerous major French company Chief Executive Officers were educated at the Ecole Polytechnique, for example. In view of this, *Demain la Parité* commissioned a study of the representation of women in the engineering schools and in the classes that prepare students for the corresponding competitive admission exams. Two heavily documented reports were issued under the direction of the late Huguette Delavault, a retired university professor of mathematics. The first one is on the place of women in scientific *classes préparatoires* (special classes for preparing the selective entrance examinations into *Grandes Ecoles*) (see the section below on Education and also Delavault 1997), the second one on the engineering *Grandes Ecoles* (Delavault 1998). These pioneer studies had a major impact in the academic circles and in the mass media.

Between 1998 and 2002, French Prime Minister Lionel Jospin championed *parité* (sex equality, particularly in politics). A law (2002) introduced the paternity leave of eleven days to be taken within four months following the baby's birth. In politics a law was passed requiring equal numbers of men and women's names on proportional ballot elections (candidates elected from lists) in cities with more than 3,500 inhabitants. This law was applied for the first time in March 2001 in local elections with the result that women held 47 per cent of municipal functions but only 4.7 per cent of the mayor positions and 12 per cent in the 2002 Parliament. The Prime Minister Jospin also ordered numerous reports on the position of women in French society: in the top positions of civil service (Colmou 1999); in the professions (Génisson 1999); in the economy (Majnoni d'Intignano 1999); and in decision-making (Cotta 2000). Catherine Blum's report on the availability of sex-disaggregated statistics in France (Blum 2000) concluded that such data are reasonably collected, but insufficiently analysed or publicised. In response, the Prime Minister issued a memorandum (March 2000) requiring that all published statistics should be sex-disaggregated.

During this period, the Research Directorate-General of the European Union, under the direction of Edith Cresson, Commissioner for Education and Research, turned the spotlight onto the question of women in science (European Commission 1999). The creation in 1998 of the 'Women and Science' Sector (which became a Unit in 2001) was essential to introducing this question into many European Union circles. The status-quo was elaborated in the European Technology Assessment Network (ETAN) report

(European Commission 2000) and this report was not only approved by the then Commissioner for research, Philippe Busquin, but was widely publicised throughout Europe (see the section below on Recent Developments (1999–2007)). The combined action of the Research Directorate-General (DG) and the European Member States continued, in particular through the creation of a group of civil servants from the European Union member states and the states associated to DG research programmes. This group, responsible for issues surrounding women's participation in the sciences and known as the EU Helsinki group, is still active. The mission of these civil servants was to provide sex-disaggregated statistics, to exchange good practises on the Women and Science issue and to initiate new actions in their own countries. A first report on their progress was made to the European Parliament in December 2000. The statistical correspondents of the Helsinki group produced impressive booklets of harmonised gender-disaggregated statistics and indicators over the European Union countries: '*She Figures* 2003' (European Commission 2003) and '*She Figures* 2006' (European Commission 2006). Dg Research is pursuing this publication every third year and '*She Figures* 2009' (European Commission 2009) is the last issue up to now.

Gender Studies

While the Northern European (especially Scandinavian) countries were highly efficient in ameliorating conditions for women in science, France has not been able, despite a strong mobilisation during the 1980s, to improve the situation. One reasonable explanation may be found in the different philosophies that subtend actions in favour of women in science in the two regions. Scandinavian and English political philosophy recognises two sexes and genders, the unequal social construction built around them, and the necessity to take specific measures in favour of women. The claimed universality of rights, which characterises French law, situates France in a quite different philosophical tradition. Gender issues are thus analysed within different frameworks beginning in the 1970s.

In France, materialist theoreticians like Colette Guillaumin (Guillaumin 1992), Christine Delphy (Delphy 1998–2001) and Nicole-Claude Mathieu (Mathieu 1991) introduced concepts like '*classe sociale de sexe*' (sex social class), and '*rapports sociaux de sexe*' (social sex relations), that did not take into account a feminine specificity relating to females' reproductive capacities. On the contrary, they rejected any reference to naturalism and claim that the hierarchy between the two sexes is historical and social, thus an arbitrary, construct. The Marxist notions of struggle between sex classes and oppression were used instead. Women symbolised as a social class dominated by that of men mainly to perform reproductive work. '*Il n'y a de femmes que pour autant qu'un rapport de force inégalitaire fait de l'oppression et de l'exploitation d'un groupe social la condition du pouvoir de l'autre*' (Women exist only as far as an unequal power relation transforms the oppression and exploitation of a social group into the condition of the other group's power) one could read in the first issue of the journal *Questions féministes* (No. 1, p.19).

The theoretical orientations inherent in the concept '*rapports sociaux de sexe*' (social sex relations), was used and applied in domains like sociology, anthropology, and history. Women's history developed autonomously and on different epistemological assumptions inside the academic institutions, with the support of Rita Thalmann and Michelle Perrot (cf. Perrot's assessment in the book recently published by Delphine Gardey and Ilana Löwy 2000, pp. 59–73). Feminist study groups were founded at various universities, the first one being *Centre d'études féminines de l'Université de Provence* (CEFUP) (Centre of feminine studies of the Provence university) in Aix-en-Provence, followed by *Groupe d'études féministes* (GEF) (Group of feminist studies) in Paris 7, *Centre Lyonnais d'études féministes* (CLEF) (Lyons centre of feminist studies) in Lyon 2, *Groupe de recherche interdisciplinaire d'étude des femmes* (GRIEF) (Group of interdisciplinary research on women studies) in Toulouse, and so on. These groups were quite active in organising conferences and publishing journals or bulletins (for instance Daune-Richard, Hurtig and Pichevin 1989).

The natural sciences were, however, outside the realm of feminist critique as it developed in close contact to the women's movement in France. Informal groups began, however, to meet towards the end of the 1970's (Peiffer 2000) and to reflect on their situation as women in the sciences. The first meeting was prompted by the publication of *Le fait féminin*[2] (Sullerot and Thibaut 1978), where sociologist Evelyne Sullerot, arguing against Simone de Beauvoir, held that women are, indeed, born women with a different anatomical fate. Furthermore, women researchers in the trade unions (Syndicat national des chercheurs scientifiques, SNCS) (National Trade-union of the scientific researchers) compiled statistics concerning women and their scientific carriers. They also investigated women's lived experiences within the various

[2] This book was translated into nine languages, including English.

scientific professions (Commission des femmes du SNCS 1981) [Commission of SNCS women]. Several projects grew out of these first gatherings (cf. (*Pénélope* 1981). An especially original one was without doubt the seminar entitled Limits/Frontiers (1980–1988) held in Paris. This women-only seminar was to be interdisciplinary, drawing participants from widely diverse fields from literature to the natural sciences; it was also to take place outside the boundaries of conventional academic institutions. The understanding of interdisciplinary work inside research institutions was at that time far too narrow to support such an enterprise. During its eight-year run, this seminar allowed women from various feminist and research traditions the freedom to test hypotheses and cast theories sometimes even in language not accepted by traditional institutions (cf. Séminaire Limites-Frontières and Rouch 1986).

In addition to Marxism, the dominant intellectual discourse in 1970s France was psychoanalysis, dominated by Jacques Lacan's attempt to formalise its concepts. Rooted in that culture, the founding members built the Limits/Frontiers seminar on the following assumption: the sciences, as well as other intellectual ventures, are nourished with images, pictures, and metaphors. These images might not be the same for men and for women, given their different social inscriptions and given their different places in the hierarchical order between the two sex classes. Experiencing oppression or exerting domination does not necessarily lead to the same construction of knowledge. Science done from the bottom up, from the 'grassroots' (to use the metaphor of the time), could differ significantly from the official science. Thus the question of sexist bias in the academic science was an important one to explore. Does the gendered identity of the scientist play a role in the production of knowledge? The emphasis was more on the individual scientist, on the results produced, than on the social organisation of science, its rhetoric or its values. But, ultimately not choosing between a materialist or an essentialist approach to sex differences, the seminar has proved unable to build a coherent theoretical framework to the gender in science question. It remained a reservoir of ideas, undogmatic, and open to various, even contradicting approaches.

Nearly at the same time, some attempts were made to introduce a feminist standpoint and/or gender problematic into the university curricula. Toward the end of 1982, circa 800 women gathered together at the University of Toulouse in order to make a first assessment of the results obtained and to organise women's studies. The proceedings of this conference show the wide range of topics discussed and the variety of theoretical and methodological approaches (AFFER 1984. cf. in particular the sections: '*Femmes, sciences et techniques*' (Women, science and technology), pp. 705–734; '*Critique féministe des sciences: problématique de la différence des sexes*' (Feminist critique of science: the issue of sex difference), pp. 735–824; '*Sciences exactes-sciences humaines: Epistémologie des formalismes*' (Exact sciences- human sciences: epistemology of formalisms) pp. 835–877). The dominant opinion was that gender studies could not be left to a handful of experts, but should be intimately linked to feminist action and women's struggles. The aim of such research was not simply to study facts, but also to change them. Thus the social and the political dimension of feminist research or research on women was stressed. As a result, the *Centre national de la recherche scientifique* (CNRS) (National centre for scientific research) financed a program on women's and feminist research '*Action thématique programmée*' (Programmed thematic action, a targeted funding) No. 6, 1984–1989, (CNRS 1989–90) for an overview of the research done). The various attempts to perpetuate the program failed for a series of complex reasons internal to CNRS and because of a changed political context, when the socialist government was replaced by a more conservative one after the president elections of 1995.

Nevertheless several research groups were created, especially in the human and the social sciences, some of which are still active, like *SIMONE-SAGESSE* (*groupe de recherche pluridisciplinaire SAvoirs, GEnre et rapports Sociaux de Sexe,* pluridisciplinary research group Knowledges, Gender and Social Sex Relations) in Toulouse, *Groupe d'étude sur la division sociale et sexuelle du travail* (GEDISST) (Study group on the social and sexual work division) and *Centre d'enseignement, de documentation et de recherches pour les études féministes* (CEDREF) (Centre of teaching, information and research for feminist studies) in Paris, etc. University positions were obtained (at the level of associate professor) in Rennes, Paris and Toulouse, journals like *Clio, Les cahiers du genre (anciennement du GEDISST)* (The Review on Gender, previously named the Review of GEDISST), *Les cahiers du CEDREF* (The Review of CREDEF), etc. were published and at the initiative of some individuals, seminars were held at the *Ecole des Hautes Ètudes en sciences sociales* (EHESS) (School of Higher Studies in Social Sciences). This timid institutionalisation made no real attempt to change the disciplinary structure of the sciences concerned or to reorganise their architecture in order to introduce new disciplines such as gender studies.

The gender question in science has thus opened a new methodological perspective questioning the epistemological foundations of the established, but did not lead to an autonomous field of knowledge in France. Gender analysis yielded important results in the human and social sciences (Collin 1992; Fraisse, Sissa, Balibar, Rousseau-Dujardin, Badiou, David-Menard and Tort, 1991; *Les cahiers du GRIF* (*Groupe de recherche et d'information féministe*, Group of feminist research and information – Brussels, 1990), especially in understandings of women's work, but did not really touch upon the natural sciences, whose gender neutrality and objectivity were often reasserted. France did not even join this movement, which is today well established in the history of science, and recovers the scientific contributions of women in the past. While in the United States the gender studies of science came of age in the 1980s and '90s, French research and its institutions proved too rigid. The growth of a rich and interesting field of research, growing out of the bottom of feminist struggles, was stunted before even beginning. The reasons of this failure are not clear, but are partly due to the lack of institutionalisation, partly due to the rejection by French academic institutions of certain 'radical' positions and finally partly due to the isolation of the single researchers in their specialised disciplines. To offset these pressures, a national association *Association nationale d'études féministes* (ANEF) (National association of feminist researches) was created in 1989. It publishes a *Bulletin* three times per year and a thematic issue once a year.[3]

Yet, after the publication in the late 1990s of a remarkable book by Michèle Le Doeuff, *Le sexe du savoir* (The sex of knowledge) (1998), initiatives have shown up to introduce gender issues in philosophy, history and science sociology. These are mainly based on the fundamental theoretical texts by women authors, written in English, which are not much known in France and not translated into French, with the exception of Evelyn Fox Keller. Several seminars also have taken place at Centre Koyré, University Paris 7, EHESS and École des Mines (Engineering school for Mining Crafts, one of the top French Engineering schools) to sensitise and attract the students and the researchers to this new field. Some meetings or colloquia also have lead to a few publications (Gardey and Löwy 2000, Fougeyrollas and al. 2003, Löwy and Rouch, 2003; Rouch et al. 2005) and has initiated several translations (such as Donna Haraway's collection of essays in the anthology by Allard, Gardey and Magnan 2007). But the academic fortress of the history and philosophy of science still remains firmly closed to the alternative epistemologies of the stand point or of the situated knowledge, marginally tolerating the critical analyses

[3] For further information: ANEF, 34 rue du Professeur Martin, F-31100 Toulouse.

of the construction of the sex and gender differences (Vidal 2006). It is rather in some other fields of the work, such as education, health and sociology that advances have taken take place, particularlry researches on the relationship between gender and crafts or industry techniques (Chabaud and Gardey. 2002, Laufer, Marry and Maruani. 2003, Marry 2004, Cacouault and Gardey 2005), contraceptive and the reproductive techniques (colloquium organised by Centre Beaubourg '*Des femmes et des hommes: le genre de la reproduction*' (Of women and men : gender of reproduction) 2007, Löwy 2006). Many of the productions related to these researches still remain scattered, under the shape of papers or files, in the reviews, less known and spread. Even if a large fraction of the scientific community, in particular history and sociology of science, begins to take seriously the field opened by the gender issues, the institutional support essential for its development is still lacking.

The Scientific Employment Pattern of France: A Few Points

Before delving into the details concerning the status of women scientists in France, let us begin with the data given in Chapter Two of the ETAN report (European Commission 2000) and the findings of the '*She Figures 2006*' (European Commission 2006), comparing the European countries. According to these data, France occupies a relatively good (although not satisfactory) position in the ranking of the countries. In decreasing order of percentages of women among university professors: in 2007 the global (all disciplines) data on French faculty gives 18 per cent women among professors compared to 40 per cent women at the lower rank (see Table 11.1). The present situation for women scientists in France results from the general conditions for the educated working women. The employment is not determined by any affirmative action or 'quotas', which are in general not appreciated, with the

Table 11.1 *Percentage of women among French faculty (2006–2007), by rank and main group of disciplines*

	Law	Arts and Literature	Science	Health	Total
MdCs	40%	52%	31%	50%	40%
Professors	19%	31%	13%	14%	18%

Source: Authors. Constructed on the basis of Ministère de l'éducation nationale, 2007b data.

fear that a woman hired to fill a quota would be considered less qualified than a regular hire (Hermann 1997).

The Societal System in France

Although French women's wages are on average 19 per cent lower than those of men and women represent only 35 per cent of the executives, generally speaking working conditions for women in France are rather favourable, and it is considered normal for a mother of young children to work. An infant-care system, even if it is not perfect and varies by town and region, does exist. The *école maternelle* (public pre-school system for children from three to six years of age) is free, open (except on school holidays) from morning to late afternoon, and attended by more than 90 per cent of all the children. In addition, the law permits parental leave until the child is three years old (unpaid, but the position is kept for the parent), an option used mostly by mothers, and tax deductions are available for child care costs (until the child is six years old). These programs for working parents, although not yet ideal, do not exist in all European countries.

Employment in French Universities and Research Institutions

Several recent publications summarize the main sex-disaggregated data on French research (Livre blanc [White book] 2002; *Mission pour la parité dans la recherche et l'enseignement supérieur* [Mission for Parity in Research and Higher Education], 2005). First, only the general trends will be discussed and subsequently the situation of women and men in French research staff will be analysed. Similar to several other countries, French research and development is split into the public and private sectors. In 2005, there were approximately the same number of researchers in the civil public (139,333 persons) and in private institutions (113,704 persons) (*Ministère de l'enseignement supérieur et de la recherche* [Ministry of Higher Education and Research] 2007). In public arena are universities (faculty members: 55,9373 in 2005, see the section below on University Faculty) and different research institutions, the largest one being the *Centre national de la recherche scientifique* [National Centre for Scientific Research] (CNRS]) (11,650 researchers in 2005, see in the section below on Research Institutes). These are subject to public regulations. Governmental industrial and commercial structures, such as the Atomic Energy Commission (Commissariat à l'Energie Atomique [CEA]), are under private regulations and will not be discussed here. The scientists in French universities and research institutions are in general civil servants or state-employees. After a short post-doctoral period, they are hired for the permanent positions between age 30 and 35, depending on the discipline. In the last several years, the government has provided incentives for state researchers to join the industrial sector, but this has not produced major effects. Since in general it is easier to be promoted to the rank of university professor than to that of *directeur de recherche* (research director, i.e. senior tenured researcher), it is frequent to start one's career at CNRS and to end it as a university professor. Generally speaking, in all professions, geographical mobility is low in France, being not in the traditions: people keep the culture of lifelong personal relationship, one usually remains employee in the same company for a whole career, and in a couple man and woman usually both work. Although mobility in academia is generally praised, in reality it is recommended only when a candidate is being promoted to professor.

The Current Trends and Figures

The latest *She Figures* 2009 (European Commission 2009) data shows that 18 per cent of the university professors in France are women. Though this percentage is not high, it ranks among the top European countries. The famous pioneer report by the European Union, better known as ETAN report (European Commission 2000), introduced the typical career of females in academia as 'scissors diagram'. It shows the percentage of females is higher as students, but goes down as one moves upward the hierarchy, while the career graph of male shows an upward trend. For example in France, females constitute 57 per cent of the total number of students (*Ministère de l'éducation nationale* [Ministry of National Education] 2007a); their number drops at the post-graduate and subsequent levels and finally only 18 per cent of the professors are female. This kind of 'scissors diagram' is not a cohort study, which follows the same persons during the course of their studies and careers, but rather a snapshot at a given year of the sex percentages at the different levels of the academic ladder. It reflects 'the leaky pipeline' effect, a phrase introduced in the mid-1990s that expresses the erosion of women's percentage when climbing the academic ladder.

It is sometimes enlightening to build relevant indicators from gender-disaggregated statistics. Chapter 8 of the ETAN report (European Commission 2000), for example, is devoted to 'Gender Statistics in Science: Measuring

Inequality'. Table 8.1 of this chapter (Chapter, 8, ETAN report) indicates the number of men who would have to vacate their positions in order to achieve equal sex distribution among professors in France: for the year 1998, this amounted to 1,576 in physics and chemistry, 1,372 in mathematics, 496 in biology, medicine, and 1,124 in the humanities. This thought experiment merely indicates the scale of the problem (this type of replacement is, of course, preposterous). Another indicator is related to the likelihood of promotion. In French universities where there are two professorial ranks (*maître de conferences* [assistant professor] and professor), one can calculate a woman's (a man's) likelihood of promotion from the current distribution of the sexes by rank, which is very discipline-dependent (see below the next section Developments in the Past Twenty-Seven Years in the Main Groups of Disciplines). For example, statistics show that women are more likely to be promoted to the rank of professor in Law than in Biology and Biochemistry. In Law currently 37 per cent of the men are professors, in comparison with 17 per cent of women; in Biology and Biochemistry 37 per cent of the men have achieved this rank, but only 13 per cent of the women, even though women are well represented at the lower level. The indicator, 'the male advantage', quantifying the *better* likelihood *of men* to be promoted is introduced in the next section.

University faculty

We first present French procedures for hiring and promotion, which may help the reader to define the equivalence with the university ranks in other countries. In France, after having defended his or her PhD (and in some disciplines having carried out one- or two-years of post-doctoral research), a scientist wishing to become faculty member applies to become a '*maître de conférences*' (MdC) at the National Council of Universities **Conseil national des universités** [CNU]). Once accepted, the person has the privilege of applying for any of the MdC positions offered by various universities. If successful, the person becomes a MdC.

To become a professor in a French university, the MdC must first defend his or her habilitation, a national diploma awarded by a university. This diploma certifies that the person is a highly qualified researcher, displaying an original approach in a scientific domain, an autonomous strategy of scientific research, and the ability to guide young researchers. The 'habilitated' candidate then applies to be considered for the position of '*professeur*' to the National Council of Universities. Once the candidate is approved, he or she may apply to positions

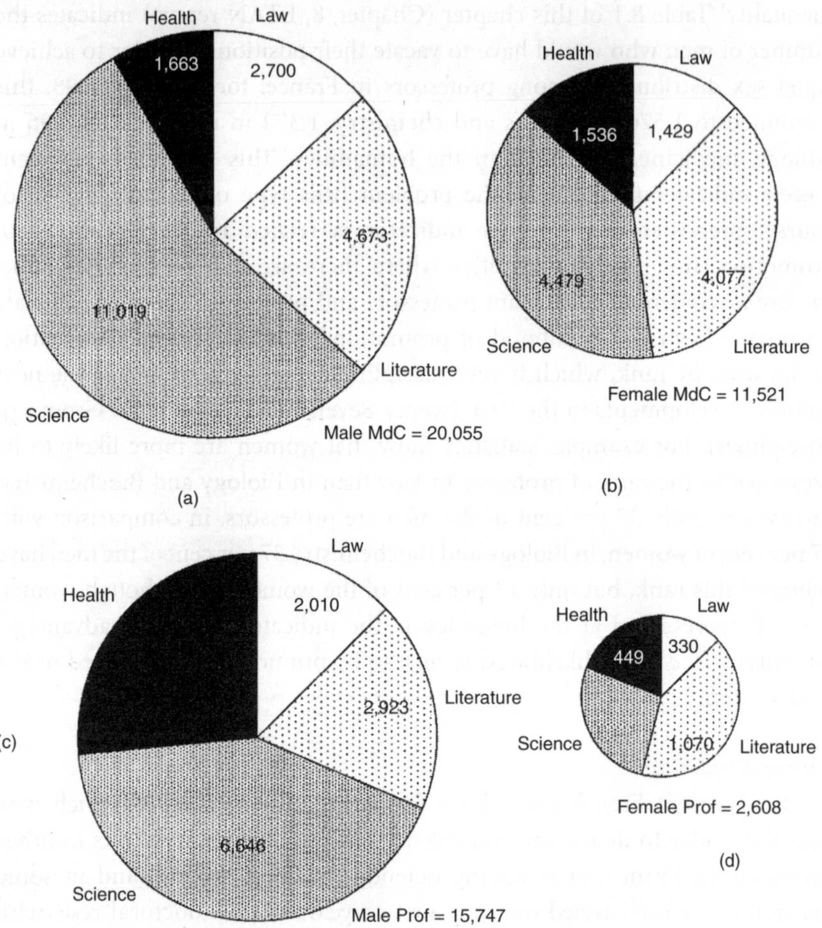

Figure 11.1 French faculty: numbers according to gender and rank in the four groups of disciplines

Source: Authors. Constructed on the basis of *Ministère de l'éducation nationale*, 2007b data.

of *professeur* at universities with public openings. In Law, Politics, Economy, Management, professors must have passed a competitive national examination.

University disciplines in France fall into four main groups: Law (and Economy), Literature (and Human Sciences), Science, and Health (Health includes Medicine and Odontology, whose faculties have a specific status, and Pharmacy). Statistics are gathered according to these groupings. A more detailed analysis of situation of women by detailed disciplines is given in Figure 11.1 which represents presents the total number of MdCs and professors by sex for the four main groups of disciplines in 2006–2007. For simplicity, we only discuss the two academic main rankings, the *'maître de conférences'* (MdC) and *'professeur'* ones.

As can be deduced from these data, Science MdCs (men and women) represent 46 per cent of the total MdC staff, Science professors (men and women) are 39 per cent of the total number of professors. The corresponding women percentages are listed in Table 11.1.

For MdCs, the lowest percentage of women is in Science (31 per cent); equality between the sexes is obtained for Health and Literature. For professors, all the women percentages are much lower than the corresponding ones for MdCs, the lowest being in Science (13 per cent women); the percentage of women among the professors is also very low for Health (14 per cent), although sex equality is obtained for MdCs. As for average age of MdCs or professors, it is not very gender-dependent.

Even though France is a centralised country, the universities retain some autonomy. Hence, it is interesting to compare the representation of women faculty in different universities. For instance, in Paris and its neighbouring areas, the percentage of female faculty is much above the national average. By contrast, the major universities in the countryside have lower percentages of women.

Two conclusions can drawn from the statistics presented above. First, the pool of qualified women for top positions is larger for Health and Literature but smaller elsewhere, particularly in Science. Second, across the board the percentage of women in the rank of professor is much lower than in the rank of MdC: this means that it is more difficult for a woman than for a man to be promoted (see the next two sections).

Developments in the Past Twenty-seven Years in the Main Groups Of Disciplines

First note that, thanks to the methodical and obstinate work of the late Huguette Delavault, we are able list sex disaggregated data from 1981, a time when this information was not widely publicised. Then, over the last ten years the statistical office of the French ministry of education has published a yearly Note of Information on the teaching staff of the universities, using the same scheme and categories, so that data can be compared over such a very long period of time (for example for 2007–2008 : Note d'information 08–25 of August 2008).[4]

[4] http://www.education.gouv.fr/cid21784/les-personnels-enseignants-de-l-enseignement-superieur.html.

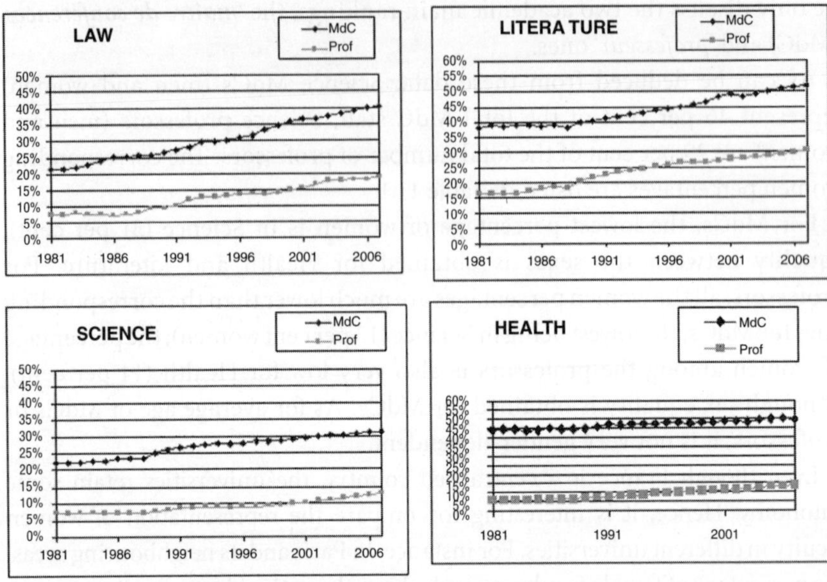

Figure 11.2 Evolution of women percentage for the MdC and professor university ranks in the four groups of disciplines between 1981 and 2007
Source: Authors. Constructed on the basis of *Ministère de l'éducation nationale*, 2007b data.

During the past 27 years, the largest increase in the number of MdCs appears in Law (the number of Law MdCs was multiplied by 4.7 over this period) and the lowest in Health (the number of Health MdCs was multiplied by 1.6 over this period), this slower increase being due to a limitation of the number of students ('numerus clausus'). The same pattern is found for professors. Has this increase permitted an increase in the percentage of women? Yes, indeed. The evolution of the women percentage in the four *main groups* of disciplines is given in Figure 11.2 for the rank of MdC and that of professor (Delavault, Boukhobza, Hermann and Konrad. 2002 and yearly Notes d'Information of the ministry of National Education for the more recent period).[5]

For the MdC rank, this percentage has remained nearly constant for Health, where it has always been close to 50 per cent; in Literature after a constant value in the 1980s, it has steadily increased since and is above 50 per cent. In Law and Science the percentage of women among the MdCs had the same value of 22 per cent in 1981, it now reaches 40 per cent in Law, but 31 per cent

[5] See also footnote 5.

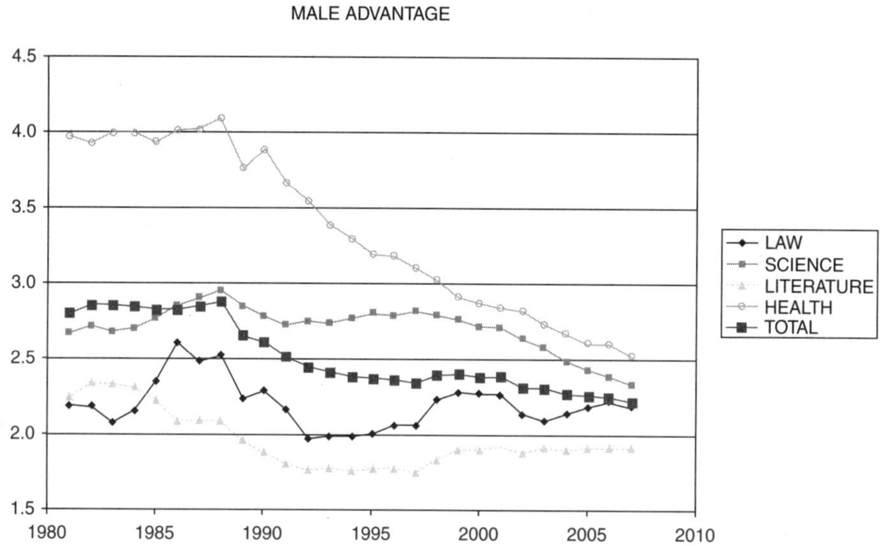

Figure 11.3 Comparison of the chances of promotion for women and men faculty, in terms of ratio during 1981 and 2007

Source: Authors. Constructed on the basis of *Ministère de l'éducation nationale*, 2007b data.
Note: The plotted 'male advantage' is obtained by the following procedure: first, the ratio of the number of professors to the total number (MdCs and professors) is calculated for men ('men's ratio'), then for women ('women's ratio'), and then the men's ratio is divided by the women's ratio. The figure evidences that this final ratio ('men/women ratio') is now ranging between 2 and 2.5 according to the discipline

in Science. For the professor rank, in the case of Science and Health, it is only around 2000 that the percentage went over 10 percent; for Law and Literature there has been a significant increase during the past 27 years.

In order to compare the difficulties of the promotion for men and women and indicator, 'the male advantage', can be defined in the following manner: first, the ratio of the number of professors to the total number (MdCs and professors) is calculated separately for men and women, and then the men's ratio is divided by the women's ratio. Figure 11.3 shows that this final ratio ('male advantage') is currently between 2.2 and 2.5 in all groups of disciplines.

This indicates that it is approximately two times more likely for a man to be promoted than for a woman. The smaller this ratio, the better likelihood of promotion for women. The likelihood for a woman to be promoted depends on the group of disciplines and is highest for Literature and lowest for Health.

Table 11.2 *Percentage of women among the researchers of CNRS (31/12/06), by scientific department and the corresponding masculine advantage*

	Math	Physics	Engineering	STIC	SdU	Chemistry	Life sciences	SHS	Total
Percentage	17%	18%	23%	20%	26%	31%	39%	44%	31.5%
Masculine advantage	1.04	1.41	1.55	1.75	1.43	1.52	1.74	1.50	1.6

Note: STIC: Sciences and Technologies of Information and Communication
SdU: Sciences of the Universe; SHS : Human and Social Sciences
Source: CNRS 2007.

Versus time, apart from Law where the variation of the male advantage is erratic, this indicator is decreasing, expressing that women's promotion likelihood is slowly improving.

Research institutes

There are many research institutes in France, the largest one being the Centre National de la Recherche Scientifique (National Centre for Scientific Research) (CNRS), which covers all disciplines except Law. It is interesting to apply the same analysis to the CNRS researchers as for university faculty members. At the CNRS here are two main ranks: the lower rank is *chargé de recherche* (CR) (tenured junior researcher), the higher one being *directeur de recherche* (DR) (research director, i.e. tenured senior researcher). They exactly correspond respectively to the academic ranks of MdC and professor. The disciplines at CNRS are grouped into scientific departments: we rather present the data according to scientific fields corresponding to the university disciplines, for sake of comparison. The percentages of women in these fields are given in Table 11.2.

The analysis for CNRS shows qualitatively the same features as for the universities. Nevertheless, the chance to be promoted is slightly better for women, as now the 'men/women ratio' is smaller (global value for all disciplines: 1.6), i.e. it is 'only' less than twice easier for a man to be promoted than for a woman. It seems that these differences exist for other public research institutions as well: for example, the National Institute of Health and Medical Research (INSERM) also evidences a ratio smaller than that of the universities. The reason why institutes behave in a different way than universities and permit female scientists to be promoted more easily is not known.

Table 11.3 *Percentage of women among French university staff in 2006–2007, by discipline and rank*

	Professors (PR)				Assistant Professors (MdC)						
	Men	Women	Total	% Women	Men	Women	Total	% Women	Men ratio (PR/ Total)	Women ratio (PR/ Total)	Male advantage
Law and political science	911	308	1,219	25.3%	1,260	1,089	2,349	46.4%	42.0%	22.0%	1.90
Economic Sciences and management	758	145	903	16.1%	1,601	1,085	2,686	40.4%	32.1%	11.8%	2.73
Law	1,669	453	2,122	21.3%	2,861	2,174	5,035	43.2%	36.8%	17.2%	2.14
Literature	508	301	809	37.2%	582	985	1,567	62.9%	46.6%	23.4%	1.99
Languages	538	405	943	42.9%	1,116	1,767	2,883	61.3%	32.5%	18.6%	1.74
Human Sciences	634	227	861	26.4%	1,143	1,042	2,185	47.7%	35.7%	17.9%	1.99
History-Geography	746	225	971	23.2%	1,058	816	1,874	43.5%	41.4%	21.6%	1.91
Interdisciplinary group	238	94	332	28.3%	530	519	1,049	49.5%	31.0%	15.3%	2.02
Science and Techniques of Physical and Sport Activities	95	27	122	22.1%	374	190	564	33.7%	20.3%	12.4%	1.63
Theology	26	5	31	16.1%	17	7	24	29.2%	60.5%	41.7%	1.45
Literature	2,802	1,301	4,103 0	31.7%	4,843	5,415	1,0258	52.8%	36.7%	19.4%	1.89

contd.

Table 11.3 *Contd.*

Mathematics	983	111	1,094	10.1%	1,591	563	2,154	26.1%	38.2%	16.5%	2.32
Computer science	653	157	810	19.4%	1,712	617	2,329	26.5%	27.6%	20.3%	1.36
Physics	822	94	916	10.3%	1,162	441	1,603	27.5%	41.4%	17.6%	2.36
Chemistry	863	167	1,030	16.2%	1,301	917	2,218	41.3%	39.9%	15.4%	2.59
Geosciences	358	41	399	10.3%	588	276	864	31.9%	37.8%	12.9%	2.93
Mechanics, Mechanical Engineering, Computer Engineering, Energetics	1,868	176	2,044	8.6%	3,628	920	4,548	20.2%	34.0%	16.1%	2.12
Biology & Biochemistry	881	234	1,115	21.0%	1,517	1,606	3,123	51.4%	36.7%	12.7%	2.89
Sciences & Techniques	6,428	980	7,408	13.2%	11,499	5,340	16,839	31.7%	35.9%	15.5%	2.31
Pharmacy	431	164	595	27.6%	479	723	1,202	60.1%	47.4%	18.5%	2.56
Medicine	3,256	389	3,645	10.7%	867	833	1,700	49.0%	79.0%	31.8%	2.48
Odontology	82	28	110	25.5%	306	102	408	25.0%	21.1%	21.5%	0.98
Total all disciplines	14,729	3,353	18,082	18.5%	20,786	14,550	35,336	41.2%	41.5%	18.7%	2.21

Source: Authors. Created on the basis of Ministere de l'enseignement superieur et de la recherche 2007 data

Gender differences by disciplines in universities

It is interesting to study the differences within a group of disciplines, as is done in Table 11.3.

Within science too, there are variations (see Bussac and Hermann 1998 for the case of Physics). We can examine the variations in some of the science disciplines. For MdCs there is sex equality for Biology and Biochemistry with 51.4 per cent of women, but in comparison there are 41.3 per cent women in Chemistry, 27.5 per cent in physics and only 20.2 per cent women in mechanics (listed as 'Mechanics, Mechanical Engineering, Computer Engineering, Energetics'). For professors, the percentage of women is only 21 per cent in Biology and Biochemistry and 8.6 per cent in Mechanics. In other scientific disciplines the percentage of women professors ranges between 10 and 19 per cent. Thus, except in Biology and Biochemistry, it remains difficult for women to begin a career in science; once in the first academic rank, it is not easy for them to be promoted, especially in Biology and Biochemistry. When looking at the same analysis by disciplines for the CNRS (Table 11.4, upper line), one comes to the same ranking between scientific fields, all percentages being lower (CNRS 2006).

Thus the chances for promotion for men and for women differ according to discipline both in the universities and at the CNRS. However, it is not in disciplines where women are the most numerous in the lowest rank that they have the greatest chance of being promoted, as shown in the universities for biology, biochemistry, and pharmacy, or at the CNRS for life sciences.

French appointment committees

The university nominations and promotions are decided by the *Conseil national des universities* (CNU) (National Council of Universities), where

Table 11.4 *Percentage of women among the researchers of CNRS (31/12/06), by scientific department and the corresponding percentages in the Comité national [National Committee]*

Percentages	Math	Physics	Engineering	STIC	SdU	Chemistry	Life sciences	SHS	Total
Women researchers	17%	18%	23%	20%	26%	31%	39%	44%	31.5%
Women in Committee	30%	18%	17%	24%	26%	25%	38%	37%	30%

Note: STIC: Sciences and Technologies of Information and Communication
SdU: Sciences of the Universe; SHS : Human and Social Sciences

both academic ranks are represented. For the last few years, efforts have been made to reach a percentage of women representatives slightly above that of the pool of corresponding faculty voters. Indeed, women are under-represented among the members elected from the union lists: only the relatively higher percentage of women nominated by the ministry has allowed this high women representation.

The global (all disciplines) percentages of women, in CNU in 2007, are, for the MdCs members, 45.9 per cent among the elected members and 57.4 per cent among the members designated by the ministry, that is a global (elected+designated) percentage of women of 50 per cent (*Ministère de l'éducation nationale* 2007c) (Ministry of national education). The corresponding figures for the professors representatives are respectively 25.9 per cent among the elected and 47.2 per cent among the designated members, that is a global (elected+designated) percentage of women of 33 per cent women. This has to be compared to the percentage of women in the same ranks in academia, i.e. 40 per cent among the MdCs and 18 per cent among the professors. The women representation in the CNU has improved since the previous 2003 council, where the total percentages of women members were 46.9 per cent for MdCs and 23.3 per cent for professors.

For research institutes, taking the example of the CNRS, there also exists a committee, the so-called *Comité national* (National Committee), which is in charge of appointments and promotions. The representation of women is not so good there, and it is easier for female scientists to be promoted at CNRS than in the university. This shows that the correlation between the appointment and promotion of women and the percentage of women present in the committees is not obvious.

Academic careers

Defending a PhD is the first step for an academic career, but less than one half of PhDs enter into that career (*Observatoire de l'emploi scientifique* [Observatory of the scientific employment] 2007). It is difficult to find sex-disaggregated statistics showing where the PhD graduates go after university. One can notice that the percentage of women receiving a PhD in Fundamental sciences and their applications has not increased significantly over the last years (27.5 per cent in 2006–2007) (*Ministère de l'éducation nationale* 2007 a). This fact helps explain the small number of women MdCs in disciplines such as Mathematics, Physics.

As noted above, in order to be promoted to the rank of professor, in France one needs first to be approved by the CNU. One notices that the percentage of women approved by the CNU is always smaller that the percentage of women among the *maîtres de conférences*. This is a reason for the small number of women professors. Many more quantitative and qualitative studies would be necessary to understand why, starting with a large percentage of women as university students, such a small percentage of professors are female (de Cheveigné 2008; Louvel 2007).

Education

There has been an interest in the scientific and technical education of girls for more than two decades. Several agreements were signed between the Ministry of National Education and the Ministry for Women's Rights. One of the main objectives of the December 20th, 1984 Agreement was 'the improvement of girls' training and professional integration'. The main priority of the September 14, 1989 Agreement was the need for girls to widen their choice of training to include industrial training more frequently. In these frameworks a considerable number of actions was implemented to push girls to focus their careers in scientific and technical sectors. In particular a 'Prize for Feminine Scientific and Technical Vocation" is awarded each year by the regional deputies of the Ministry of Education and those of Women's Rights to each of 650 girls choosing traditionally masculine studies and careers, presenting their projects the year before entering higher education. In spite of its rather modest amount (1000 Euros), this prize is well publicised in many regions and offers a moral support to girls from modest families choosing scientific or technical careers.

As for French higher education in general, the system is dual: i.e. it is split into universities and *Grandes Ecoles* for Engineering, Agriculture, Business and Administration. These dual education systems meet only at the PhD level. Access to top positions in French economy, science or politics is achieved through the Grandes Ecoles, both the scientific ones and those for Business or Administration. To prepare for the selective exam to enter the Grandes Ecoles, the students spend two years in special classes, *classes préparatoires* (preparatory classes). In classes preparing for access to business schools, female and male students are present in equal numbers. In the scientific classes, the overall percentage of women is 29.5 per cent, with around 25 per cent for those centered on maths and physics, around 10 per cent in more technological areas.

Technical Higher Education

In the *Instituts Universaires de Technologie* (University Institutes for Technology) (IUT) the students receive a two-year education to become technicians. In the secondary domain IUTs, women represent less than 22 per cent of the students (*Ministère de l'éducation nationale* 2007a). They are less than 10 per cent in electrical engineering, industrial engineering, mechanical engineering, thermal engineering, networks and telecommunications, computer sciences, and mechanical engineering.

University Students

Women now represent 57 per cent of university students and 47 per cent of the students preparing PhDs (*Ministère de l'éducation nationale*, 2007a). If the percentage of women students has increased over the last two decades in medicine and odontology (currently 60 per cent women students), pharmacy (67 per cent women students) and in nature and life sciences (58 per cent women students), it has remained low in Fundamental sciences and their applications (27.3 per cent women students), which includes mathematics, physics, chemistry and engineering. It is even lower in engineering: for example, in electronics and electrical engineering, at the licence level (first three years), women represent only 6.5 per cent of all students, in comparison with 38 per cent for all scientific disciplines.

Top-ranked Grandes Ecoles: a co-education failure story

In French society, the situation of *Grandes Ecoles* is mythic. Many families dream of having a son entering the *Ecole Polytechnique* (for engineering) or the *Ecole Nationale d'Administration* (for high state administration), but up to now very few of them have invested such hopes in their daughters. The *Ecoles*, besides offering privileged conditions of studies, provide a very useful 'old boys' network that persists over a lifetime. The best students, in science for example, follow this path, which lead many of the scientific students to an administrative career. Apart from biology the percentage of women on science tracks in the *Grandes Ecoles* is currently between 15 and 30 per cent, in comparison to 36 per cent of women in science in the universities (a particularly low percentage is the 10 per cent of women students in the *Ecoles Nationales Supérieures des Arts et Métiers* (Higher Education Institutes for Industrial Art and Design), the top engineering school for mechanical engineering, power engineering, and industrial engineering in France.

A non-official ranking exists among the *Grandes Ecoles*: for research in mathematics, physics and chemistry, the *Ecole Normale Supérieure de la rue d'Ulm*, Paris, (ENS) (see the section Historical Background) and the *Ecole Polytechnique* ('X') are considered the top ones. The students graduating from ENS mainly do research or teach at the university level; those graduating from X enter industry and top state technical administration (*Corps de l'Etat*), but a non-negligible fraction of them defend a PhD and continue in research. These renowned schools were both founded two hundred years ago.

As for women students, *Ecole Normale Supérieure de Jeunes Filles* (the corresponding women's school to *Ecole Normale Supérieure de la rue d'Ulm*) was founded in 1881 to educate future women teachers for the girls' secondary public schools (see in the section Historical Background the part on Milestones in Women's Education). It became co-educational in 1986 by merging into *Ecole Normale Supérieure de la rue d'Ulm*.

At that time, the selective entrance examinations for *Ecole Normale Supérieure de la rue d'Ulm* (men) and *Ecole Normale Supérieure de Jeunes Filles* (women) were separate, with different rankings, even though some courses were common. The number of positions in science for a given year was similar in the two schools: out of the total for these two schools approximately 40 per cent of the positions were attributed to women. Although not formulated this way, it represented a kind of 'quota' as the female percentage was much higher than in the candidates' pool. The alumnae of *Ecole Normale Supérieure de Jeunes Filles* who studied there in the mid-1960s are now mostly university professors or research directors at Centre National *de la Recherche Scientifique*, they thus followed careers similar to those of their male colleagues of ENS Ulm.

Since the decision in 1986 to go co-ed, the number of girls who passed the entrance examination for the *Ecole Normale Supérieure de la rue d'Ulm* dropped dramatically for mathematics, and may happen to be null some years in certain disciplines. In physics the percentage of female students is now approximately 15 per cent and it corresponds almost exactly to the proportion of girls in the classes preparing for the selective examination (see Adda 1993–1996; Delon 1995; Ferrand, Imbert and Marry 1999). This new situation is clearly dramatic in mathematics, but also problematic in physics where women were over-represented before co-education. Yet the women who went through such studies have pursued normal careers and have held regular scientific jobs. This is an example of a co-educational decision which did not reach its target; indeed, a generation of women researchers is lost. Previously, when women students entered *Ecole Normale Supérieure de Jeunes Filles*, their ambition was typically to teach in a secondary school, until their professors convinced a good number of them that research and teaching at the university level would be an exciting job within their reach.

Note that is only about ten years after the decision of co-education of the *Ecoles Normales Supérieures* in the mid-1980s, that analyses and discussions on the new situation of women students there really began. Up to now, the only measure, taken in 2001, has been a memorandum of the directors of the scientific *Ecoles Normales Supérieures* to the correctors of the written (anonymous) problems for the selective entrance examination. They were asked particularly to evaluate the in-depth answers to the questions: several preliminary studies had shown that girls have a tendency to answer questions more in detail, losing time and marks as compared to boys, who follow a better strategy.

Whatever the difficulties resulting from co-education in this particular case, there is no discussion in France on returning to separate education for women and men students, in the context of universality of rights that prevails in France (see section on Gender Studies). The law on parity in politics (Picq 2002), voted in 1999, which required a modification of the French constitution and quotas are in general not accepted. It should be noted that in Scandinavian countries, where affirmative actions have been taken, in particular to create women professors positions, not all were successful.

Ecole Polytechnique, perhaps because its military status, was among the last engineering schools to accept women in 1972. Its women students indeed complete their military service like their male colleagues, follow the same curriculum, become state engineers or civil engineers, and get approximately the same salary as their male colleagues at the beginning of their career. They are privileged in the French system, since the women graduating from less prestigious schools have more handicaps. The very first year that women students were accepted, a woman ranked first (this happened for the second time in 1995). Since then, the number of women entering *Ecole Polytechnique* has slowly increased. It jumped in 1994 and 1995, when it exceeded 50 women students (out of 400 French students) but unfortunately has levelled out to around 55–60 per year since, with fluctuations due to such small numbers. Among the foreign students, the percentage of women students is similar.

Engineering Grandes Ecoles

The first French Engineering School to admit a woman student was *Ecole Nationale Supérieure de Chimie de Paris* (Paris Chemistry National Higher School, a Chemistry engineering school) in 1916. In 1925, a women-only engineering school, the *Ecole Polytechnique Féminine* (Women Polytechnique School), was founded. It went co-educational just before its 50th anniversary; women are now 40 per cent of its students. One of most renowned French Engineering schools, *Ecole Centrale des Arts et Manufactures* (Higher Education Institute for Industrial Art and Design), admitted its first woman student in 1920, that is, 52 years before *Ecole Polytechnique.*

The periodical surveys completed by the *Conseil National des Ingénieurs et Scientifiques de France* (CNISF) (National Council of French Engineers and Scientists) include gender categories, and the particular situation of French women engineers is analysed by the French association of women engineers (Association Française des *Femmes Ingénieures,* 2005). In 2005, women received 25 per cent of engineering degrees; 16 per cent of active French engineers were women, 50 per cent of the women engineers were under thirty years, in comparison to 25 per cent of the men engineers. Out of 100 women engineers 12 per cent (in contrast to 6 per cent men) work in agronomy, 6 per cent (in contrast to 11 per cent men) in electricity and electro-technique.

Recent Developments (1999–2007)

Recent developments concerning women in science and engineering are related not only to the general French political situation, but also to the importance that the European Union has placed on this issue since 1998 (see section The Role of Politics). The context of European competitivity and of a declining student enrollment for science and technology (Gago 2004; OECD 2005) has brought a new interest on the situation of women in science.

Politics

The issue of political 'parité' has raised general awareness about the situation of women in top positions. After A.M. Colmou's recommendations (Colmou, 1999), a 'steering Committee for the equal access of women and men to the top positions of French civil services' was established by the

Ministry of Civil Service in November 2000 for a duration of five years. This Committee was to oversee equity issues in general, that also concern university teaching staff and scientists of the public research institutions, who are civil servants. It produced three extensive reports, the last one (Le Pors and Milewski, 2005) presented detailed information on higher education and public research.

Scientific education of women

In February 2000, the Minister for Employment and Solidarity, the Minister of National Education, Research and Technology, the Minister of Agriculture and Fishing, the Deputy Minister in charge of School Teaching and the Junior Minister for Women's Rights and Professional Training signed an Agreement for the promotion of equal opportunities for girls and boys at school, and for men and women in continuing education. This revitalised and widened the scopes of the previous agreements (1984 and 1989). This agreement was extended in 2002 to the Ministries of Justice, Equipment, Transportation and Lodging as well as the Ministry of Culture and Communication.

This agreement was then renewed in 2006 for five years. It was then signed by the Ministry of Employment, Social Cohesion and Housing, the Ministry of National Education, Higher Education and Research, the Ministry of Justice, the Ministry of Transportation, the Ministry of Equipment, Tourism and Sea Affairs, the Ministry of Agriculture and Fishing, the Ministry of Culture and Communication, the Deputy Ministry of Social Cohesion and Parity and the Deputy Ministry of Higher Education and Research.

In January 2006, the Deputy Minister for Higher Education and Research François Goulard installed the Committee for Professional Equality between Women and Men in Higher Education and Research. In its first report in December 2006, the Committee proposed practical measures aiming at developing the attractiveness of scientific careers for women, acting against the 'glass ceiling' and taking a better account of maternity and parenthood in general in scientific and technical careers. In 2007, the High Council of science and technology, the mission of which is to provide advices to the President of the French Republic, issued a report on the declining interest for scientific studies and professions. Particular remarks and recommendations on the situation of girls and boys in science and on that of women in scientific jobs were expressed.

Actions in universities

From 1997 until 2002, the Directorate for Higher Education at the Ministry of Education provided permanent support for the problems related to the situation of women students in science and for those of women in academia. For a better understanding of the situation and its origins, a statistical and sociological report on the place of women in French universities was ordered by the Directorate. It was issued in March 2000 and has been widely diffused (400 copies). This document had a large impact: the National Council on Higher Education and Research (CNESER) asked for a presentation of these findings and decided to establish a working group interested in the situation of women in universities and aimed at proposing practical measures for improvement. In addition to these measures, since 2001 universities have been invited to propose women-friendly programmes. A collection of gender-disaggregated data was also realized in every university and published on the ministry of education website.

The interest for gender issues in universities decreased somewhat after 2002, as other considerations, and other problems like suburban disorder retained more the attention of politicians than the situation of women in French science. However, as can be seen from the data on Figures 11.3 and 11.4, a slow but constant progress has been realized in academia, including over the last years.

Creation of missions devoted to the Women in Science issue

Inside the ministry of research, the *Mission pour la parité en sciences et technologies* (Mission for Gender Equality in Science and Technology), then entitled *Mission pour la parité dans la recherche et l'enseignement supérieur* (Mission for Parity in research and higher education) and now called *Mission de la parité et la lutte contre les discriminations* (Mission of Gender Equality and Struggle against Discriminations)[6] was created in 2001.

Its aim is to initiate programs proposed by the minister and to check the gender relevance in any new proposals and politics of the ministry. For several years it benefited from partial funding of the European Social Fund of the European Commission. It has supported several studies that produced White books on women scientists in public (Livre Blanc 2002) and private research,

[6] http://www.recherche.gouv.fr/pid20161/mission-parite.html

in French research institutions (*Mission pour la parité dans la recherche et l'enseignement supérieur* 2005). Its main types of actions are:

(a) incitation, which aims at attracting more girls into scientific and technical studies and professions;
(b) support, allow and help women scientists and their associations to promote actions in favour of scientific and technical careers;
(c) recognition, through scientific prizes, in particular the Irène Joliot-Curie yearly awards, which turn the spotlight onto careers of exemplary women and favour their access to decision-making positions.

As for CNRS, the committee 'Disciplines, professions, careers and gender: the place of women at CNRS', presided by the General Director of CNRS, was also created in 2001. A few months later the 'Mission for the place of women in CNRS' was established.[7] It has opened several fronts of statistical activities aimed at showcasing the situation of women in CNRS and the factors that affect their careers. It has initiated multidisciplinary workshops on 'Sexes and gender in scientific work', that take place in different regions of France and are designed to bring young researchers and engineers of both sexes together to share viewpoints on gender-related questions for the purpose of education and debate. A book *Women in the history of the CNRS* (Kaspi 2004) presents a collection of monographs of notable women scientists since the CNRS creation in 1939.

Associations

Several French women's associations already existed before 2000, related in some way to the Women and Science issue: *Association Française des Femmes Diplômées des Universités* (French Association of University Women),[8] *Association Française des Femmes Ingénieurs* (French Association of Women Engineers),[9] Association *femmes et mathématiques* (*Women and mathematics*),[10] In December 2000 the association *Femmes et Sciences* (Women and Science)[11] was created, the objectives of which are both to promote scientific education for girls and to improve the situation of women scientists.

7 http://www.cnrs.fr/mpdf/
8 http ://www.affdu.fr
9 http://www.femmes-ingenieurs.org
10 http ://www. femmes-et-maths.fr
11 http://femmesetsciences.fr

The last three associations act together as role models in secondary schools to promote scientific and technical studies and career for boys and girls. They also do a permanent watch of the nominations in public committees, which are often not gender-balanced... The association *Pour la parité dans les métiers scientifiques et techniques* [For parity in scientific and technical professions], founded in 2002, pursues similar objectives in the Grenoble region.[12]

The associations *Femmes Ingénieurs, femmes et mathématiques* and *Femmes et Sciences* are members of the European Platform of Women Scientists[13] which aims at giving a voice to European Women Scientists, at making women scientists better understand the role they can play in the European research policy debate and the way to fully benefit from these opportunities by bundling their powers and forces.

Conclusion

This short paper does not intend to fully describe the complex situation of women scientists in France. Yet it tries to give a cursory outline of the situation: a rather favourable social system has created the possibility for women to work, in particular in science. Quite a number of women educated in women-only *Ecoles Normales Supérieures* led to the tradition of the presence of women scientists in universities and research institutes. Since the decision of co-education in these elite schools, their number of women students has dropped, but women can now be found on other educational paths, including engineering schools.

In recent years, following the concern about the insufficient representation of women in French politics and in conjunction with the numerous European union initiatives for women in science, new attention has been brought to the issue of women in science both at the governmental and institutional levels. Some slow progress has been obtained in the universities, the situation in CNRS remaining terribly static.

From a political point of view, in spite of official positions now constantly mentioning the situation of women scientists, the every day practical actions very often ignore this issue. A constant watch is necessary, to ensure a proper women representation in decision-making committees. At the European Union level, merging of the Women and Science of Directorate General

[12] http://blog.parite-sciences.org/

[13] http ://www.epws.org

Research unit into the 'Scientific culture and gender issues' unit[14] decreased the visibility of the women concern; in the current research projects of 7th Framework Programme (FP), like the Networks of Excellence, the Gender Actions Plans are now optional, whereas they were compulsory in the 6th FP. In France, a modification of the status of researchers, with creation of short-term contracts for young scientists in their thirties, is now under discussion: this would reduce the stability of researchers' positions and would not help young women scientists to combine a career and a family.

So the general situation of French women scientists combines continuous slow progress with threats. Hopefully progress will dominate and all the various efforts described in this paper will result in concrete actions that will permit men and women to contribute equally to the science of tomorrow.

Acknowledgements

Claudine Hermann (CH) would like to pay a tribute to the late Huguette Delavault who taught her so much in the field of Women and Science and provided her a great deal of original data. She would like to thank the members of the European Union Helsinki Group on Women and Science with whom she served as one of the two French representatives (1999–2006) and her friends at Association Femmes et Sciences. Françoise Cyrot-Lackmann (FCL) extends warm thanks to the team that worked with her (2001–2002) when she served as the Director of the Mission for Gender Equality in Science and Technology at the French Ministry of Research. Special thanks are due to Peiffer and H. Rouch for their contribution on the sub-section on gender studies. They are grateful to M. Lynch for a careful reading of the manuscript.

This paper is the updated version of the article published in *Science in Context* 15(4), 529–556 (2002) and we thank Cambridge University Press for allowing its new publication.

References

Adda, J. 1993. 'Is Gender a Relevant Variable for Mathematics Education ? The Case of the French Situation', in Gender and Mathematics Education, an ICMI study in Stiftsgarden Akersberg, Höör, Sweden, edited by B. Grevholm, G. Hanna, Lund: Lund University Press, 197–203.

[14] http://ec.europa.eu/research/science-society/index.cfm?fuseaction=public.topic&id=27

Adda J. 1996. 'Is Gender a Relevant Variable for Mathematics Education ? The French case', Towards gender equity in mathematics education, An ICMI study, edited by G. Hanna, New ICMI studies series, vol.3, Kluwer Academic Publishers, Dordrecht, 197.

AFFER (association *Femmes, féminisme et recherches*) [Women, feminisme and researches association] 1984. ed., Actes du colloque national "Femmes, féminisme et recherches" (Toulouse, 17–19 décembre 1982), Toulouse. The colloquium took place in 1982 and the proceedings were published in 1984.

Allard L., Gardey D. and Magnan N. 2007, 'Manifeste cyborg et autres essais. Sciences – Fictions – Féminismes', Anthologie d'œuvres de Donna Haraway établie par Laurence Allard, Delphine Gardey et Nathalie Magnan, Paris : Editions Exils.

Association Française des Femmes Ingénieures 2005. Les femmes ingénieurs en France – Résultats issus de la 15° enquête socio-économique (année 2002) du Conseil National des Ingénieurs et Scientifiques de France (CNISF) exploités et présentés par l'Association Française des Femmes Ingénieurs. Paris.

Blum C. 2000, 'Les situations des homes et des femmes: statistiques pertinentes' rapport pour le service des droits des femmes du ministère de l'emploi et de la solidarité, Paris, La documentation française.

Bussac M.-N. and Hermann C. 1998. 'Femmes en physique', Bulletin de la Société Française de Physique volume 114 pages 26–27 Paris.

Chabaud-Rychter, D. and Gardey, D. 2002. dir. L'engendrement des choses. Des hommes, des femmes et des techniques. Paris: Éditions des Archives Contemporaines.

Couacault, M. and Gardey, D. 2005. dir. 'Sciences, recherche et genre', dossier in Travail, Genre et Société, n° 14 pages 27–128.

CNRS 1989–90. Action Thématique Programmée n°6, Recherches sur les femmes et recherches féministes. Présentation des travaux, vol. 1: 1984–1987; vol. 2: 1986–1989, Paris : éditions du CNRS.

CNRS 2007. Bilan social 2006, Direction des Ressources Humaines, Observatoire des Métiers et de l'Emploi Scientifique, Paris, Centre National de la Recherche Scientifique. http://www.dgdr.cnrs.fr/drh/publi/bilan-social/pdf/bilan-social-2006.pdf, accessed on September 19th 2009.

Collin, F. 1992. dir. Le sexe des sciences. Les femmes en plus, éd. Autrement, série Sciences en société n°6, Paris.

Colmou, A.-M. 1999. L'encadrement supérieur de la fonction publique: vers l'égalité entre les hommes et les femmes. Quels obstacles? quelles solutions. Paris: La Documentation Française.

Commission femmes du SNCS 1981. La recherche des femmes (Enquête, réflexions sur les femmes chercheurs au CNRS), Paris.

Cotta, M. 2000. Femmes dans les lieux de décision, avis du Conseil Economique et Social sur le rapport présenté par Madame Michèle Cotta au nom de la section du travail, séance des 19 et 20 décembre 2000, Journal Officiel de la République Française Année 2000 n°18, Jeudi 28 décembre 2000, Paris: Les éditions du Journal Officiel.

de Cheveigné, S. 2008. The Career Paths of Women (and Men) in French Research, Accepted for publication in *Social Studies of Science*, Manuscript # SSS06018

Daune-Richard, A.-M., Hurtig, M.-C., Pichevin, M.-F. 1989. dir., Catégorisation de sexe et constructions scientifiques, Petite collection CEFUP, Université de Provence.

Delavault, H. 1997. Vers la parité dans les instances de décision? La place des filles dans une filière de formation des cadres. 1.Du lycée aux grandes écoles scientifiques.

http://www.int-evry.fr/demain-la-parite/lyceeauxgrandesecoles.pdf

Delavault, H. 1998. Vers la parité dans les instances de décision ? La place des filles dans une filière de formation des cadres 2. Les écoles d'ingénieurs.

http://www.int-evry.fr/demain-la-parite/grandesecoles.htm

Delavault, H., N. Boukhobza, C. Hermann and C. Konrad 2002. Les enseignantes-chercheuses à l'université - demain la parité ?. Paris: L'Harmattan

Delon, F. 1995. 'The French experience: the effect of de-segregation', Equity in mathematics education: influences of feminism and culture, edited by P. Rogers and G. Kaiser, the Falmer Press, Ch.17, p.141.

Delphy Christine (1998–2001) *L'ennemi principal*, 2 vols. Vol. 1 Economie politique du patriarcat, Vol. 2 Penser le genre. Paris, Syllepse.

European Commission 1999. Women and science: Proceedings of the conference, Brussels, April 28–29 1998. Luxembourg: Office for Official Publications of the European Communities, European Commission; see in particular the testimony of Huguette Delavault in III.5 on her own experience and on the French situation.

European Commission 2000. Science policies in the European Union-Promoting excellence through mainstreaming gender equality. Luxembourg: Office for Official Publications of the European Communities, European Commission

European Commission 2003. Women and Science – Statistics and Indicators – She Figures 2003. Luxembourg: Office for Official Publications of the European Communities, European Commission, ISBN 92-894-8229-X.

European Commission 2006. Women and Science - Statistics and Indicators – She Figures 2006. Luxembourg: Office for Official Publications of the European Communities, European Commission, ISBN 92–79-01566–4.

European Commission 2009. She Figures 2009- Statistics and Indicators on Gender Equality in Science – Luxembourg: Publications Office of the European Union, ISBN 978-92-79-11388-8.

Fave-Bonnet, M.-F. 1996. 'Women in Educational management', *European Journal of Education*, vol.31, N°4, 1996, 389.

Feldman, J. and Morelle A. 1994. '"Les femmes savantes" in France: under the glass ceiling', World yearbook of Education 1994. The Gender Gap in Higher Education, Eds S. Stiver Lie, L. Malik and D. Harris (Series Editor). London: Kogan Page, pp. 56–69.

Ferrand, M., Imbert, F. and Marry, C. 1999. L'excellence scolaire,: une affaire de famille. Le cas des normaliennes et normaliens scientifiques, Paris : Bibliothèque de l'éducation, L'Harmattan.

Fougeyrollas, D., Rouch, H., Zaidman, C. 2003. ed. Science et genre. L'activité scientifique des femmes. États-Unis, Grande-Bretagne, France. Cahiers du Cedref/Université Paris 7.

Fraisse, G., Sissa, G., Balibar, F., Rousseau-Dujardin, J., Badiou, A., David-Menard, M., Tort, M. 1991. L'exercice du savoir et la différence des sexes. Paris: L'Harmattan.

Gago, J.M. (dir.), 2004. Increasing Human Resources for Science and Technology in Europe, Report for the Conference "Europe Needs More Scientists", Brussels, 2/04/2004 http://ec.europa.eu/research/conferences/2004/sciprof/pdf/gago.pdf

Gardey, D., Löwy, I. 2000. dir., L'invention du naturel. Les sciences et la fabrication du féminin et du masculin. Paris : Editions des Archives contemporaines.

Génisson, C. 1999. Davantage de mixité professionnelle pour plus d'égalité entre les hommes et les femmes, Paris: La documentation française.

Guillaumin, C. 1992. Sexe, race et pratique du pouvoir. L'idée de nature. Paris : côté-femmes.

Hermann C. 1997. 'Women scientists in France', Council of Europe, Parlementary Assembly, Committee on Science and Technology, Hearing on the role of women in the field of science and technology (Paris, 26 November 1997) AS/Science (1997) 29.

Hulin N. 2002. Les femmes et l'enseignement scientifique, collection Science, histoire et société, Paris : Presses universitaires de France.

Hurtig, M.-C., Kail, M. and Rouch, H. 1991. ed. Sexe et genre. De la hiérarchie entre les sexes. Paris : éditions du CNRS.

Kaspi, A. and G. Ramunni. 2004, Les femmes dans l'histoire du CNRS, Paris: Mission pour la place des femmes du CNRS.

Laufer, J., Marry, C., Maruani, M. 2003. Le travail du genre. Paris: La Découverte

Le Doeuff, M. 1998. Le sexe du savoir. Paris : Aubier.

Le Pors A. and Milewski, F. 2005. Vouloir l'égalité, Troisième rapport du Comité de pilotage pour l'égal accès des femmes et des hommes aux emplois supérieurs des fonctions publiques. Paris : Collection des rapports officiels, rapport au ministre de la Fonction publique, La documentation française.

Les cahiers du GRIF (1990) 'Savoir et différence des sexes' No. 45, Paris.

Livre blanc 2002. Les femmes dans la recherche française, mars 2002. Paris: ministère de la recherche.

Löwy, I. and Rouch, H. 2003. 'La distinction entre sexe et genre. Une histoire entre biologie et culture', Les cahiers du genre, n° 34.

Löwy, I. 2006. L'emprise du genre. Masculinité, féminité, inégalité. Paris: La Dispute.

Louvel, S. 2007. La résistible ascension des femmes dans le monde académique: le cas des universités françaises, Colloque international de MAGE 'Genre, sciences et recherche : regards franco-allemands', Darmstadt, 31 mai et 1er juin 2007 http://halshs.archives-ouvertes.fr/halshs-00185535/fr/

Majnoni d'Intignano, B. 1999. Egalité entre les hommes et les femmes: aspects économiques, Paris : la Documentation Française.

Marry, C. 2004. Les femmes ingénieurs. Une révolution respectueuse. Paris: Belin

Mathieu, N.-C. 1991. L'anatomie politique. Catégorisations et idéologies du sexe. Paris: côté-femmes.

Ministère de l'enseignement supérieur et de la recherche 2007. Etat des lieux: les femmes dans la recherche

Ministère de l'éducation nationale 2007a. Repères et références statistiques sur les enseignements, la formation et la recherche [AERS 2007], Direction de l'Evaluation, de la Prospective et de la Performance, ISBN 978-11-095421-3, ISSN 1635–9089SBN 978-11-095421

Ministère de l'éducation nationale 2007b. Note d'information 07.46 décembre, Les personnels enseignants de l'enseignement supérieur 2006–2007, Direction de l'Evaluation, de la Prospective et de la Performance.

Ministère de l'éducation nationale 2007c. Elections au CNU 2007, Eléments statistiques, mise à jour du 10 décembre 2007, DGRH A1–1

Ministère de l'éducation nationale 2007d. Démographie des personnels enseignants affectés dans l'enseignement supérieur au cours de l'année 2006–2007, Direction générale des ressources humaines, DGRH A1–1

Mission pour la parité dans la recherche et l'enseignement supérieur [Mission for parity in research and higher education] 2005. Femmes dans les organismes de recherche, Ministère de l'éducation nationale, de l'enseignement supérieur et de la recherche, ministère délégué à l'enseignement supérieur et à la recherche, décembre 2005.

Montreynaud F. [1992] 2000. Le XXème siècle des femmes, Paris : Editions Nathan

Observatoire de l'emploi scientifique 2007. L'état des lieux de l'emploi scientifique en France, Février 2007, ministère de l'éducation nationale, de l'enseignement supérieur et de la recherche, ministère délégué à l'enseignement supérieur et à la recherche http://www.recherche.gouv.fr/cid5765/l-etat-des-lieux-emploi-scientifique-france.html

OECD, 2005. Conference on Declining Student Enrolment in Science and Technology, Amsterdam.

http://www.oecd.org/document/52/0,3343,en_2649_34319_35398132_1_1_1_1,00.html

Peiffer, J. 2000. 'Les débuts de la critique féministe des sciences en France (1978–1988)'; L'invention du naturel. Les sciences et la fabrication du féminin et du masculin. ed by Gardey, D. and Löwy I., 75–86. Paris : éditions des Archives contemporaines.

Pénélope, pour l'histoire des femmes 4, 1981. 'Les femmes et la science'. Paris: Université Paris 7.

Picq, F. 2002. 'Parité, la nouvelle 'exception française'', Modern and Contemporary France 10(1): 13.

Préfecture d'Ile-de-France 1995. Les femmes en Ile-de-France, faits et chiffres, délégation régionale aux Droits des Femmes d'Ile-de-France, direction régionale d'Ile de-France de l'INSEE.

Rouch, H. 1986. 'Une expérience d'interdisciplinarité hors institution: le séminaire Limites-Frontières'. L'université au féminin. Perspectives universitaires 3. Montréal :AUPELF, p.333–341.

Rouch, H., Dorlin, E., Fougeyrollas, D. 2005. dir. Le corps entre sexe et genre. Paris: L'Harmattan

Sullerot S. And Thibaut O. 1978, "Le fait féminin", ouvrage collectif sous la direction d'Evelyne Sullerot, avec la collaboration d'Odette Thibaut, préface André Lwoff, prix Nobel. Paris : Fayard

Vidal, C. 2006. dir. Féminin-Masculin. Mythes et idéologies. Paris: Belin Séminaire Limites-Frontières, 1980–1988, complete collection of the contributions to be consulted at the Bibliothèque Marguerite Durand, Paris, and at the Centre Lyonnais d'Etudes Féministes, Lyon 2 University.

CHAPTER TWELVE

Women and Science

Issues and Perspectives in the Indian Context

Arpita Subhash

THE analyses of science from sociological, philosophical, and historical perspectives show that it is socially and culturally constructed in a large measure, and the norms and values of society and scientists reflect in the representation of knowledge.[1] Ideologies of gender, nature, and science that arose in the seventeenth century supported the increasing split between a man's and women's worlds and between the public and private spheres; and supported the exclusion of women from intellectual pursuits such as science. According to feminists, sociologists of scientific knowledge have been slower to recognise that science and technology reflect and reinforce the interests of patriarchy (Harding, 1986). Various studies and surveys done in sociology of science have substantiated that women have neither received sufficient opportunities to be a part of science nor have they got the due credits for their contribution towards the progress of science (Eisenhart and Finkel, 1988; Rose, 1994; Wenneras and Wold, 1997). Women's studies in the 1970s seeking to understand the roots of women's subordination pointed out biology, gender socialisation, economy, sexual division of work, and so on, as various reasons for it. In

[1] This view is clearly evident from the perspectives relating to social construction of science (Knorr Karin D., 1981; Latour, 1979). This debate even prolongs now in the formal social studies of science.

the 1980s, feminist theorising increasingly questioned the androgynous model of human nature. It aimed at recovering women's culture and critiqued masculine ideology by pointing to the interconnections between women's subordination and the destruction of the environment (Poonacha, 2003). They also highlighted the fallacy of gender-neutrality of science and brought out the gendered norms within the culture and practice of science which act to the detriment of women in general, and women scientists in particular. The gendered nature of science not only excludes women, it also endangers the role of science and technologyin working for sustainable development (Harding, 1996). Despite the significant increases in the number of women in science, women scientists face a cluster of impediments to their ascent in science, irrespective of national or social system (Etzkowitz and Kemelgor, 2001).

This chapter explores some of the important perspectives and data on gender and science, and relates and draws comparisons with the 'state of the art' literature on gender and science in India.

An Overview of Women, Gender, and Science

The literature on women and science in Western nations has been extensive, and ranges from discussions about women in science to philosophical analyses of the gendered nature of science itself (Schiebinger, 1987). These works focus on various aspects of representation of women in science and technology like historical studies of women in science (Keller, 1983; Sayre, 1975) analyses of position and status of women in science (Brainard and Carlin, 1998; Wenneras and Wold, 1997), and so on. Feminist sociologists have analysed constraints on the education, attainment of scientific degrees and careers, and participation and rewards of the contemporary women in the conduct of scientific work (Brainard and Carlin, 1998; Eisenhart and Finkel, 1988; Silverman, 1997; Wenneras and Wold, 1997; Fox, 1998). Some scholars analysed and sometimes reworked those studies and experiments which show bias against women – their bodies and their capabilities (Bleier, 1986; Fausto-Sterling, 1989; Klein and Rowland, 1989). Feminist intervention, however, have not occurred uniformly across the sciences. Biology and medical sciences have been the most successful areas from the point of feminist critique and the impact it has been able to make even on the content of these sciences. But physical science and mathematics have by and large resisted feminist analysis. However, male bias in these subjects too has started falling apart.[2] Feminist scholars have also objected to some of the current practices of science, namely, reductionism (Hubbard, 1990; Spanier, 1995), de-contextualising the object (Hubbard, 1990; Shiva, 2001) and objectivity (Haraway, 1988) and advocated for more holistic approach. Few scholars have tried to categorise the range of studies in this subject (Harding, 1986; Keller, 1982). From initial stock-taking on women scientists and their positions and problems in science organisations, new areas of inquiry have evolved in feminist critique of science. These more radical approaches on gender and science which try to explore the questions like how science as organisation, as institution, and as industry[3] has dealt with women, and to what extent, gender factor influences scientific knowledge production.

Beyond such efforts, by the 1990s, feminists' critique moved on to the reconstruction of science which has led to questions like 'do women do science differently?' or 'Can science be different?' Feminists answer more or less in affirmative to these questions (Harding, 1991; Shiva, 1993). However the hypothesis that women will do science differently still needs testing. Harding (1991) is cautions against the assumption that any one method can encompass the varieties of possible types of knowledge and experience. Keller (1995), too, is not certain that a different science is practiced by women or that a gender-free science would be a more humane science. She points that no change can be introduced in the practice of science until feminists can articulate an adequate response to the question of how 'nature' interacts with 'culture' in the production of scientific knowledge and until they find an adequate way of integrating the impact of multiple social and political forces, psychological predispositions, and experimental constraints, on the growth of science. Keller specifically says that gender ideology is but one aspect of the constitutive role of language, culture, and ideology in the construction of science, and strengths of feminists theory need now to be integrated with the strengths of other kind of scholarship in these areas.[4]

[2] Scholars opine that even among various science disciplines divisions like masculine and feminine are present. For example, Martin (1988) records that most mathematicians are men, and math, like the rest of natural sciences, is perceived to be masculine – a subject for those who are rational, emotionally detached, instrumental, and competitive, and is accorded a high status.

[3] Science is little different from many institutions and industries when it comes to patterns of control, employment, and profit. Few women work in very high positions, and this is especially true of women in less developed nations.

[4] E.F. Keller, 'The Origin, History, and Politics of the Subject Called "Gender and Science": A First Person Account', in Jasanoff, S. et al. (eds.), (1995), *Handbook of Science and Technology Studies*, London: Sage Publications.

Mapping Women in Science in India: a Quantitative Perspective

While there are significant number of studies on various aspects of gender and science originating from the Western countries, it is relatively a neglected area of research in India. This is not to say that gender studies are not undertaken in India. Various scholars and institutions devote their time and energy in studying the status and position of women in India and exploring ways through social research for their betterment. Yet status of women in professions particularly in science has still not drawn the attention it deserves in social studies relating to science and technology.[5] There are only few reports and studies on gender and science and they touch upon only random issues. Further, these studies are scattered and no systematic mapping of women in science in social studies is available.

The question of women in science in India stands out because of the almost complete lack of empirical research specifically on women scientists. Formation of the Indian Women Scientists' Association (IWSA) more than 26 years ago in 1973 in Mumbai helped as the first source of data collection on women scientists. Later, the survey of women scientists at BARC (Bhabha Atomic Research Centre) by Begum and Balaraman (1975) revealed that due to the socialisation of women and the lack of childcare facilities women have to work harder than men (Subrahmanyan, 1998). Family background and support for pursuing science careers was found to be of utmost importance (Chakravarthy, Jain, Nagpaul, Chawala and Mehta 1991; Mukhopadhyay, 1994; Parikh and Sukhatme, 1992; Rama Rao and Shenoy, 1998). Studies also show that women scientists in India have fully internalised their gender roles and give great importance to family. Moreover, they do not perceive that there is any other option to it (Chakravarthy, Jain, Nagpaul, Chawala and Mehta 1991; Parikh and Sukhatme, 1992; Rama Rao and Shenoy, 1998). There also exists a view that most families are reluctant to allow their daughters to pursue science degrees, especially in applied, male-dominated sciences as apart from high investment, it might lead to potential marriageability problems (Mukhopadhyay, 1996). However, other studies have not found support for this (Parikh and Sukhatme, 1992). Gupta and Sharma (2001), in their study on women academic scientists asserted that women scientists not only have

[5] I am aware of the important studies being carried out at what has came to be known as women studies departments and groups at various universities and institutions in India.

a dual burden of job and family, but they also face some unique problems as women in both formal and informal organisation of work leading to an additional source of stress. Hence they face triple burden.

Studies done on women in medical profession highlighted their disproportionate concentration in certain specialties (Bhargava, 1983) and discrimination in service conditions (Balasubramanyan, 1985). Chakravarthy (1986) studied work performance of men and women scientists and found that women tend to be engaged in 'pure' research rather than in administration. This means they are less involved in the decision-making processes in an institution. Also, the research work done by them is more a compilation, collection, and review work, rather than analytical and creative work. Studies also show that women scientists had to work more to attain returns comparable to men scientists (Begum and Balaraman, 1975; Parikh and Sukhatme, 1992; Subrahmanyan, 1998).

The issue of discrimination of women in science is complex involving several factors and it is seen that only recent generations of Indian women are willing to acknowledge discrimination.[6] The conservative nature of society in India forces loneliness and professional exclusion of women scientists from informal groups, committees, and science societies which impinges on their contribution and productivity (Chakravarthy, et al.1991; Subrahmanyan, 1998; Sur, 2001). Bal (2002) opines that discrimination against women practitioners of science may not be due to conscious efforts from men colleagues. However, cultural upbringing and values do contribute to the discrimination observed. 'Favourable' peer review is crucial in natural sciences for grant approval, selection to faculty positions, manuscript publication and funding. Indian attitudes and cultural upbringing where easy mixes of the two sexes is not socially encouraged does contribute to this discomfort, and men are more comfortable with men as peers rather than having women as peers. Hence, one can explain the attempts to keep women out of professional groups and committees where members are nominated (Bal, 2002).

Study by Madheswaran and Shroff (2000) shows that discrimination against women is more due to pre-market discriminatory practice in the family with respect to education, health, and nutrition and lesser due to discrimination in the market place. Again, even if there is no obvious disparity between the salary

[6] Krishnaraj (1991) opines that discrimination, formal, or informal, is not as straightforward as it may appear, moreover, protective legislation for women and the rhetoric of legal equality tend to act as a veil hiding structural blocks that women face.

earned by men and women occupying the same position in an institution, there is imbalance in exercise of authority and in decision-making.

The decade of the 1990s saw the emergence of more detailed studies on women in science highlighting various dimensions of the question. The government of India, through the department of science and technology, also funded few studies on women professional scientists and technologists and their status and problems with the aim of incorporating it into their policy for development.[7] These national level studies on women scientists, engineers, and agricultural scientists found that women's representation in science subjects is increasing. Also, their prime employers are educational institutions and government.

The study on social status of women scientists and engineers vis-à-vis men scientists found that socio-economic background of women scientists was better than men and the status of women's spouses was better than men's spouses (Jaiswal, 1993). Regarding professional status of women scientists it was found that majority of them were at a lower level of professional hierarchy; women had a higher level of job dissatisfaction and a lower level of job commitment than men and there was latent discrimination against women, both at organisational level and in relationships.

The study by Mukhopadhyay (1994) looked into the role of patrifocal family structure[8] and ideology in women's under-representation in science and engineering education. Her study showed that patrifocal family model and its caste and class situation leads to gender-differentiated educational goals. Mukhopadhyay (1996) also talks of an ongoing tension between macro-structurally generated pressures that increase the desirability of education for women and micro-structurally generated pressures that constrain women's education in order to preserve a patrifocal cultural model of family.

[7] Major studies being *Women Scientists in R&D and Academic Organisations* (1991) by Radha Chakravarthy et al., NISTADS, New Delhi; *Women Engineers in India* (1992) by P.P. Parikh and S.P. Sukhatme, IIT, Bombay and *Professional Women in Agriculture* (1998) by D. Rama Rao and N.S. Shenoy, NAARM, Hyderabad. These studies formed major foundation stones for other empirical studies done on women scientists in India.

[8] Patrifocality refers to the kinship and family structure ideology that gives precedence to men over women. Among its characteristics are the subordination of individual goals to collective family welfare; structural features (patrilineality, patrilocality). These reinforce the centrality of sons and the peripheral status of daughters; gender-differentiated family responsibilities; regulation of female sexuality to maintain the purity of the patriline and family honour through arranged marriages and restricted male-female interactions; and female behaviour which emphasises obedience, self-sacrifice, adaptability, nurturance, restraint, and traits conducive to family harmony (Mukhopadhyay and Seymour, 1994).

Some studies also looked at challenges of making school and curriculum more gender friendly. Effects of single sex classroom teaching are not yet conclusively proved by various studies. While Chakravarthy, Jain, Nagpaul, Chawala and Mehta (1991) as well as Parikh and Sukhatme (1992) argue against it, few other studies (Dale, 2002) find it beneficial to women by providing them more sense of empowerment and giving them more opportunities for interaction with teacher and among peers, thus, facilitating better learning and higher grades.[9] Swaminathan (1991) argues that strategies and resocialisation practices for female children cannot by themselves increase the representation of women in science. This a because their concentration in certain subjects and absence from other 'hard' subjects is itself rooted in and forms part of an ideology that attempts to characterise as natural what are in fact mainly social and cultural distinctions. She also challenges the notion that sex-role conditioning can be reversed or removed by appropriate state intervention.[10] Similarly Bhog (2002) points out to the controversy generated by the National Curriculum Framework (NCF) introduced by the government on 14 November 2000, which according to her analysis, points to a move towards ensuring that women learn to play out their 'traditional' social roles as good mothers, wives, and daughters within the family and the nation.

Various other studies focused on issues of women in science careers. Few studies looked into the difference between scientific productivity of male scientists vis-à-vis their female counterparts (Basu, 1997; Chakravarthy, et al. 1991; Gupta et al., 1999; Kumar, 2001). While Chakravarthy et al.'s study (1991) showed that women scientists' productivity is marginally lower than that of men scientists, other studies did not show difference in productivity. On the issue of professional growth in career it was found that women scientists have fewer training opportunities in India and abroad. Women scientists have to do better than men to get equal professional recognition. They are not provided with adequate opportunities to participate in professional activities outside their institutions and outside their cities. These factors are also influence productivity of a scientist.

[9] However, arguments of Chakravarthy, et al.; and Parikh and Sukhatme are based more on their conviction. Inquiry of this question in a comprehensive manner was beyond the scope of their study.

[10] She exposes the inherent sexist bias in the *Report of the Working Group on Personnel Policies for Bringing Greater Involvement of Women in Science and Technology*, 1981 and argues that recommendations provided in it, namely, restructuring of courses, changes in textbooks, and so on, would not help in improving status of women as these recommendations are premised on women's prime responsibility being that of bearing and rearing of children.

Women scientists indicated marriage, childbirth, and health as reasons for interruption in their academic and professional careers. However, marriage and motherhood was not found to be hampering women scientists' productivity (Chakravarthy et al., 1991). Low salaries and the lack of suitable accommodation facilities are often cited as major reasons because of which respondents have had to refuse jobs. Most of the women scientists consider it to be uneconomical to have separate establishments at different place from their family and consequently refuse promotions (Chakravarthy, Jain, Nagpaul, Chawala and Mehta 1991; Parikh and Sukhatme, 1992; Rao and Shenoy, 1998). All the studies stressed on the professional attitude of the women in science profession and that both women and men scientists have the same orientation towards professional goals. The professional goals are not gender specific whereas family goals are gender specific (Chakravarthy, Jain, Nagpaul, Chawala and Mehta 1991; Parikh and Sukhatme, 1992; Rao and Shenoy, 1998). Few studies also give account of exploitation of women as targets of scientific investigation for reasons far removed from science (Bal, 2002). However, very few studies talked about sexual harassment and other professional harassment of women scientists. The study by Rao and Shenoy (1998) stated that about 22.3 per cent respondents indicated facing such problem.[11]

Thus, according to Krishnaraj (1991), there has been a change in the manner of looking at the issues and problems of women in the science – what was earlier a 'feminist concern' that looked at how women could accommodate themselves within the system in now turning into sharper 'feminist perspective'. Institutional barriers to the progress of women in science and research are beginning to attract the attention of scholars, but such studies have not yet designed to capture the ethos of the institutions or the nature of science in these institutions.

A comparison of Western and Indian literature on gender and science shows that Indian perspective on gender and science is more of a liberal tradition (except perhaps some eco-feminist critiques of science which fall under radical school), where as studies in the West belongs to a much wider range of liberal to radical perspectives. Western literature on gender and science deals with empirical studies exploring status and position of women in science, methodology of science, and the gender bias therein, to the content of science. Indian studies on gender and science deal more with women in science and

[11] Unfortunately, the researchers considered such incidents as marginal and opined that a congenial atmosphere/work condition existed for women at the place of work, in most organisations.

are mostly in the empirical tradition pointing out to various obstacles, status, and performance of women scientists in India. The review studies on gender bias in various research projects like those done by Anne Fausto-Sterling, Bonnie Spanier, and Helen Zweifel (Lederman and Bartsch, 2001) are also non-existent in the Indian context.

However, slight change is noted in studies that have come out in recent years and studies on androcentric nature of organisations of science (Subrahmanyan, 1998),[12] or bias in government policies on women in science and technology (Chakravarthy, 1991; Mukhopadhyay, 1996; Swaminathan, 1991) have started appearing. Scholarly attention on the topic of gender and science in India is increasing in recent years and studies on women *in* science as well as science and technology for women are also increasing.

Low Representation of Women in Science

The most basic and invariably the first question asked is related to the statistics on the female students in science and scientific professions. Almost all the studies done in India in the field of Gender and Science highlight the fast improving yet minority position of women in science in India (Chakravarthy, Jain, Nagpaul, Chawala and Mehta 1991; Parikh, 1992; Rama Rao, 1998, Research and Development Statistics 2005). The quantitative profile of women scientific/engineering personnel in India over the years as given in government reports and *Manpower Profile of India Yearbooks* (1993–2008) brought out by the Institute of Applied Manpower Research and *Research and Development Statistics* by the Department of Science and Technology, University Grants Commission's Annual Reports (various years), Census Reports of Government of India, and so on, are discussed below.

Women in Science Education

Data regarding the enrollment of girls in general higher education in science faculties at graduate and post-graduate level is given in Table 12.1. During the year 2001, women formed 37.38 per cent of the total enrolment in general science faculty at the postgraduate level and 38.47 per cent at the graduation level.

It is seen that the percentage of women enrolled in higher education and percentage of women enrolled in general science faculties are comparable

[12] Subrahmanyan (1998) opines that despite family support, the manner in which research in the science in structured in India is highly 'male' and hence has proved to be a barrier to their advancement.

Table 12.1 *Enrollment in Higher Education in General Science Faculty by Level and Sex*

Year	All India							
	Level							
	Post Graduation				Graduation			
			Total	Girls %		Total	Girls %
1971	Boys	29,713	37,806	21.4	403,266	495,285	18.58
	Girls	8,093			92,019		
1981	Boys	37,255	52,809	29.45	353,052	489,415	27.86
	Girls	15,554			136,363		
1991	Boys	47,542	71,891	33.87	475,861	756,896	37.13
	Girls	24,349			281,035		
2000	Boys	91,611	147,075	37.71	881,379	1,408,673	37.43
	Girls	55,464			527,294		
2001	Boys	91,611	148,393	37.38	891,715	1,449,244	38.47
	Girls	55,464			557,529		

Source: Manpower Profile India Yearbook, 2000, 2001, and 2008, Institute of Applied Manpower Research, New Delhi, India.

the figure is around 37 per cent. Again, in general science subjects, women enrollment is on rise over the years. Also, as enrollment of women in level of graduation and post graduation is comparable, it indicates that the chances of continuing in research increases for women who acquire the first degree in science and complete graduate education.

Enrollment of women in different science disciplines like medicine, engineering, agriculture, and veterinary science at graduation level can be seen from Table 12.2.

Table 12.2 *Percentage of Women Enrolled in Professional Science Disciplines at Graduation Level*

Year	Medicine	Engg./Tech./ Architecture	Agriculture/ Forestry	Veterinary Science
1971	20.18	0.97	0.22	1.18
1981	23.33	4.26	3.47	3.81
1987	31.48	7.16*	6.41	8.20
1999	33.68	14.87	18.93	11.57
2000	33.42	16.21	19.09	13.40

Note: * 1986 data

Source: Manpower Profile India Yearbook 2000 and 2001, Institute of Applied Manpower Research, New Delhi, India.

It can be seen that medicine has always acquired a place of preference among women and their relative presence in medicine is good as compared to other science subjects. The increase in women enrollment in engineering and agriculture has been spectacular over the years and it can be said that now more and more women are turning to technical and professional degrees. Yet, except medicine, the percentage of enrollment of women in all these professional science subjects is still below 20 per cent.

Women scientists prefer medical sciences, life sciences, chemistry, and physics and have now started entering areas of logic, mathematics, earth sciences, and so on. Amongst women engineers, Chakravarthy et al. (1991) notice the change in preference for industrial-mechanical engineering, electronics, computers, aeronautical, space and nuclear engineering, and so on.

Here it would be interesting to know the data on enrollment of girls in various faculties other than science subject and see the trend over the last decade. Comparative trend of various subjects over the years is given in Table 12.3.

It is seen that the enrollment of women is increasing sharply in all the subjects except education. Education was, and to a larger extent, still is today

Table 12.3 *Enrollment of Girls as Percentage of Total Enrollment in University Education by Faculty*

	All India					
Year	Arts	Science	Commerce	Education	Engg./Tech.	Medicine**
1960–61	18.6	–	1.1	32.5	0.8	20.4
1970–71	33.5	18.5	2.8	37.3	1.0	21.3
1980–81	37.5	27.9	15.2	46.7	4.6	23.8
1990–91	39.8	36.8	24.0	44.2*	10.9*	34.3*
1995–96	41.5	35.5	29.0	41.2*	14.2*	34.5*
1999–2000	44.9	37.4	34.0	42.6*	16.2*	37.8*
2000–01	38.0	38.0	35.6	42.8*	22.3*	40.6*
2001–02	43.8	39.1	38.7	43.5*	24.9*	40.6*
2002–03	45.4	40.2	36.7	50.6*	22.2*	44.7*
2003–04	45.4	39.8	36.7	52.1*	23.1*	46.3*
2004–05	45.7	39.7	36.9	51.1	23.3	46.5
2005–06	45.8	40.0	37.0	51.4	23.4	46.6

Notes: * Only for degree level, not postgraduate
** Excludes dentistry, public health, nursing, midwifery, and pharmacy.
Source: Manpower Profile India Yearbook 2000, 2001, 2008, Institute of Applied Manpower Research, New Delhi, India; University Grants Commission (UGC).

considered to be most apt subject for women as it is compatible with other responsibilities of women as a mother and wife. About 55 per cent women enroll themselves in faculty of education which is double the percentage share of all students enrolled in the faculty of education. But whether the stereotyped image of teaching as a female profession is changing, needs in-depth inquiry. The other trend is that while faculties like science and medicine show only a small increase in women enrollment over the last decade (an increase of one and three percentage points respectively), the increase in women enrollment in arts and engineering faculties is to the tune of 5 percentage points. Commerce faculty showed the largest increase of women enrollment over the decade of 1990s, registering an increase of 10 percentage points, from 24 per cent of enrollment in 1990 to 34 per cent during 1990–2000.

Thus, above data show that the stereotype profile of women in higher studies is breaking up and more women are taking up hard sciences and professional courses like engineering, veterinary science, and forest sciences. However, a note of caution could be expressed here. Women literacy is still very low in India with a little over half of the women being literates (54.16 per cent, Census 2001). Again, the low numbers of women in academia and science can be a simple reflection of lower female literacy in the Indian context and/or a result of more complex factors than just literacy alone. Women's participation in science subjects is also dependent on the opportunities available to them for successful science career. Thus, the following section on 'Women in Science Careers' deals with the presence of women in science careers.

Women in Science Careers

The survey of degree holders and technical personnel, Census 1981 was the last comprehensive survey done exclusively for science and technical personnel. From this survey, the total number of employed and self-employed women scientists and engineers in India in 1981 was worked out to be 9.67 per cent of the total. It was also seen that while 25 per cent of science qualified women were unemployed and trying for job, as much as 15 per cent were unemployed and not even trying for jobs. This compared poorly with that of men where the percentages were 2.1 and 0.74 respectively. There also exists high wastage ratio of women science and technology (S&T) personnel with only 43.6 per cent of them being employed, while nearly 70 per cent of male S&T personnel

Table 12.4 *Level of Qualification of Women and Total Research and Development Personnel by Sector*

Quali		All India								
		Year								
		1988			1996			2005		
		Total persons	Women	Women %	Total persons	Women	Women %	Total persons	Women	Women %
PhD	Inst.Sec	8,289	586	7.07	14,350	2,165	15.08	15,627	1,989	12.73
	Indust.Sec	2,048	78	3.81	2,489	241	9.68	4,772	703	14.73
	Total	10,327	664	6.43	16,839	2,406	14.29	20,399	2,692	13.20
PG	Inst. Sec	14,346	1,335	9.31	14,333	2,044	14.26	24,392	2,572	10.54
	Indust.Sec	6,054	392	6.48	6,796	717	10.55	19,975	3,061	15.32
	Total	20,400	1,727	8.47	21,129	2,761	13.07	44,367	5,633	12.70
Grad	Inst. Sec	7,643	544	7.12	8,879	807	9.09	13,330	1,655	12.42
	Indust.Sec	11,403	683	5.99	9,746	821	8.42	21,840	2,893	13.24
	Total	19,046	1,227	6.44	18,625	1,628	8.74	35,170	4,548	12.93
Dipl	Inst. Sec	2,489	243	9.76	5,010*	453*	9.04	3,264	525	16.08
	Indust.Sec	4,072	142	3.49	6,457*	303*	4.7	6,503	489	7.52
	Total	6,561	385	5.87	11,467*	756*	6.6	9,767	1,014	10.38
Other	Inst. Sec	8,024	645	8.04	-	-	-	2,202	192	8.72
	Indust.Sec	7,311	245	3.35	-	-	-	4,270	291	6.82
	Total	15,335	890	5.80	-	-	-	6,472	488	7.54
Total	Inst. Sec	40,791	3,353	8.22	42,572	5,469	12.85	58,815	6,939	11.80
	Indust. Sec	30,888	1,540	4.99	25,488	2,082	8.17	57,360	7,437	12.96
	Total	71,679	4,893	6.83	68,060	7,551	11.09	116,175	14,375	12.37

Notes: As on 1st April, excluding higher education sector
* Includes others also
Quali – Qualification; Inst.Sec-Institutional sector; Indus.sec – Industrial Sector; PG – PostGraduation; Grad – graduation; Dipl. – Diplom
Source: Manpower Profile India Yearbook, Institute of Applied Manpower Research, New Delhi, India.

are employed. Similarly, in the case of women agriculture graduates in the country only 34.4 per cent were found to be working (Rama Rao and Shenoy, 1998). The study on women scientists also showed that higher one goes up the ladder of scientific occupational status and hierarchy, the number of women becomes fewer. Also, the share of women in managerial and higher leadership positions has increased at a lower rate than the growth of number of women with science qualifications (Chakravarthy et.al. 1991).

According to DST data, the number of women engaged in R&D and joint sector during 1988 did not exceed 4,900 of the total 72,000 R&D personnel, that is, 6.8 per cent of the total. The latest data shows a significant increase over the years (R&D Statistics, 2007–08). There were 19,707 females directly engaged in R&D activities, which form only 15.7 per cent of the total R&D personnel engaged in R&D work. Information related to field of specialisation and qualifications were available for 14,375 candidates engaged in R&D activities (see Table 12.4). The analyses of this data showed that engineering and technology shared 39.0 per cent, natural science 29.9 per cent, medical sciences 14.8 per cent, agricultural sciences 10.3 per cent, and social sciences 6.0 per cent. By level of qualifications they comprised of 18.7 per cent PhDs, 39.2 per cent postgraduates, 31.6 per cent graduates, and 10.5 per cent diploma holders and other qualifications. In the case of institutional sector 73.9 per cent of total females possessing PhDs were deployed on R&D activities, whereas, for the industrial sector this was 26.1 per cent. Of the total women R&D personnel, 57.9 per cent, were having postgraduate and above qualifications. Of the personnel employed in R&D establishments, 15.6 per cent, were women. The data of R&D personnel by gender, discipline, and sector of employment can be seen from the Table 12.5. Here also comparison of three years, 1992, 1996 and 2005, is being made.

Table 12.4 shows the level of qualification of women and that of total research and development personnel by sector during three years 1988, 1996, and 2005. It should be noted that Table 12.6 gives scientists in only some specific disciplines and Table 12.4 might include scientists from other disciplines apart from the broad fields covered in Table 12.5.

Thus, it is seen that the number of women scientists is increasing over the years and their percentage presence among R&D personnel has increased from 6.83 per cent in 1988, and 11.09 per cent in 1996 to 12.37 per cent in 2005. Even though the percentage of women personnel has declined in the institutional sector from 1996 to 2005, the absolute number of women in the

Table 12.5 *Research and Development Personnel by Gender, Sector, and Discipline in 1992, 1996, and 2005*

Discipline	Year	Institutional Sector			Industrial Sector			Total		
		Total	Fem	%	Total	Fem	%	Total	Fem	%
Natural Sciences	1992	15,080	1,942	12.9	10,107	857	8.48	25,187	2,799	11.1
	1996	10,522	1,582	15.0	7,488	642	8.57	18,010	2,224	12.3
	2005	18,824	1,721	9.14	15,780	2,572	16.30	34,604	4,293	12.41
Agricultural Science	1992	11,139	878	7.88	736	29	3.94	11,875	907	7.64
	1996	13,185	1,993	15.1	761	30	3.94	13,946	2,023	14.5
	2005	11,528	1,280	11.11	2,496	203	8.13	14,024	1,483	10.58
Engineering/Technology	1992	19,265	1,384	7.18	21,993	1,139	5.18	41,258	2,523	6.11
	1996	16,066	1,225	7.62	14,830	973	6.56	30,896	2,198	7.11
	2005	24,675	2,757	11.17	30,612	2,856	9.33	55,287	5,613	10.15
Medical Sciences	1992	1,944	496	25.5	1,791	520	29.0	3,735	1,016	27.2
	1996	1,367	471	34.5	1,195	328	27.4	2,562	799	31.1
	2005	2,270	771	33.96	7,194	1,352	18.79	9,464	2,123	22.43
Total (All Sciences)	1992	47,428	4,700	9.91	34,627	2,545	7.35	82,055	7,245	8.83
	1996	41,140	5,271	12.8	24,274	1,973	8.13	65,414	7,244	11.1
	2005	57,297	6,939	12.11	56,028	7,437	13.27	113,379	14,375	12.68

Source: Department of Science and Technology, Research and Development Statistics 2000–2001, and 2007–08.

Table 12.6 *Full Time Equivalent of Manpower Employed in R&D by Sector, Activity and Sex as on 1 April 1998*

Establishment	R&D Activity		Auxiliary Activity		Administrative Activity		Total	
	Persons	Female	Persons	Female	Persons	Female	Persons	Female
(A) Institutional Sector								
Major Scientific Agencies	33,513	3,840	50,016	5,025	45,220	7,535	12,8749	16,400
Central Govt. Ministry/ Dept.	7,101	846	7,502	899	10,424	1,547	25,027	3,292
State Govt.	19,841	1,614	21,838	1,836	43,435	6,466	85,114	9,916
Total (A)	60,455	6,300	79,356	7,760	99,079	15,548	238,890	29,608
(B) Industrial Sector								
Public Sector/ Joint sector	9,246	800	6,454	234	3,567	517	19,267	1,551
Private sector	25,727	2,601	14,846	2,163	9,662	1,750	50,235	6,514
Total (B)	34,973	3,401	21,300	2,397	13,229	2,267	69,502	8,065
Total (A+B)	95,428	9,701	100,656	10,157	112,308	17,815	308,392	37,673

Source: Research and Development Statistics, DST, 2000–01.

category has still increased. This indicates the slow but steady increase in representation of women in science careers.

Another point, which comes forth from Table 12.4, is the presence of highly qualified women R&D personnel. More than half of women R&D personnel have education level of post-graduation and above. In year 1988, 48.8 per cent women R&D personnel were holding postgraduate or PhD degrees and in the year 1996, this proportion further increased to 68.4 per cent. In the year 2005, as much as 57.91 per cent women out of total women R&D personnel were postgraduates and doctorates. For comparison, men postgraduates and PhD degree holders constitute around 55.44 per cent of total male R&D personnel in the year 2005.

It was also found that the industrial sector has comparatively higher proportion of personnel engaged in R&D activities and low proportion of personnel engaged in administrative activities compared to the institutional sector. Table 12.6 gives the break up of all 30,8392 personnel employed in R&D establishments according to sector, activity, and sex in the year 1998.

Table 12.7 *Percentage of Female Employees in R&D Establishments by Sector and Activity, 1998*

Establishment	R&D Activity	Auxiliary Activity	Administrative Activity	Total
(A) Institutional Sector				
Major Scientific Agencies	11.5	10.0	16.7	12.7
Central Govt. Ministry/Dept.	11.9	12.0	14.8	13.2
State Govt.	8.1	8.4	14.9	16.7
Total (A)	10.4	9.8	15.7	12.4
(B) Industrial Sector				
Public Sector/ Joint sector	8.7	3.6	14.5	8.1
Private sector	10.1	14.6	18.1	13.0
Total (B)	9.7	11.2	17.1	11.6
Total (A+B)	10.2	10.1	15.9	12.2

Note: All figures in percentage
Research and Development Statistics, DST, 2000–01.

Calculating the percentage of women employees in each category of Table 12.6, shown in Table 12.7, it is found that the percentage of women employed in R&D establishments is only 12.2 per cent (Table 12.7) which is a slight increase over the 1996 figure of 11 per cent. Till 1992 their presence was below 10 per cent of the total employment while in 1988 they formed just 6.3 per cent of the total workforce in R&D organisations.

However there are some improvements in terms of the employment of women in the R&D establishments during the last few years.

Institutional sector and government organisations are major employers of women. Of the total 19,707 women working in R&D establishments, 62.26 per cent work in the institutional sector. Women engaged in R&D activities were 19,707 in numbers, which accounted for 15.55 per cent of total R&D personnel engaged in direct R&D work. Similarly, out of total personnel engaged in auxiliary activities, 15.71 per cent were women but they formed around 18.32 per cent of total administrative and non-technical staff. Majority of women are engaged in administrative and non-technical activities (67.72 per cent) while for men this percentage is 63.26 per cent. Also, it was seen that while only 32.28 per cent of total women employees get to do direct R&D works, 36.73 per cent men out of total men employed are engaged in direct R&D work.

Table 12.8 *Percentage of Female Employees in R&D Establishments by Sector and Activity, 2005*

Establishment	R&D Activity	Auxiliary Activity	Administrative Activity	Total
(A) Institutional Sector				
Major Scientific Agencies	16.89	12.72	21.02	16.92
Central Govt. Ministry/Dept.	20.98	21.28	21.97	21.52
State Govt.	12.65	12.66	16.54	14.75
Total (A)	16.28	14.31	19.43	16.99
(B) Industrial Sector				
Public Sector/ Joint Sector	10.03	9.11	19.54	10.82
Private Sector	15.46	32.18	44.48	20.68
Total (B)	14.48	25.34	38.44	18.64
Total (A+B)	15.55	15.71	20.42	17.32

Note: All figures in percentage
Source: Research and Development Statistics, DST.

Women scientist's representation in certain major R&D agencies of India in the year 2002 is presented in the Table 12.9.

It is seen that there exists wide variance in women scientist's participation in these agencies. Among these agencies, the Indian Council for Medical Research (ICMR) has highest rate of participation. This may be because of high rate of enrollment and success of women in medical science discipline in comparison to other fields of science and technology. The fact that women have entered late into science professions is confirmed by presence of more women scientists among the younger age group. For example, in the Council of Scientific and Industrial Research (CSIR), ratio of men to women scientists is 8:1, while the ratio in younger age group is 7:1 (CSIR, 1997); similarly in ICAR, in the highest age group (more than 56 years) only about 5 per cent of the scientists are women while in the youngest age group (less than thirty years), their proportion exceeds 22 per cent.[13]

The CSIR data of 1996 shows that majority of women scientists are at lower levels. Only a little above 10 per cent of total women scientists are at Scientist

[13] Unpublished census of agricultural scientists at NCAP (National Centre for Agricultural Economics & Policy Research) by Prof. D.N. Jha, Professor, NCAP. Information through personal communication.

Table 12.9 *Women Scientists in Some R&D Agencies in India in 2002*

R&D Agencies	Total Scientists	Women Scientists	Women %
CSIR*	5528	595	10.76
ICMR	615	168	27.31
ICAR**	4542	540	11.88
ISRO	11057	1056	9.55
DBT & Autonomous Institutions	456	121	26.53
DST	147	18	12.24

Notes: * 1996 data
** Unpublished census of agricultural scientists at NCAP (National Centre for Agricultural Economics & Policy Research) by Prof. D.N. Jha, NCAP
Source: author, on the basis of Research and Development Statistics, DST, NCAP, ICMR and CSIR.

Table 12.10 *Membership of the Women Scientists in World Academies*

US national academies	6.2%
Royal Swedish Academy of Sciences	5.5%
Third World Academy of Sciences	3.9%
Royal Society of London	3.6%
Chinese Academy of Sciences	5.1%
Indian National Sciences Academy	3.1%

Source: ETAN (1999), European Commission

EII and above levels. At Scientist F level, women formed only 0.6 per cent of total scientists and no women scientist was found at level G (CSIR, 1997: 20). Similarly, in BARC, very few women have risen to the top grades, just three out of total 502 women scientists and engineers (Rastogi, 2002).

From the above discussion it can be seen that across the spectrum the enrollment of women is increasing at all levels of education over the years. The gap between the enrollment rates of women at primary, secondary, and higher education is narrowing down. At all levels of higher education, namely, graduation, post-graduation, and research the proportion of women enrollment is comparable to each other. This point to the trend of retention of women in education at all levels, especially in higher education.

The proportion of women in higher education and that of women in science disciplines is same. Similarly, percentage share of women and men opting for general science courses, with respect to their absolute numbers in higher education respectively is nearly equal (UGC, 2001). This shows that no particular aversion for science subjects exists among women students in India and their preference for science courses is the same as that of men. This also indicates that if impetus is given to women education and as more women enroll for higher education, representation of women in science subjects would also increase. However, this should not be understood as lower importance of positive role models for women in science, improved pedagogy and gender-sensitive curriculum for girl students to generate and retain their interest in science, scholarships and other support structures for higher education in science and most importantly more opportunities for women in science careers.

The slightly higher proportion of women at the research level in comparison to undergraduate and graduate level indicates that the chances of women who acquire the first degree and complete graduate education continuing in research increases. It is likely that the flexible time schedule and a longer time span of six to seven years for doctoral degree can be adjusted with marriage and raising a family. It also provides an avenue for getting back into education and career after a break (Chanana, 2000).

The proportion of women as career scientists has also increased over the years but this increase is very slow and it has remained almost stagnant for more than a decade now. Also, the data shows that there exist high wastage ratio of women S&T personnel and wide gap exists between percentage of women studying science and percentage of women doing science.

The reason for low participation of women in science is believed to be three pronged. Family's attitude to women's education – particularly in challenging fields such as science; women's commitment and confidence in themselves; and society's (especially employer's) attitude towards providing opportunities to women have been cited as important areas of focus for increasing the participation of women in science. Lack of women science/engineering faculty and of chances to interact with women faculty have also been cited as discouraging to young women in science majors.

Few studies found gender prejudice, discouragement from family and perceived incompatibility of work in science careers with that of familial responsibilities to be important reasons for less representation of women in science (Chakravarthy, et al. 1991; Parikh and Sukhatme, 1992; Rama Rao and Shenoy, 1998). Study by Mukhopadhyay (1996) in India shows that, in contrast to 'internal', 'self-selection' models, families rather than students are key decision-makers. The decisions are guided by collective family goals rather than student desires or aptitudes. Both girls and boys are socialised to subordinate personal desires to collective goals and 'adapt to' and even 'prefer' choices made for them.

Studies on classroom settings (Sadker and Sadker, 1991) and experiences of women students show that various things affect women's performance: teacher's expectation from student; the stereotypes propagated by them in day-to-day lives; and the expectation from women students that they would not speak up. Thus part of the blame lies with conventional pedagogical approaches. Alternative approaches like emphasising practical applications, collaborative problem solving and group work can definitely make science more understandable and appealing to students, especially girls.

Studies have also found that women scientists seem to be concentrated in lower positions in comparison to men and have a lesser involvement in identification of research problems, formulation of hypothesis, and interpretation of results. They are more often engaged in scientifically unproductive work or routine activities like data collection and literature review and are more isolated from communications. Female science role model must be visible, accessible, and successful in order to positively affect female science graduates motivation and determination (Chakravarthy, 1992). The serious problem of lack of mentors for women in science and technology professions has also been highlighted by few studies. There are very few women at the top, and those who are there insist on mentoring women and men equally as they do not want to be seen favouring women.

However, in spite of power and gender inequality in the market, many women prefer to work outside the home in an attempt to improve their social position, rather than be subject to family-based dependency and coercion. A large number of women continue to work in gendered homes and work sites, and there is an ongoing struggle on their part to challenge the embedded patriarchal relations and existing structural inequalities, where women constantly seek to balance work and domestic responsibilities (Kelkar, Shrestha and Veena, 2002).

Conclusion

From a societal perspective, gender inequality in science represents an under-utilisation of a significant proportion of society's human capital. From the perspective of science, the unequal representation of women has implications for the production of scientific knowledge. This may not only have epistemological consequences, such as distortions of knowledge claims of science in general, but also results in substantive gap in our knowledge of the world we inhabit.

The statement 'No data, no visibility; No visibility, no priority' from the *World Science Report 1996*, UNESCO (Koblitz, 1996), aptly sums up the condition of women scientists in India. It was found that the available information on women scientists, their status and their position in the respective work sphere is not adequate. Moreover, agencies responsible for compiling and publishing such statistics work in a lackadaisical manner. For example, during 1991, postgraduate degree holders and technical personnel survey was undertaken along with census and the data are being processed by CSIR in 2001(R&D Statistics, 2001). This underlines the urgency for proper actions to be taken with regard to collection, documentation, and dissemination of data reflecting the participation, status, and contribution of women in education and career. Similarly gender-wise breakup on the percentage of professors in Indian universities is lacking.

This study suggests various areas in gender and science studies where further inquiry could be fruitful. Studies on school and classroom atmosphere in Indian settings, teaching methods, and alternative approaches especially in the context of science would be very useful. Very few studies brought out the perception and participation of women scientists in women's movement in India. Scholars feel that women's movement have a very important role in creating favourable situations for change in conditions, yet women scientists are seen to be distant from it. Thus, further study on this issue is required. It is said that invisibility of women from science in many disciplines leads to the absence of research interest and consequent lack of information on many problem areas of particular concern to women's lives (Chakravarthy et al., 1991). However, the fact that women scientists do take up different topics of research is a question which needs to be investigated through studies. There has been a change in women's attitudes towards career and personal life over the generations (Parikh and Sukhatme, 1992; Subrahmanyan, 1998). However, longitudinal study on this topic comparing more generations of scientists is desirable. Study of workplace and scientific organisations and their congeniality to women with respect to sexual harassment and/or professional harassment (which is inflicted on women scientists because they are women) is also another important area for study. Similarly, the need for comparative study between various research agencies – private laboratories, central nodal institutes, engineering and medical institutes, central universities, and other marginal universities is called for.

The Government of India realises the importance of women's education and imparting high level skills to its population including women. Special scholarships and awards have been instituted to attract students in general and women in particular to the science and technology stream, yet more needs to be done. Over all sensitivity to women's issues in all areas of administration needs to be sharpened. For example, recently published report 'IT Manpower challenge and response: Interim report of the task force on HRD in IT' by the Department of Secondary Education and Higher Education, Ministry of Human Resource Development (HRD) do not contain even a single word on women in information technology (IT) while it discusses various strategies on how to increase quantity and quality of manpower.

After globalisation, it has become important for the government to provide facilities and support to private sector. Thus it becomes the duty of the government to see that the flexibility that the private sector has in fixing compensation and other perks are not used in a discriminate and random manner, hurting the principle of equity and gender sensitivity. Apart from this, earmarking more funds and resources for women's education in general and science education in particular is yet to get higher priority in government policies.

Equality in access to science is not only a social and ethical requirement for human development, but also a necessity for realising the full potential of scientific communities worldwide and for orienting scientific progress towards meeting the needs of humankind. Hence, the difficulties encountered by women, constituting over half of the population in the world, in entering, pursuing, and advancing in a career in the sciences and in participating in decision-making in science and technology should be addressed urgently.

References

Abbott, Alison (1998). 'Plea for Female Academics in Germany', *Nature*, 393, p. 402.

Ahmad, K. (1979). 'Equity and Women's Higher Education', *Journal of Higher Education*, 5(1): 231–49.

Arun, Shoba and T.G. Arun (2002). 'ICTs, Gender and Development: Women in Software Production in Kerala', *Journal of International Development*, 14(1): 3–50.

Bal, Vineeta (2002). 'Gendered Science: Women as Practitioners and as Targets of Research', *Economic and Political Weekly*, 37(52): 5163–5167.

Balasubramanyan, V. (1985). 'Biology and Gender Bias: Some Issues in Discrimination against Women in the Workplace', *Economic and Political Weekly*, 20(20): 876–77.

Barinaga, M. (1994). 'Surprises across the Cultural Divide', *Science*, 263: 1468–72.

Basu, A. (1997). 'Contemporary Indian Women in Science: A Scientometric Study'. Unpublished paper, NISTADS, New Delhi, as quoted in N. Kumar, (2001). 'Gender and Stratification in Science: Empirical Study in Indian Setting', *Indian Journal of Gender Studies,* 8(1): 51–67.

Begum, Roshan J. and Kamala Balaraman (1975). 'Contemporary Social Problems of Women Engineers and Scientists in India'. Paper presented at the Fourth International Conference of Women Scientists and Engineers, Poland.

Bhargava, G. (1983). 'Sex Stereotyping and Sex-Congruency: Components in the Sex Role Definition of Medical Specialities in India', *Social Science Medicine*, 17, (15): 1012–26.

Bhog, Dipta (2002). 'Gender and Curriculum', *Economic and Political Weekly*, 37(17): 1638–1642.

Bleier, Ruth (Ed) (1986) *Feminist Approaches to Science*. Pergamon, New York.

Brainard, S. G. & Carlin, L. (1998). 'A Longitudinal Study of Undergraduate Women in Engineering and Science,' *Journal of Engineering Education*, 87(4): 369–375

Brian, Martin (1988). 'Mathematics and Social Interests', *Search*, 19(4): 209–214.

Chakravarthy, Radha (1986). 'Productivity of Indian Women Scientists', *Productivity*, 27(3): 259–69.

Chakravarthy, Radha, Ashok Jain, P.S. Nagpaul, Anju Chawla and Geeta Mehta. 1991. 'Women Scientists in R&D and Academic Organisations'. Report submitted to Department of Science and Technology, Government of India, National Institute of Science, Technology and Development Studies, New Delhi.

—— (1992). *Enhancing Women's Participation in S&T: Report on a Science Camp of School Girls*. WISTAR, UNESCO & NISTADS, New Delhi.

Chanana, Karuna (2000). 'Treading the Hallowed Halls: Women in Higher Education in India', *Economic and Political Weekly*, 35(12): 1012–1022.

Dale, Baker (2002). 'Good Intentions: An Experiment in Middle School Single-Sex Science and Mathematics Classrooms with High Minority Enrollment,' *Journal of Women and Minorities in Science and Engineering*, 8(1): 1–23.

Eisenhart M. and E. Finkel (1988). *Women's Science: Learning and Succeeding from the Margins*. Chicago: The University of Chicago Press.

ETAN (European Technology Assessment Network Report) (1999). *Improving the Human Research Potential and the Socio-economic Base*, European Commission, Available online at http://www.cordis.lu/rtd2002/science-society/women.htm (retrieved on 15 January 2003).

Etzkowitz and Kemelgor (2001). 'Gender Inequality in Science: A Universal Condition?' *Minerva*, 39(2): 153–174.

Fausto-Sterling, Anne (1989). 'Life in XY Corral', *Women's Studies International Forum*, 12(3): 319–331.

Feminism and Science: In memory of Ruth Bleier (1989). *Women's Studies International Forum*,12(3): 249–388.

Fox, Mary Frank (1998) 'Women in Science and Engineering: Theory, Practice, and Policy in Programs', *Signs: Journal of Women in Culture and Society* 24(1): 210–23

Fuchs, Stefan J. von Stebut and J. Allmendinger (2001). 'Gender, Science, and Scientific Organizations in Germany', *Minerva*, 39(2): 175–201.

Gender Analysis and the History of Technology Special Issue (1997). *Technology and Culture*, 38(1).

Gupta, B.M., Suresh Kumar and B.S. Aggarwal (1999). 'A Comparison of Productivity of Male and Female Scientists of CSIR', *Scientometrics*, 45(2):269–89.

Gupta, N. and A.K. Sharma (2001). 'Triple Burden among Women Scientist in India: A Sociological Study of Women Faculty at some Reputed Centres of Higher Learning and Research', *Social Action*, 51(4): 394–416.

Hacker, Sall (1989). *Pleasure, Power and Technology: Some Tales of Gender, Engineering and the Cooperative Workplace*. London: Unwin Hyman.

Hammond, Tracy and Jan Hammond (2002). 'Gender-Based Under-representation in Computer Science and Related Disciplines'. *32nd ASEE/IEEE Frontiers in Education Conference*, Boston, MA.

Haraway, Donna (1988). 'Situated Knowledges: The Science Question in Feminism and the Privilege of Partial Perspectives', *Feminist Studies*, 14(3):575–599.

Harding, Jan (1996). *Science and Technology – A Man's World? S&T in the Commonwealth 1999/2000*. London: Kensington Publications.

Harding, Sandra (1986). *The Science Question in Feminism*. Milton Keynes: Open University Press.

Harding, Sandra (1991) *Whose Science: Whose Knowledge? Thinking from Women's Lives*. Cornell University Press, Ithaca.

Heyman, Gail, Martyna, Bryn and Bhatiya, Sangeeta, (2002). Gender and Achievement-Related Beliefs among Engineering Students. *Journal of Women and Minorities in Science and Engineering*, Vol. 8, No.1: 41–52.

Higher Education Statistics Agency (2000). *Annual Report*. United Kingdom: Higher Education Statistics Agency

Holloway, M. (1993). 'A Lab of Her Own', *Scientific American*, 269(5): 94–103.

Hubbard, Ruth (1990). *The Politics of Women's Biology*. New Brunswick and London: Rutgers University Press.

Jaiswal, R.P. (1993). *Professional status of Women: A Comparative Study of Women in Science and Technology*. Jaipur: Rawat Publications.

Jasanoff, S., G.E. Markle, J.C. Petersen and T. Pitch (1995). *Handbook of Science and Technology Studies*. London: Sage Publications.

Jha,D.N Unpublished census of agricultural scientists at NCAP (National Centre for Agricultural Economics & Policy Research), New Delhi

Kelkar, G., Girija Shrestha and Veena N. (2002). 'IT Industry and Women's Agency: Explorations in Bangalore and Delhi, India', *Gender, Technology and Development Journal Special Issue*, 6(1).

Keller, E.F. (1982). 'Feminism and Science', *Signs: Journal of Women in Culture and Society*, 7(3).

Keller, Evelyn Fox. (1983) Feminism as an Analytic Tool for the Study of Science. *Academe*, 69(5): 15–21.

—— (1995). 'The Origin, History, and Politics of the Subject Called "Gender and Science": A First Person Account', in S. Jasanoff et al. (eds.), *Handbook of Science and Technology Studies*, pp. 80–94. London: Sage Publications.

Klein, R. and R. Rowland (1989). 'Hormonal Cocktails: Women as Test-Sites for Fertility Drugs' *Women's Studies International Forum*, 12(3).

Knorr-Cetina, K. (1981). *The Manufacture of Knowledge: An Essay on the Constructivist and Contextual Nature of Science*. New York: Pergamon.

Koblitz, A. Hibner (1996). Global Perspectives: Challenges in Interpreting Data. *World Science Report–1996*, UNESCO.

Krishnaraj, M. (1991). *Women and Science*. Bombay: Himalaya Publishing House.

Kumar, Neelam (2001). 'Gender and Stratification in Science: An Empirical Study in the Indian Setting', *Indian Journal of Gender Studies*, 8(1): 51–67.

Latour, Bruno and Steve Woolgar (1979) *Laboratory Life: the Social Construction of Scientific Facts*, Sage, Los Angeles, USA.

Lederman, M. and I. Bartsch (2001). *The Gender and Science Reader*. London: Routledge.

Madheswaran, S. and S. Shroff (2000). 'Education, Employment and Earnings for Scientific and Technical Workforce in India: Gender Issues', *The Indian Journal of Labour Economics*, 43(1): 127–37.

Manpower Profile India Yearbook 2000; 2001 and; 2008, Institute of Applied Manpower Research, New Delhi, India.

Mukhopadhyay, C. (1996). 'The Cultural Context of Gendered Science: An Indian Case Study'. Papers from GASAT 8, Ahemedabad, India. Available online at http://www.wigsat.org/gasat/papers1/27.txt (retrieved on 20 January 2003).

Mukhopadhyay, C. and S. Seymour (1994). *Women, Education and Family Structure in India*. Colorado: Westview Press.

Mukhopadhyay, Carol C. 1994. 'Family Structure and Indian Women's Participation in Science and Engineering', in Carol C. Mukhopadhyay and Susan Seymour (eds.), *Women, Education and Family Structure in India*, Colorado: Westview Press: 103–132.

Mulkay, M. J. (1977). 'Sociology of the Scientific Research Community', in Ina Spiegel-Rosing and Derek de Solla Price (eds.), *Science, Technology and Society: A Cross-disciplinary Perspective*, pp. 93–148. London: Sage Publications.

Parikh,P.P and Sukhatme, S.P (1992) Women Engineers in India, Report submitted to Department of Science and Technology, New Delhi.

Poonacha, Veena (2003). 'Women's Studies in Indian Universities', *Economic and Political Weekly*, 38(26): 2653–2658.

Rastogi, R. (2002). 'Achievements of Women Scientists and Technologists in BARC'. Paper presented at the Conference of Women Scientists and Technologists: Role in National Development', DBT, 8–9 March, New Delhi.

Rose, Hilary (1994). *Love, Power and Knowledge: Towards a Feminist Transformation of the Sciences*. Cambridge: Polity Press.

Sadker, David and Myra Sadker (1991). 'Sexism in American Education: The Hidden Curriculum', in Leslie R. Wolfe (ed.), *Women, Work, and School: Occupational Segregation and the Role of Education*. Colorado: Westview Press, pp.57–76.

Sayre, Anne. 1975. *Rosalind Franklin and DNA*. New York: W.W. Norton and Company.

Sclove, R.E. (1995). *Democracy and Technology*. London: Guilford Press.

Schiebinger, L. (1987). 'The History and Philosophy of Women in Science: A Review Essay', *Signs: Journal of Women in Culture and Society*, 12(2): 305–32.

Shiva Vandana (2001). 'Democratizing Biology: Reinventing Biology from a Feminist, Ecological, and Third World Perspective', in M. Lederman and I. Bartsch (eds.), *The Gender and Science Reader*. London: Routledge.

Silverman, Edward R. (1997). 'NSF Employment Study Confirms Issues Facing Women, Minorities', *The Scientist*, (April 14).

Spanier, Bonnie (1995). *Im/Partial Science: Gender Ideology in Molecular Biology*. Bloomington and Indianapolis: Indianapolis University Press.

Subrahmanyan, Lalita (1998). *Women Scientists in the Third World: The Indian Experience*. New Delhi: Sage Publications.

Sur, Abha (2001). 'Dispersed Radiance: Women Scientists in C. V. Raman's Laboratory', *Meridians: Feminism, Race, Transnationalism*, 1(2): 95–127.

Swaminathan, Padmini (1981). *Report of the Working Group on Personnel Policies for Bringing Greater Involvement of Women in Science and Technology*.

—— (1991). 'S&T for Women: A Critique of Policy', *Economic and Political Weekly*, 26(1): 59–63.

Wenneras, C. and A. Wold (1997). 'Nepotism and Sexism in Peer Review', *Nature*, 387: 341–43.

Government Reports

Chakravarthy, R., Ashok Jain, P.S. Nagpaul, Anju Chawla, and Geeta Mehta (1991). *Women Scientists in R&D and Academic Organisations*. DST Project Report, NISTADS, New Delhi.

CSIR (Council for Scientific and Industrial Research) (1997). *Profile of CSIR Scientists*. Human Resource Development Group, New Delhi.

DSHE (Department of Secondary and Higher Education) (2001). *Annual Report 2000–2001*. Ministry of Human Resource Development, Government of India.

—— (2002). *Annual Report 2001–2002*. Ministry of Human Resource Development, Government of India.

IAMR (Institute of Applied Manpower Research) (1993). *Manpower Profile India Yearbook*. New Delhi: IAMR.

—— (1995). *Manpower Profile India Yearbook*. New Delhi: IAMR.

—— (1997). *Manpower Profile India Yearbook*. New Delhi: IAMR.

—— (1999). *Manpower Profile India Yearbook*. New Delhi: IAMR.

—— (2001). *Manpower Profile India Yearbook.* New Delhi: IAMR.

Parikh, P.P. and S.P. Sukhatme (1992). *Women Engineers in India.* DST Project Report, IIT, Bombay.

Rama Rao, D. and N.S. Shenoy (1998). *Professional Women in Agriculture.* DST Project Report, The National Academy of Agricultural Research Management (NAARM), Hyderabad.

Report of the Working Group on Personnel Policies for bringing greater involvement of women in Science and Technology, 1981, Ministry of Social Welfare, Government of India.

Research and Development Statistics. 2000–2001, Department of Science and Technology, Government of India.

Research and Development Statistics, 2007–2008, Department of Science and Technology, Government of India.

University Grants Commission (UGC) (1995). *Annual Report 1994–1995.* New Delhi: UGC.

—— (1996). *Annual Report 1995–1996.* New Delhi: UGC.

—— (1997). *Annual Report 1996–1997.*New Delhi: UGC.

—— (2001) *Annual Report 2000–2001.* New Delhi: UGC.

—— (2006) *Annual Report 2005–2006.* New Delhi: UGC.

Conclusion

The Persistent Patterns!

Neelam Kumar

> Historical participation by women in science and technology has been persistent but with inconsistent patterns because of the social, economic, and intellectual obstacles that have stood in their way.
>
> *(Kohlstedt, 2004: 1)*

What Kohlstedt (2004) opines about the history of women in science continues to be true even in the twenty-first century. Modern science was born as an exclusively masculine activity. By excluding women during its professionalisation, the world of science resulted in the association of 'man' and 'scientist'. Changes did occur, but science remains gendered and social and cultural barriers do stand in the way to women's participation in (and within) science and technology. There has been enormous increase with more women studying science and getting engineering education all over the world. But is there a proportional increase in the scientific professions and faculty appointments, especially at the higher echelons? Although the gender gap is slowly diminishing, increase in the absolute numbers or proportion of women receiving graduation, postgraduation or doctorates do not tell the entire story! Globally, women account for slightly more than one-quarter of the researchers. In 121 countries with available data, women represent slightly more than one-quarter of researchers (29 per cent). Women continue to be clustered in lower academic positions and relatively fewer at the higher ranks. According to the reports by the National Science Foundation in the US, around 19 per cent of the professorship was held by women in the year 2006. Women hold a larger share of instructor and assistant professor positions (42 per cent) than of associate (34 per cent) or full professor (19 per cent) positions (NSB, 2008). Females who enjoy full recognition of their scientific achievements are still very rare. At the top fifty US universities, the proportion of female full professorships in math-intensive fields ranges from 3 per cent to 15 per cent (NSB, 2008; National Science Foundation, 2005, 2006). Women obtain nearly 30 per cent of the doctorates in chemistry, but as one goes up the ladder of prestige and seniority, the less encouraging are the numbers (Cavallaro et al., 2007). Beyond the postdoctoral level, women scientists have slower rates of promotions and less recognition through awards and hold fewer departmental chairs relative to the eligible pool (National Academy of Sciences, 2007). The situation in non-academic venues is also not encouraging, where women leave science, engineering, and technology jobs at twice the rate of men, although this figure includes not only jobholders with PhD degrees but also those with bachelor's and master's degrees (Belkin, 2008).

Is Europe different? European Commission declares that there are still too few women in science in Europe. According to the latest survey on Statistics and Indicators on Gender Equality in Science by the European Commission (*She Figures*, 2009), women in scientific research remain a minority, accounting for 30 per cent of researchers. Despite the number of female researchers growing faster than that of men and an increase in the proportion of female PhDs, the under-representation of women in scientific disciplines and careers remains a serious challenge in Europe! Over the period, there has been an increase in the overall number of female researchers in almost all fields of science in the European Union (EU): the highest growth rates have been recorded in the fields of the medical sciences, the humanities, engineering and technology and in the social sciences. Yet, *She Figures* (2009) shows that women still account for only 37 per cent of all researchers in the higher education sector, 39 per cent in the government sector, and 19 per cent in the business enterprise sector. Career progression of women and men in the higher education sector confirms a pattern of 'vertical segregation', whereby the majority of women in academia are to be found in the lower hierarchical positions. Women account for 59 per cent of all graduates, but only 18 per cent of full professors in Europe are women. The under-representation of women is even more striking in the field of

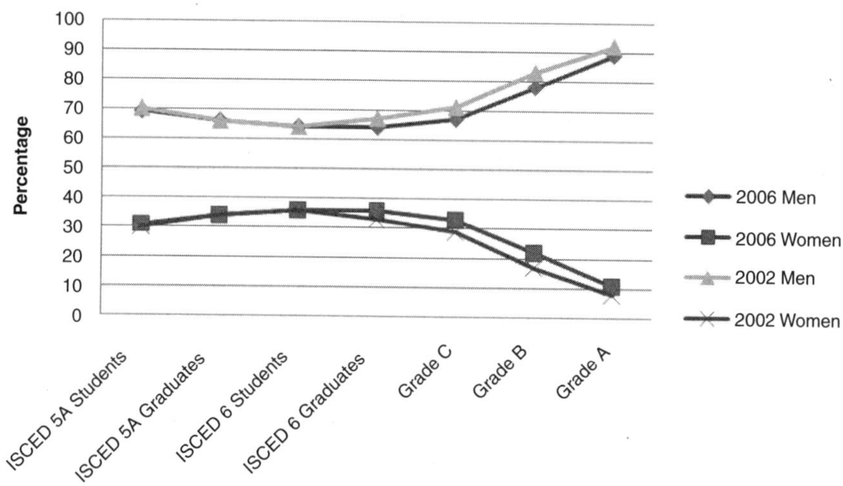

Figure E.1: Proportions of Women and Men in a Typical Academic Career in Science and Engineering, Students and Academic Staff in European Union, 2002/2006

Source: She Figures, 2009, European Union.

Notes: Definition of grades:

A: The single highest grade/post at which research is normally conducted.

B: Researchers working in positions not as senior as top position (A) but more senior than newly qualified PhD holders.

C: The first grade/post into which a newly qualified PhD graduate would normally be recruited.

ISCED 5A: Tertiary programmes to provide sufficient qualifications to enter into advanced research programmes & professions with high skills requirements.

ISCED 6: Tertiary programmes which lead to an advanced research qualification (PhD).

science and engineering, where only 11 per cent of professors are female. The report also reveals that on average only 13 per cent of institutions in the higher education sector are headed by women. Figure E.1, which became popularly known as 'Scissors Diagram' still captures women's position in science in European countries.

Asia has been discussed recently for its rising science and technology strength in a report by National Science Foundation, the United States of America (NSF, 2007). Can we assume that this reflects both female as well as male equally?[1] The data from Asian countries are not always as complete as one could wish. In Asia, women constituted only 18 per cent of researchers, but there is considerable heterogeneity (UNESCO, 2009).

[1] The report does not analyse the data using the gender dimension.

South Asia had the lowest rate of 18 per cent. Less than 30 per cent of researchers were female in the Asian Arab States (21 per cent), as well as in Japan (13 per cent) and the Republic of Korea (15 per cent). South East Asia reported a high share of female researchers at 40 per cent. Most Central Asian countries reported gender parity (around 50 per cent). Along with the expansion of higher education in China in the recent years, the number of women staff and faculty also has shown significant increase. Women represented 43 per cent of the total workforce in the year 2004 and 19.3 percent of the total professors were female (Yezhu, 2007). The situation in Asian context is not so well documented as it is in the case of United States or Europe.

Table: *Women's share of the total number of researchers (headcount)*

Region/country or territory	Women's share of the total number of researchers (headcount)*
Europe	32%
Latin America & Caribbean	46%
Other Europe	42%
Asia	15%
Central Asia	50%
Arab states in Asia	18%
South Asia	12%
South East Asia	42%
Africa	29%
World total	27%

Source: UNESCO Institute for Statistics, May 2006.
Note: *estimates

Critique that has been addressed at various times in social studies of science and technology, as well as in feminist science and technology studies (Bray, 2007; MacKenzie and Wajcman, 1999; Wajcman and Anh, 2007), that there are very few studies of gender and technology relations in non-Western societies. Yet, cross-cultural comparisons (or cross-national analyses) have been generally absent from gender, science and technology studies, with a few notable exceptions (for example, Traweek, 1988b). But even the few glimpses from Asian countries lead to curious and interesting pictures and paradoxes! Women are significantly under-represented in Information and Communication Technologies (ICT) in an analysis of data from 21 countries and noted a 'striking cross-national uniformity in the sex-typing of computer science programs' (Charles and Bradley, 2005). The dominant Western perception of the relationship between gender and computer science codes the latter as 'masculine', and the low number of women is seen at least partly as an effect of that coding. Malaysia, in contrast, represents a different case! Gender and computer science is more diversely coproduced Malaysia than commonly believed in Western research (Lagesen, 2007). There are large numbers of women in computer science, and computer science is not perceived as 'masculine'. The position of women in the IT industry, and in new rapidly transforming digital economies in countries such as the Philippines, Brazil, Malaysia, and Vietnam, gives hope of a more gender balanced division of labour. Women's education, and their positions in computer science departments and software employment being equivalent to those of men in case of Malaysia, for example, undoubtedly contributes to such relative optimism about gender and technology relations in developing countries (Mellström, 2009). But what is true in case of Malaysia is not the reality across the whole of Asia. Asia characterises diversity and heterogeneity in the case of gender and science. The gender stratification hypothesis maintains that such gender differences are closely related to cultural variations in opportunity structures for girls and women. Considerable cross-national variability in the gender gap has been explained by important national characteristics reflecting the status and the welfare of women (Else-Quest et al., 2010).

Gendered nature of science has epistemological as well as empirical implications. Science, which has its own ethos and norms, such as universalism, should be free from the particularistic and irrational biases. Why is science gendered even in the twenty-first century? What are the explanations? Insights across cultures in the form of participation patterns of women in science and technology, as well as the particularities of women's experiences, moderated by time, place, and the discipline, in relation to the larger cultural settings may reveal more about gender-science nexus. This volume selectively dips into the data and findings across a few nations, yet, reveals universal as well as a few diverse facts, such as vertical and horizontal gender segregation and a dramatic increase in the number of women getting into engineering in some nations. The fields of science, technology, engineering, and mathematics drive economies worldwide. The under-representation of women at the highest levels of science, technology, mathematics, and engineering is a matter of concern in the era of knowledge-based economies. What are the social, economic, political, socio-cultural, and socio-psychological barriers and why does science and technology remain 'gendered', transcending time and space?

References

Belkin,L. (2008). 'Diversity Isn't Rocket Science, Is It?' *The New York Times*, May 15, p. A-2.

Bray, Francesca (2007). 'Gender and Technology', *Annual Review of Anthropology*, 36: 37–53.

Cavallaro, M., S. Hansen, and M. Wenner (2007). 'Bridging the Gender Gap', as cited in Stephen J Ceci, Wendy M. Williams, and Susan M. Barnett (2009). 'Women's Underrepresentation in Science: Sociocultural and Biological Considerations', *Psychological Bulletin*, 135(2): 218–61.

Ceci, Stephen J., Wendy M. Williams, and Susan M. Barnett (2009). 'Women's Underrepresentation in Science: Sociocultural and Biological Considerations', *Psychological Bulletin*, 135(2): 218–61.

Charles, Maria, and Karen Bradley (2005). 'A Matter of Degrees: Female. Underrepresentation in Computer Science Programs Cross-nationally', in J.McGrath.

Cohoon and William C. Aspray (eds.), *Women and Information Technology: Research on the Reasons for Underrepresentation.*

Dyhouse, C. (1995). *No Distinction of Sex?* London: UCL Press.

Else-Quest, Nicole M., Janet Shibley Hyde, and Marcia C. Linn (2010). 'Cross-national Patterns of Gender Differences in Mathematics: A Meta-analysis', *Psychological Bulletin*, 136(1): 103–27.

ETAN (European Technology Assessment Network (2000). *Science Policies in the European Union: Promoting Excellence through Mainstreaming Gender Equality.* Luxembourg: Office for Official Publications of the European Communities. Available online at www.cordis.lu.

Fox, Mary Frank (2010). 'Women and Men Faculty in Academic Science and Engineering: Social-Organizational Indicators and Implications', *American Behavioral Scientist*, 53: 997–1012.

Hyde, Janet Shibley (2005). 'The Gender Similarities Hypothesis', *American Psychologist*, 60(6): 581–92.

INSA (Indian National Science Academy) (2004). *Science Career for Indian Women: An Examination of Indian Women's Access to and Retention in Scientific Careers.* New Delhi: Indian National Science Academy. Available online at www.insaindia.org.

Keller, E.F. (1995). 'The Origin, History, and Politics of the Subject Called "Gender and Science"', in S. Jasanoff, G. Markle, J.C. Peterson and T. Pinch (eds), *Handbook of Science and Technology Studies*, pp. 80–94. Thousand Oaks, CA: Sage Publications.

Kohlstedt, Gregory Sally (2004). 'Sustaining Gains: Reflections on Women in Science and Technology in 20th-century United States', *NWSA Journal*, 16(1): 1–26.

Lagesen, Vivian Anette (2007). 'A Cyberfeminist Utopia? Perceptions of Gender and Computer Science among Malaysian Women Computer Science Students and Faculty', *Science,Technology,& HumanValues*, 33(1): 5–27.

Leggon, Cheryl B. (2010). 'Diversifying Science and Engineering Faculties: Intersections of Race, Ethnicity, and Gender', *American Behavioral Scientist*, 53: 1013–28.

Mackenzie, D., and J. Wajcman (eds.) (1999). *The Social Shaping of Technology*. Maidenhead, UK: Open University Press.

Mellström, Ulf (2009). 'The Intersection of Gender, Race and Cultural Boundaries, or Why is Computer Science in Malaysia Dominated by Women?' *Social Studies of Science*, 39: 885–907.

National Academy of Sciences (2007). *Beyond Bias and Barriers: Fulfilling the Potential of Women in Academic Science and Engineering*. Washington DC.: National Academy of Sciences.

NSB (National Science Board) (2008). *Science and Engineering Indicators 2008*. Arlington, VA: National Science Foundation.

NAE (National Academy of Engineering) (2005). *Diversity in Engineering: Managing the Workforce of the Future*. Committee on Diversity in the Engineering Workforce. National Academy of Engineering. Washington, DC: National Academies Press. Available online at www.nap.edu.

NSF (National Science Foundation) (2007). *Asia's Rising Science and Technology strength: Comparative indicators of Asia, the European Union, and the United States*. Arlington, VA.: National Science Foundation (NSF).

—— (2009). *Women, Minorities, and Persons with Disabilities in Science and Engineering: 2009*, NSF 09-305. Arlington, VA; Division of Science Resources Statistics, NSF). Available online at http://www.nsf.gov/statistics/wmpd/.

NECUSE (New England Consortium for Undergraduate Science Education) (1996). *New England Consortium for Undergraduate Science Education: A Guide for Faculty*. UK: Brown University, NECUSE.

Normile, D. (2005). 'Gender Equity: Japan Mulls Workforce Goals for Women', *Science Magazine*, 308. Available online at www.sciencemag.org.

She Figures (2009). *Statistics and Indicators on Gender Equality in Science*. Luxembourg: Publications Office of the European Union.

Sodei, T. (2005). 'A Comparative Study of the Research Conditions of Women Scientists and the Present States of Women's/gender Studies in Asia countries Toward the Sustainable Development'. The fifth conference of the Science Council of Asia (SCA), Hanoi, 11–13 May. Available online at www.scj.go.jp/en.

Sonnert, G., M.F. Fox, and K. Adkins (2007). 'Undergraduate Women in Science and Engineering: Effects of Faculty, Fields, and Institutions over Time', *Social Science Quarterly*, 88: 1333–56.

Thomas, K. (1990). *Gender and Subject in Higher Education*. Buckingham: Open University Press.

Traweek, Sharon (1988a). 'Feminist Perspectives on Science Studies', *Science, Technology and Human Values*, 13: 250–53.

—— (1988b). *Beamtimes and Lifetimes: The World Of High Energy Physics*. Cambridge: Harvard University Press.

UNESCO (United Nations Educational, Scientific and Cultural Organization) (2009). A Global Perspective on Research and Development. UNESCO Institute for Statistics Fact Sheet, No. 2, UNESCO.

Yezhu, ZHAO (2007). An Analysis of the Status of Female Faculty in Chinese Higher Education, *Frontiers of Education in China*, 2(3): 415–29.

Wajcman, J. and Le Anh (2007). 'The Gender Relations of Software Work in Vietnam', *Gender, Technology and Development*, 11(1): 1–26.

Bibliography

Abir-Am, Penina G., and Dorinda Outram, (eds.) (1987) *Uneasy Careers and Intimate Lives: Women in Science*, 1787–1979. Rutgers University Press, New Brunswick.

Acker, J. (1990) Hierarchies, Jobs, Bodies: A Theory of Gendered Organisations. *Gender and Society*, 4(2): 139–158.

Ackerman, P. L. (2006) Cognitive Sex Differences and Mathematics and Science Achievement. *American Psychologist*, 61(7): 722–723.

Amano, Masako (1997) Women in Higher Education, *Higher Education,* 34(2): 215–235.

Ambrose, Susan A, Kristin L. Dunlke, Barbara B, Lazarus, Indira Nair and Deborah A. Harkus (eds.) (1998) *Journeys of Women in Science and Engineering.* Temple University Press, Philadelphia.

Andersen, H. (2001) The Norm of Universalism in Sciences. Social Origin and Gender of Researchers in Denmark. *Scientometrics*, 50 (2): 255–272.

Astin, H.S. (1991) Citation Classics: Women's and Men's Perceptions of Their Contributions to Science. In H. Zuckerman, J.R. Cole and J.R. Bruer, eds. *The Outer Circle: Women in the Scientific Community*, W.W. Norton and Company, NY, pp. 57–76.

Bagilhole, B (1993) How to Keep a Good Woman Down: An Investigation of the Role of Institutional Factors in the Process of Discrimination against Women Academics. *British Journal of Sociology of Education* 14(3): 261–274.

Bailyn, L. (2004) Time in Careers – Careers in Time. *Human Relations*, 57(12): 1507–1521.

Bailyn, L. (2006) *Breaking the Mold: Redesigning Work for Productive and Satisfying Lives*. Cornell University Press, Ithaca, NY.

Bailyn, L., and Harrington, M. (2004) Redesigning Work for Work-family Integration. *Community, Work and Family*, 7(2): 197–208.

Barber, Leslie A. (1995) U.S. Women in Science and Engineering, 1960–1990: Progress Toward Equity? *Journal of Higher Education*, 66(2): 213–235.

Barinaga, M. (1993) Is There a Female 'Style' in Science? *Science*, 260(5106): 384–391.

Barnes, Mary (1988) Mathematics: A Barrier for Women? In Barbara Caine, E.A. Grosz, and Marie de Lepervanche (eds.) *Crossing Boundaries: Feminisms and the Critique of Knowledges*, Allen and Unwin: Sydney, pp. 28–42.

Baron-Cohen, S. (2006) Sex Differences in Mind: Keeping Science distinct from Social Policy. In S. J. Ceci and W. M. Williams (eds.), *Why aren't More Women in Science? Top researchers debate the evidence.* American Psychological Association, Washington, DC, pp. 159–172.

Barr, Jean and Lynda Birke (1998) *Common Science? Women, Science and Knowledge*, Indiana University Press, Indiana.

Bart, J. (Ed) (2000) *Women Succeeding in the Sciences: Theories and Practices Across Disciplines*, Purdue University Press, West Lafayette.

Bayer, A. E. and Astin, H. S. (1975). Sex Differentials in the Academic Reward System. *Science*, 188(4190): 796–802.

Benbow, C. P. (1988) Sex Differences in Mathematical Reasoning Ability in Intellectually Talented Preadolescents: Their Nature, Effects, and Possible Causes. *Behavioral and Brain Sciences*, 11(2): 169–182.

Benbow, C. P. (1992) Academic Achievement in Math and Science of Students between 13 and 23: Are There Differences among Students in the Top 1% of Math Ability? *Journal of Educational Psychology*, 84 (1): 51–61.

Benbow, C. P., Lubinski, D., Shea, D. L., and Eftekhari-Sanjani, H. (2000) Sex Differences in Mathematical Reasoning Ability: Their Status 20 years Later. *Psychological Science*, 11(6): 474–480.

Benbow, C. P., and Stanley, J. C. (1980) Sex Differences in Mathematical Ability: Fact or Artifact? *Science*, 210(4475): 1262–1264.

Benjamin, Marina, (Ed). (1991) *Science and Sensibility: Gender and Scientific Inquiry.* Basil Blackwell, Cambridge, MA.

Benjamin, Marina, (Ed) (1993) *A Question of Identity: Women, Science, and Literature.* Rutgers University Press, New Brunswick.

Benston, Margaret Low (1992) 'Women's Voices/Men's Voices: Technology as Language'. In Gill Kirkup, Laurie Smith Keller, (eds.) *Inventing Women. Science, Technology and Gender*, Cambridge: Polity Press, pp. 33–41.

Ben-Zeev, T., Duncan, S., and Forbes, C. (2005) Stereotypes and mathematics performance. In J. Campbell (ed.), *Handbook of Mathematical Cognition* Psychology Press, New York, pp. 235–249.

Birke, Lynda (1986) *Women, Feminism, and Biology: The Feminist Challenge*, Methuen New York.

Birke, Lynda. (1991) Science, Feminism and Animal Natures I: Extending the Boundaries. *Women's Studies International Forum.* 14(5): 443–449.

Bleier, Ruth. (Autumn 1988) The Cultural Price of Social Exclusion: Gender and Science, *NWSA Journal,* 1(1): 7–19.

Bleier, Ruth. (1988): A Decade of Feminist Critiques in the Natural Sciences. Ed.by Judith Walzer Leavitt and Linda Gordon. *Signs: Journal of Women in Culture and Society,* 14(1): 182–195.

Bleier, Ruth (Ed) (1986) *Feminist Approaches to Science.* Pergamon, New York.

Bleier, Ruth (1984) *Science and Gender: A Critique of Biology and Its Theories on Women.* Pergamon, New York.

Blickenstaff, J. C. (2005) Women and Science Careers: Leaky Pipeline or Gender Filter? *Gender and Education,* 17(4): 369–386.

Bloom, N. and Van Reenen, J. (2006). Management Practices, work-life Balance, and Productivity: A Review of Some Recent Evidence. *Oxford Review of Economic Policy,* 22(4): 457–482.

Borthwick, Meredith (1984) *The Changing Condition of Women in Bengal: 1850–1905,* Princeton University Press, Princeton.

Bray, Francesca (1997) *Technology and Gender: Fabrics of Power in Late Imperial China,* University of California Press, Berkeley.

Briscoe, Anne M., and Sheila M. Pfafflin. (1979). Expanding the Role of Women in the Sciences. *Annals of the New York Academy of Sciences,* 323: 1–5

Britton, D. M. (2000) The Epistemology of the Gendered Organisation. *Gender and Society,* 14(3): 418–434.

Brush Lorelei, R. Susan, F. Chipman Donna and M. Wilson (1985) *Women and Mathematics: Balancing the Equation.* Lawrence Erlbaum Associates. Hillsdale, NJ.

Brush, Stephen G (1991) Women in Science and Engineering. *American Scientist,* 79(5): 404–419.

Brush, Stephen G (1995) Women, Science, and Universities, *Bulletin of Science, Technology, and Society,* 15(4): 205–214.

Burton Antoinette, (1994) *Burdens of History: British Feminists, Indian Women, and Imperial Culture 1865–1915.* University of North Carolina Press, Chapel Hill.

Burton Antoinette (2003) *Dwelling in the Archive: Women Writing House, Home, and History in Late Colonial India.* Oxford University Press, New York.

Butler, J (1900) *Gender Trouble: Feminism and the Subversion of Identity.* Routledge, New York.

Byrne, Eileen (1993) *Women and Science: The Snark Syndrome.* The Falmer Press, Washington, DC.

Campion, P., and W. Shrum (2004) Gender and science in development: Women scientists in Ghana, Kenya and India. *Science, Technology and Human Values,* 29(4): 459–85.

Cancian, Francesca M. (1992) Feminist Science: Methodologies that Challenge Inequality. *Gender and Society,* 6(4): 623–642.

Ceci, S. J., and Williams, W. M. (2007). *Why Aren't More Women in Science? Top Researchers Debate the Evidence.* American Psychological Association: Washington, DC.

Chambers, D. W. (1983) Stereotypic Images of the Scientist: The Draw-a-Scientist Test. *Science Education,* 67 (2): 255–65.

Chanana, K. (2000). Treading the Hallowed Halls: Women in Higher Education. *Economic and Political Weekly,* 35 (12): 1012–22.

Chanana, K. (2001) *Interrogating Women's Education: Bounded Visions, expanding Horizons*, Rawat Publications, New Delhi.

Chapman, Olive (1993) Women's Voice and the Learning of Mathematics. *Journal of Gender Studies,* 2(2): 206–222.

Charles, Maria, and Bradley, Karen. (2006). A Matter of Degrees: Female Underrepresentation in Computer Science Programs Cross-nationally. In J.McGrath Cohoon and William C. Aspray (eds.), *Women and Information Technology: Research on the Reasons for Underrepresentation,* pp. 183–204, MIT press, MIT Cambridge, MA.

Chaudhuri, Maitrayee (2004) *Feminism in India*, Kali for Women, New Delhi.

Clifton, T. J., and Shepard, E. (2004). Work and Family Programs and Productivity: Estimates Applying a Production Function Model. *International Journal of Manpower,* 25(8): 714–728.

Cockburn, Cynthia and Susan Ormrod (1993) *Gender and Technology in the Making,* Thousand Oaks, CA, Sage.

Cole, Jonathan R (1979) *Fair Science: Women in the Scientific Community,* Free Press, New York.

Cole, J. R. (1987) Women in Science. In D. N. Jackson and P. Rushton (eds.), *Scientific Excellence: Origins and Assessment,* Sage, Beverly Hills, London, pp 359–375.

Cole, J. R. and Cole S. (1973) *Social Stratification in Science*, The University of Chicago Press, Chicago.

Cole, J. R., and Singer, B. (1991) A Theory of Limited Differences: Explaining the Productivity Puzzle in Science. In H. Zuckerman, J. R. Cole, and J. T. Bruer, (eds.) *The Outer Circle: Women in the Scientific Community,* W. W. Norton and Company, New York, pp 277–323.

Cole, J. R. and Zuckerman, H. (1984) The Productivity Puzzle: Persistence and Change in Patterns of Publication of Men and Women Scientist. In P. Maehr and M. W. Steinkamp (eds.), *Advances in Motivation and Achievement,* JAI Press, Greenwich, pp. 217–257.

Cole, J. R. and Zuckerman, H. (1987) Marriage, Motherhood, and Research Performance, Science *Scientific American,* 256(2): 119–125.

Condron, Linda. (1993) Women and Technology: Feminist Perspectives. *Bulletin of Science, Technology, and Society,* 13(3): 139–141.

Conefrey, T. (2000) Laboratory Talk and Women's Retention Rates in Science. *Journal of Women and Minorities in Science and Engineering,* 6 (3): 251–64.

Connell, R. W. (2005) A Really Good Husband: Work/life Balance, Gender Equity and Social Change. *Australian Journal of Social Issues*, 40(3): 369–383.

Cowan, Ruth Schwartz (1979) From Virginia Dare to Virginia Slims: Women and Technology in American Life. *Technology and Culture,* 20(1): 51–63.

Cowan, Ruth Schwartz (1996) Technology is to Science as Female is to Male: Musing on the History and Character of our Discipline. *Technology and Culture,* 37(3): 572–582.

Creager, Angela, Elizabeth Lunbeck, Londa Schiebinger (2001) *Feminism in Twentieth-Century Science, Technology and Medicine,* University of Chicago Press, Chicago.

Crompton, R., Lewis, S., and Lyonette, C. (2007) *Women, Men, Work and Family in Europe,* London, Palgrave.

Davis, Cinda Sue, Angela B. Ginorio, Carol S. Hollinshead, Barbara B. Lazarus, Paula M. Rayman & associates. (1996) *The Equity Equation: Fostering the Advancement of Women in the Sciences, Mathematics, and Engineering,* Jossey-Bass, San Francisco.

Delamont, Sara (1987) Three Blind Spots? A Comment on the Sociology of Science by a Puzzled Outsider. *Social Studies of Science,* 17 (1): 163–170.

Delamont, Sara (2002), Hypathia's Revenge: Feminist Perspectives in SandTS. *Social Science Studies,* 32 (1): 167–174.

De Souza, Eunice (2003) *Women's Voices: Selections from Nineteenth and Early Twentieth Century Indian Writing in English,* Oxford University Press, New Delhi.

DiGuiseppe, Sarah R. and Sara F.A. Pedersen. (1996) Challenges to Equity in Science Education. *AWIS Magazine,* 25(5): 19–28.

Dikkers, J. S. E., Geurts, S., Dulk, L. d., Peper, B., and Kompier, M. (2004). Relations among Work-home Culture, the Utilization of Work-home Arrangements, and Work-home Interference. *International Journal of Stress Management,* 11(4): 323–345.

Dumais, Lucie. (1992) Impact of the Participation of Women in Science: On Rethinking the Place of Women, Especially in Occupational Health. *Women and Health,* 18:3: 11–25.

Eccles, J. S. (2006) Where Are All of the Women? In S. J. Ceci and W. M. Williams (eds.), *Why Aren't More Women in Science? Top Researchers Debate the Evidence,* pp. 199–210, American Psychological Association, Washington, DC.

Eisenberg, Anne. (1992) Women and the Discourse of Science. *Scientific American,* 267(1): 122.

Else-Quest, Nicole M.; Hyde J.S.; Linn, Marcia C. (2010) Cross-National Patterns of Gender Differences in Mathematics: A Meta-Analysis. *Psychological Bulletin,* 136 (1): 103–127

Ernst, Waltraud (1986) Psychiatry and Colonialism. *Bulletin of the Society for the Social History of Medicine,* 39: 27–31

ETAN (2000) Science policies in the European Union: Promoting excellence through mainstreaming gender equality – A report from the ETAN (European Technology Assessment Network), *Expert Working Group on Women and Science,* European Communities, Brussels.

Etzkowitz, Henry, Carol Kemelgor, and Michael Neuschatz. (1994) The Paradox of Critical Mass for Women in Science. *Science,* 266(5182): 51–54.

Etzkowitz, Henry and Carol Kemelgor (2001) Gender Inequality in Science: A Universal Condition? *Minerva*, 39 (2): 153–174.

Etzkowitz Henry, Carol Kemelgor, Brian Uzzi (2000). *Athena Unbound: The Advancement of Women in Science and Technology.* Cambridge University Press, Cambridge.

Etzkowitz H., Stefan Fuchs, Namrata Gupta, Carol Kemelgor and Marina Ranga (2008): The Coming Gender Revolution in Science. In E. J. Hackett, O. Amsterdamska, M. Lynch and J. Wajcman (eds.): *The Handbook of Science and Technology Studies.* MIT Press, Cambridge, pp. 403–428.

Evetts, Julia (1996) *Gender and Career in Science and Engineering.* Taylor and Francis, Bristol.

Fara, Patricia (2004). *Pandora's Breeches: Women, Science and Power in the Enlightenment.* Pimlico, London.

Farber, S. (1977) The Earnings and Promotion of Women Faculty: Comment. *The American Economic Review,* 67(2): 199–200.

Faulkner, Wendy (2000) 'Dualisms, Hierarchies and Gender in Engineering', *Social Studies of Science,* 30(5): 759–92.

Faulkner, Wendy (2007)`Nuts and Bolts and People': Gender-Troubled Engineering Identities, *Social Studies of Science*, 37(3): 331–356.

Fausto-Sterling, A. (1987) Society Writes Biology, Biology Constructs Gender. *Daedalus,* 116(4), pp. 61–76.

Fausto-Sterling, A. (1991) Race, Gender and Science. *Transformations* 2:2 pp. 4–12.

Fausto-Sterling, A. (1992) Building Two-Way Streets: The Case of Feminism and Science. *National Women's Studies Association Journal,* 4(3): 336–349.

Fausto-Sterling, A. (1992) *Myths of Gender: Biological Theories about Women and Men,* 2nd edition (with two new chapters). Basic Books, New York.

Fausto-Sterling, A. (2000) *Sexing the Body: Gender Politics and the Construction of Sexuality,* Basic Books, New York.

Fausto-Sterling, A. (2003) Science matters, culture matters. *Perspectives in Biology and Medicine,* 46(1): 109–124.

Findlen, Paula (1993) Science as a Career in Enlightenment Italy: The Strategies of Laura Bassi. *Isis,* 84(3): 441–469.

Fox, Mary Frank (1998) 'Women in Science and Engineering: Theory, Practice, and Policy in Programs'. *Signs: Journal of Women in Culture and Society,* 24(1): 210–23.

Fuchs, Stefan, Janina von Stebut and Jutta Allmendinger (2001): Gender, science, and scientific organizations in Germany. *Minerva,* 39(2): 175–201.

Gallagher AM, Kaufman JC, (2005). *Gender Differences in Mathematics: An Integrative Psychological Approach.* Cambridge University Press, Cambridge, UK.

Gambles, R., Lewis, S., and Rapoport, R. (2006) *The Myth of Work-Life Balance. The Challenge of Our Time. Men, Women, and Societies.* London: Wiley.

Geetha V (2002) *Gender Theorising Feminism*, Stree, Kolkata.

Gergen, Kenneth J. (1988) Feminist Critique of Science and the Challenge of Social Epistemology. in Mary McCanney Gergen (ed.) *Feminist Thought and the Structure of Knowledge*, New York University Press, New York, pp. 27–48.

Gergen, Mary McCanney, (ed.) (1988) *Feminist Thought and the Structure of Knowledge*, New York University Press, New York.

Gerhard Sonnert and Gerald Holton (1995) *Gender Differences in Science Careers: The Project Access Study*, Rutgers University Press, New Brunswick.

Gerhard Sonnert and Gerald Holton (1996) Career Patterns of Women and Men in the Sciences. *American Scientist*, 84(1): 63–71.

Glover, J (2000). *Women and Scientific Employment*, Macmillan, New York.

Glover, J. (2002). Women and Scientific Employment: Current Perspectives from the UK, *Science Studies*, 15 (1): 29–45.

Gornick, Vivian (1983) *Women in Science: Portraits from a World in Transition*. Simon and Schuster, New York.

Grosz, E.A. and Marie de L (1988) Feminism and Science. In Barbara Caine, E.A. Grosz, and Marie de Lepervanche (eds.) Crossing Boundaries: Feminisms and the Critique of Knowledges, Allen and Unwin, Sydney, pp. 5–27,

Gupta, Namrata (2007) Indian Women in Doctoral Education in Science and Engineering: A Study of Informal Milieu at the Reputed Indian Institutes of Technology, *Science Technology Human Values*, 32(5): 507–533.

Gupta, Namrata and Arun K. Sharma (2002) 'Women Academic Scientists in India'. *Social Studies of Science* 32(5/6): 901–15.

Haas L. and Hwang, H. 2007. Gender and organizational culture. Correlates of companies' responsiveness to fathers in Sweden. *Gender and Society*, 21(1): 52–57.

Haas L, Violet B., and Carolyn C. Perrucci, (eds.) (1984) *Women in Scientific and Engineering Professions*. University of Michigan Press, Ann Arbor.

Hanson, Sandra L (1996) *Lost Talent: Women in the Sciences*. Temple University Press. Philadelphia.

Haraway, Donna (1989) *Primate Visions: Gender, Race, and Nature in the World of Modern Science*. Routledge, New York.

Haraway, Donna (1991) *Simians, Cyborgs and Women: The Reinvention of Nature*, Routledge, New York.

Haraway, Donna (1991) 'A Cyborg Manifesto: Science, Technology, and Socialist-Feminism in the Late Twentieth Century,' in *Simians, Cyborgs and Women: The reinvention of Nature*, Routledge, New York, pp. 149–181.

Harding, Jan, (Ed) (1986) Perspectives on Gender and Science. Falmer Press, London.

Harding, Sandra (2006) *Science and Social Inequality: Feminist and Postcolonial Issues*, University of Illinois Press, Illinois.

Harding, Sandra (1990) Feminism and Theories of Scientific Knowledge. *Women: A Cultural Review*, 1(1): 87–98.

Harding, Sandra (1986) *The Science Question in Feminism*, Cornell University Press, Ithaca.

Harding, Sandra (1991) *Whose Science: Whose Knowledge? Thinking from Women's Lives*, Cornell University Press, Ithaca.

Harding, Sandra and Merrill B. Hintikka, (eds.) (1983) *Discovering Reality: Feminist Perspectives on Epistemology, Metaphysics, Methodology, and Philosophy of Science*, D. Reidel, Dordrecht, Holland.

Hasan, Zoya and Ritu Menon (2006) *Unequal Citizens*, Oxford University Press, New Delhi.

Hassan Farkhonda (2000) Islamic Women in Science, *Science*, 290(54896): 55–56.

Herman C. and Ellen D. (2004). Access denied: Career progression versus work-life balance amongst women network technicians. Paper presented at *Gender Perspectives in Organisations – Impact and Life Choices*. Conference Roskilde University, Denmark, January 2004.

Herzenberg Caroline L (1986) *Women Scientists from Antiquity to the Present*. Locust Hill Press, West Cornwall.

Holloway, Marguerite. (1993) A Lab of Her Own. *Scientific American*, 269(6): 94–103.

Hornig, Lilli S. (1984) Women in Science and Engineering: Why So Few? *Technology Review*,87(8): 31–41.

Horowitz, Roger (ed.) (2001) *Boys and Their Toys: Masculinity, Class, and Technology in America*, Routledge, New York.

Hyde, J. S. (1990). Meta-analysis and the Psychology of Gender Differences. *Signs: Journal of Women in Culture and Society*, 16(1): 55–73.

Hyde, J. S. (2005). The Gender Similarity Hypothesis. *American Psychologist*, 60(6): 581–592.

Hyde, J. S. (2006). Women in Science: Gender Similarities in Abilities and Sociocultural forces. In S. J. Ceci, and W. M. Williams (eds.), *Why Aren't More Women in Science? Top researchers debate the evidence*. American Psychological Association, Washington, DC, pp. 131–145.

Hyde, J. S., Lindberg, S. M., Linn, M. C., Ellis, A. B., and Williams, C. C. (2008). Gender similarities characterize math performance. *Science*, 321(5888): 494–495.

Hyde, J. S., and Mertz, J. E. (2009) Gender, Culture and Mathematics Performance. *Proceedings of the National Academy of Sciences, USA, 106(22)*, 8801–8807.

Hyde, J. S., and Linn, M. C. (2006). Gender Similarities in Mathematics and Science. *Science*, 314 (5799): 599–600.

Hyde, J.S., E. Fennema and S.J. Lamon. 1990. Gender Differences in Mathematics Performance: A meta-analysis. *Psychological Bulletin*, 107(2): 139–155.

Hyde, Janet Shibley (1990) Meta-Analysis and the Psychology of Gender Differences. *Signs: Journal of Women in Culture and Society*, 16(1): 55–73.

Iyengar Keerti (2005) How Gender Sensitive are Obstetrics and Gynaecology Textbooks? *Economic and Political Weekly*, 40(18): 1839–1845.

Jane, Duran (1997) *Philosophies of Science: Feminist Theories*. Westview Press.

Jaiswal, Rajendra Prasad (1993). *Professional Status of Women: A Comparative Study of Women and Men in Science and Technology*. Rawat Publications, Jaipur.

Jasanoff, Sheila, Gerald E. Markle, James C. Peterson, & Trevor Pinch (eds.) (1995) *Handbook of Science and Technology Studies*, Sage, Thousand Oaks, London.

Jody Bart, (ed.), (2000) *Women Succeeding in the Sciences: Theories and Practices across the Disciplines*, Purdue University Press, Purdue.

John, Mary E (1996) *Discrepant Dislocations: Feminism, Theory, and Postcolonial Histories*. University of California Press.

Jones, M. Gail, and Jack Wheatley (1988) Factors Influencing the Entry of Women into Science and Related Fields. *Science Education* 72(2): 127–142.

Jordanova, Ludmilla (1989) *Sexual Visions: Images of Gender in Science and Medicine between the Eighteenth and Twentieth Centuries,* Madison: University of Wisconsin Press.

Jordanova, Ludmilla. (1993) Gender and the Historiography of Science. The *British Journal for the History of Science,* 26(91): 469–483.

Kahle, Jane Butler, ed. (1985) *Women in Science: A Report from the Field,* Falmer Press, Philadelphia.

Kass-Simon, G., and Patricia Farnes, (eds.) (1990) *Women of Science: Righting the Record,* Indiana University Press, Bloomington.

Keller, Evelyn Fox (1982) Feminism and Science. *Signs: Journal of Women in Culture and Society,* 7(3): 589–602.

Keller, Evelyn Fox. (1983) Feminism as an Analytic Tool for the Study of Science. *Academe,* 69(5): 15–21.

Keller, Evelyn Fox. (1988) Feminist Perspectives on Science Studies. *Science, Technology, and Human Values,* 13(3/4): 235–249.

Keller, Evelyn Fox (1985) *Reflections on Science and Gender,* Yale University Press, New Haven.

Keller, Evelyn Fox (1987) Women Scientists and Feminist Critiques of Science. *Daedalus,* 116(4): 77–91.

Keller, Evelyn Fox (1983) *A Feeling for the Organism,* W. H. Freeman and Company, New York.

Keller, Evelyn Fox (1978) Gender and science, *Psychoanalysis and Contemporary Thought,* 1(30): 409–433.

Keller, Evelyn Fox. The Gender/Science System: Is Sex to Gender as Nature is to Science? In Nancy Tuna (Ed) *Feminism and Science.* Indiana University Press, Bloomington, pp. 33–44.

Keller, Evelyn Fox and Helen E. Longino, (eds.) (1996) *Feminism and Science.* Oxford University Press, New York.

Keller, Evelyn Fox and Helen Longino, (eds.) (1998) *Feminism and Science,* Oxford University Press, London.

Keller, Evelyn Fox (2001) 'The Anomaly of a Woman in Physics,' in Mary Wyer, Donna Giesman, Mary Barbercheck, Hatice Ozturk, Marta Wayne. (eds.), *Women, Science and Technology: A Reader in Feminist Science Studies,* Routledge, New York, pp. 9–16.

Kelly, Alison. (1985) The Construction of Masculine Science. *British Journal of Sociology of Education,* 6(2): 133–154.

Kimura, D. (2006) Under-representation or misrepresentation? In S. Ceci and W. Williams (eds.) *Why Aren't More Women in Science? Top Researchers Debate the Evidence.* American Psychological Association, Washington, DC, pp. 39–46.

Kirkup, Gill and Laurie Smith, Keller (eds.) (1992) *Inventing Women: Science, Technology and Gender,* Polity Press, Cambridge.

Kirkup, Gill and Laurie Smith Keller, (eds.) (1992*) Inventing Women: Science, Technology, and Gender.* Polity Press, Cambridge, MA.

Kohlstedt, S.G. (1978) In from the periphery: American Women in Science, 1830–1880. *Signs: Journal of Women in Culture and Society,* 4(1): 81–96.

Kohlstedt, S.G. (1995) Women in the History of Science: An Ambiguous Place. In: Constructing Knowledge in the History of Science (A. Thackray, ed.), *Osiris,* 2nd series, 10: 39–58.

Kohlstedt, S.G., ed. 1999. *History of Women in the Sciences: Readings from Isis.* University of Chicago Press, Chicago, IL,

Kohlstedt, S.G. (2004) Sustaining Gains: Reflections on Women in Science and Technology in 20th-century United States. *NWSA Journal* (National Women's Studies Association) 16(1): 1–26.

Kohlstedt S. G and Longino H (1997) What Do Research On Women in Science Have to Do with Each Other? *Osiris,* 12:3–15

Kohlstedt S.G. and Helen Longino (eds.) (1997)) Women, Gender and Science: New Directions, *Osiris* 12.

Kourany, Janet, A (2001) *The Gender of Science.* Upper Saddle River, Prentice Hall N.J.

Kramer, Pamela E. and Sheila Lehman. (1990) Mismeasuring Women: A Critique of Research on Computer Ability and Avoidance. *Signs: Journal of Women in Culture And Society* 16(1): 158–172.

Krishna Raj, Maithreyi (1991) *Women and Science: Selected Essays,* Himalaya Publishing House, Bombay.

Kumar, Neelam (2001) Gender Stratification in Science: An Empirical Study in the Indian Setting. *Indian Journal of Gender Studies* 8(1): 51–67.

Kumar, Neelam (2009) *Women and Science in India*: Oxford University Press, New Delhi.

Kumar, Neelam (2010) Gender Concerns in Indian Science. In Ashok Jain (ed.) Science and the Public, pp. 333–346. PHISPC series, Centre for Studies in Civilizations, New Delhi.

Kumar, Neelam (2010) *Science and Engineering in India: Gendered Equation?' In Godfroy-Genin, Anne-Sophie* (ed.) Women in Engineering and Technology Research Lit Verlag, pp. 647–660.

Kuwahara, M. (2001) Japanese Women in Science and Technology. *Minerva,* 39 (2), 203–216.

Lancaster, Roger N (2003) The Trouble with Nature: Sex in Science and Popular Culture, University of California Press, Berkeley, CA.

Laqueur, Thomas (1990). *Making Sex: Body and Gender from the Greeks to Freud* Harvard University Press, Cambridge, MA.

Laslett, Barbara, Sally Gregory Kohlstedt, Helen Longino, and Evelynn Hammonds (eds.) (1997) *Gender and Scientific Authority.* University of Chicago Press, Chicago.

Leahey, E., and G. Guo (2001). Gender differences in mathematical trajectories. *Social Forces* 80: 713–732.

Leahey, Erin (2006) Gender Differences in productivity: Research Specialization as a Missing Link, *Gender and Society,* 20(6): 754–780

Leahey, E., and Guo, G. (2001). Gender differences in math trajectories. *Social Forces*, 80, No. 2, 713–732.

Lederman Muriel and Bartsch, Ingrid *(2001) The Gender and Science Reader* Routledge, London.

Lemoine, W. (1992): The Frequency Distribution of Research Papers and Patents According Sex. The Case of CSIR, India, *Scietometrics*, Vol. 24, No. 3: 449–469.

Lerman Nina, Arwen Palmer Mohun and Ruth Oldenziel (1997) 'Versatile Tools: Gender Analysis and the History of Technology,' *Technology and Culture*, Vol. 38, No. 1, pp. 1–8.

Linn, M. C., and Hyde, J. S. (1989). Gender, Mathematics, and Science. *Educational Researcher*, 18(8): pp. 17–19+22–27

Lohan, Maria (2000). 'Constructive Tensions in Feminist Technology Studies'. *Social Studies of Science*, 30(6): 895–916.

Long, J.S (1978): Productivity and academic position in scientific career. *American Sociological Review*, 43 (6): 889–908.

Long, J. S (1992): Measures of Sex Differences in Scientific Productivity. *Social Forces*, Vol. 71, No. 1 (Sep., 1992), pp. 159–178.

Long, J. S (2001) *From Scarcity to Visibility: Gender Differences in the Careers of Doctoral Scientists and Engineers*. National Academy Press, Washington, DC.

Long J. S, Allison P. D and McGinnis (1993) Rank Advancement in Academic Careers: Sex Differences and the Effects of Productivity *American Sociological Review*, 58(5): 703–723.

Long, J. S., Allison, P. D., & McGinnis, R. (1993). Rank Advancement in Academic careers: Sex Differences and the Effects of Productivity. *American Sociological Review, 58*(5): 703–722.

Long, J. S and Fox, M. F (1995). Scientific Careers: Universalism and Particularism. *Annual Review of Sociology*, 21: 45–71.

Longino, Helen E (1992) Taking Gender Seriously in Philosophy of Science, PSA, Proceedings of The Biennial Meeting of The Philosophy of Science Association, 2: 333–340.

Longino, Helen (1996) Cognitive and Non-Cognitive Values in Science: Rethinking the Dichotomy. In Jack Nelson and Lynn Hankinson Nelson *Feminism, Science, and Philosophy of Science* (eds.) Kluwer, Dordrecht, pp 39–58.

Longino, Helen E. and Sally Gregory Kohlstedt (1997) The Women, Gender, and Science Question: What Do Research on Women in Science and Research on Gender and Science Have to Do with Each Other?, *Osiris*, 2nd Series, Vol. 12, Women, Gender, and Science: New Directions, pp. 3–15.

Longino, H.E. (2002) Subjects, Power, and Knowledge: Description and Prescription in Feminist Philosophies of Science. In J.A. Kourany, (ed.) *The Gender of Science*, Prentice Hall, Upper Saddle River, NJ, pp. 310–321.

Longino, Helen and Evelynn Hammonds (1990) Conflicts and Tensions in the Feminist Study of Gender and Science, in E. F. Keller and M. Hirsch, *Conflicts in Feminism*. pp. 164–84, Routledge, New York.

Lorber, Judith (1994) *Paradoxes of Gender*. Yale University Press, New Haven.

Lowy Ilana (1999) Gender and Science, *Gender and History,* 11:3: 514–527

McGrayne, S. (2001) *Nobel Prize Women in Science: Their Lives, Struggles, and Momentous Discoveries.* National Academies Press, Washington DC.

Massachusetts Institute of Technology (1999) *A Study on the Status of Women Faculty in Science at MIT,* http://wen.mit.edu.

McCaughey, Martha. (1993): Redirecting Feminist Critiques of Science. *Hypatia* 8:4: 72–82.

McIlwee, Judith and J Gregg Robinson (1992) *Women in Engineering: Gender, Power and Workplace Culture.* State University of New York, New York.

Mineke Bosch (2002) Women and Science in the Netherlands: A Dutch Case? *Science in Context*: 15(4): 484–527.

Mohanty, Chandra Talpade (1988) Under Western Eyes: Feminist Scholarship and Colonial Discourses. *Feminist Review, No.* 30: 61–88.

Morrow, Charlene and Perl, Teri (1998) *Notable Women in Mathematics: A Biographical Dictionary,* Greenwood Press, Westport, CT.

Mukherjee, S. (2001) 'Disciplining the Body? Health Care for Women and Children in Early Twentieth-Century Bengal', in D. Kumar (ed.), *Disease and Medicine in India: A Historical Overview,* Tulika, New Delhi, pp. 198–214.

Mukhopadhyay, Carol C (2004) A Feminist Cognitive Anthropology: The Case of women and Mathematics. *Ethos,* 32:4: 458–492.

Nagral, Sanjay (2005) Teaching Surgery: Through A Gender Lens, *Economic and Political Weekly,* 40(18): 1835–1838

National Academy of Sciences (2006). *Beyond Bias and Barriers: Fulfilling the Potential of Women in Academic Science and Engineering,* The National Academies Press, Washington, D.C.

National Science Foundation (2006) *Science and engineering indicators,* National Science Foundation, Arlington, VA.

National Science Foundation, Division of Science Resources Statistics, *Women, Minorities, and Persons with Disabilities in Science and Engineering: 2007,* NSF 07-315, Arlington, VA Available from http://www.nsf.gov/statistics/wmpd.

Nelson Lynn H and Nelson J (eds.) (1996) *Feminism, Science, and the Philosophy of Science,* Kluwer Academic Publication, Dordecht.

Nelson, Lynn Hankinson (1990) *Who Knows: From Quine To A Feminist Empiricism,* Temple University Press, Philadelphia.

Nemecek, Sasha (1997) The Furor Over Feminist Science. *Scientific American,* 276(1): 99–100.

Ng Choon Sim, Cecilia, and Rohini Hensman. (1994) Science and Technology: Friends or Enemies of Women? *Journal of Gender Studies* 3:3: 277–287.

Noble, David F (1992) *A World without Women: The Clerical Culture of Western Science.* A.A. Knopf, New York.

Nye, Robert A. (1997) 'Medicine and Science as Masculine 'Fields of Honour',' *Osiris,* 12: 60–79.

Oldenziel, Ruth (1999) *Making Technology Masculine: Men, Women, and Modern Machines in America,* 1870–1945, Amsterdam University Press, Amsterdam.

Oudshoorn, Nelly (1999) On Masculinities, Technologies, and Pain: The Testing of Male Contraceptives in the Clinic and the Media, Science, *Technology, & Human Values,* 24(2), pp. 265–89.

Osen, Lynne M (1974) *Women in Mathematics.* M.I.T. Press, Cambridge.

Outram, Dorinda (19910 'Essay Review: Fat, Gorillas and Misogyny: Women's History in Science,' *BJHS* 24: 361–7.

Dorinda Outram (1991) Fat, gorillas and misogyny: women's history in science. *The British Journal for the History of Science,* 24(3), pp 361–367.

Palomba R. ed. (2000) *Minerva's Daughters,* Franco Angeli Milano.

Palomba R., Menniti A (2003) She Figures: Women and science statistics and indicators, DG Research, Brussels.

Parikh, P.P and Sukhatme, S.P (2004), 'Women Engineers in India', *Economic and Political Weekly,* 39(2), pp. 193–201.

Pattatucci, Angela M., (ed.) (1998) *Women in Science: Meeting Challenges and Transcending Boundaries,* Sage Publications, Thousand Oaks, CA.

Phipps, A. (2008) *Women in Science, Engineering and Technology: Three Decades of UK Initiatives,* Trentham Books Ltd., London.

Pinnick, Cassandra L. (1994) Feminist Epistemology: Implications for Philosophy of Science. *Philosophy of Science* 61(4): 646–658.

Preston, Anne E. (2004) *Leaving Science: Occupational Exit from Scientific Careers* Russell Sage Foundation, Washington.

Prpic, K (2002). Gender and Productivity Differentials in Science. *Scientometrics,* 55(1): 27–58.

Ramasubban, R. (1996) Women in science: the case of India. In *World Science Report,* UNESCO, Publishing, France.

Ramaswamy, Vijaya (ed) (2003) *Re-searching Indian Women,* Manohar, New Delhi.

Ram, Kalpana and Margaret Jolly (1998) (eds.), *Maternities and Modernities: Colonial and Postcolonial Experiences in Asia and the Pacific,* Cambridge University Press.

Randel, B., Stevenson, H. W., and Witruk, E. (2000). Attitudes, Beliefs, and Mathematics Achievement of German and Japanese High School Students. *International Journal of Behavioral Development,* 24(2): 190–198.

Ray, Bharati (1995) *From the Seams of History: Essays on Indian Women,* Oxford University Press, New Delhi.

Rayman, Paula and Annabelle Brett. (1995) Women Science Majors: What Makes a Difference in Persistence after Graduation? *Journal of Higher Education,* 66(4): 388–414.

Riegle-Crumb, C. (2005). The cross-national context of the gender gap in math and science. In L. Hedges and B. Schneider (eds.), The social organization of schooling pp. 227–243. Russell Sage Foundation, New York, NY.

Reskin, B. F (1976) Sex Differences in Status Attainment in Sciences: The Case of the Postdoctoral Fellowship. *American Sociological Review,* 41(4): 597–612.

Reskin B (1978) Scientific Productivity, Sex and Location in the Institution of Science, *American journal of Sociology,* 83(5): 1235–43.

Reskin, B (1993) Sex Segregation in the Workplace. *Annual Review of Sociology*, 19: 241–70.

Reskin, B. F and Padavic, I. (1999) Sex, Race, and Ethnic inequality in United States Workplaces. In Chafetz, J.S (ed.) *Handbook of the Sociology of Gender*. Kluwer Academic/ Plenum Publishers, New York, pp. 343–374.

Richter, Derek, ed. (1982) *Women Scientists: The Road to Liberation*. Macmillan, London.

Riger, Stephanie (1992) Epistemological Debates, Feminist Voices: Science, Social Values, and the Study of Woman. *American Psychologist*, 47(6): 730–740.

Riska, Elianne and Katarina Wegar (1993) *Gender, Work, and Medicine:* Sage, Newbury Park, CA.

Rose, Hilary (1983) Hand, Brain, and Heart: A Feminist Epistemology for the Natural Sciences. *Signs: Journal of Women in Culture and Society*, 9(1): 73–90.

Rose, Hilary (1992) Gendered Reflexions on the Laboratory in Medicine, in A. Cunningham and P. Williams, (eds.) *The Laboratory Revolution in Medicine*. Cambridge Univ. Press, pp. 324–42.

Rosser, Sue V. (1984) A Call for Feminist Science. *International Journal of Women's Studies*, 7(1): 3–9.

Rosser, Sue V. (1992) Are There Feminist Methodologies Appropriate for the Natural Sciences and Do They Make a Difference? *Women's Studies International Forum*, 15(5/6): 535–550.

Rosser, Sue V. (1988) Good Science: Can It Ever Be Gender Free? *Women's Studies International Forum*, 11(1): 13–19.

Rosser, S.V (1990) *Female Friendly Science*. Pergamon Press, New York.

Rosser, Sue V (1997) The New Millennium is Now Here: Women's Studies Perspectives on Biotechnics and Reproductive Technologies. *Transformations*, 8(1): 1–19.

Rosser, Sue V., ed. (1995) *Teaching the Majority: Breaking the Gender Barrier in Science, Mathematics, and Engineering*. Teachers College Press, New York.

Rosser, S.V (2004) *The Science Glass Ceiling: Academic Women Scientists and Their Struggle to Succeed*. Routledge, New York.

Rossiter, Margaret (1982). *Women Scientists in America: Struggles and Strategies to 1940*. Johns Hopkins University Press, Baltimore.

Rossiter, Margaret W. (1974) Women Scientists in America Before 1920. *American Scientist*, 62(3): 312–323.

Rossiter, Margaret W. (1980) 'Women's Work' in Science, 1880–1910. *ISIS*, 71, (3) 381–398.

Rossiter, Margaret W (1997) Which Science, Which Women? Women, Gender, and Science, *Osiris* 12: 169–185.

Sands, Aimee (2001) 'Never Meant to Survive, a Black Woman's Journey: An Interview with Evelynn Hammonds' in Mary Wyer, Donna Giesman, Mary Barbercheck, Hatice Ozturk, Marta Wayne (eds.), *Women, Science and Technology: A Reader in Feminist Science Studies*, Routledge, New York, pp. 17–25.

Sayers, Janet (1987) Science, Sex Differences, and Feminism. In, Beth Hess and Myra Marx Ferree *Analyzing Gender: A Handbook of Social Science Research*. Sage Publications, Newbury Park, CA, pp. 68–91.

Scott, Joan. (1986). Gender: A Useful Category of Historical Analysis. *American Historical Review* 91(5), pp. 1053–1075.

Scott, Joan Wallach (1996) Gender: A Useful Category of Historical Analysis, in Joan Wallach Scott (ed.) *Feminism and History*, Oxford: University Press, London, 152–82.

Schiebinger, L (1987) The history and philosophy of women in science: A review essay. *Signs: Journal of Women in Culture and Society*, 12(2): 305–32.

Schiebinger, Londa (1987) Maria Winklelmann at the Berlin Academy: a Turning Point for Women in Science, *Isis* 78(2): 174–200.

Schiebinger, Londa (1989) *The Mind Has No Sex? Women in the Origins of Modern Science*. Harvard University Press, Cambridge.

Schiebinger, Londa (1993) *Nature's Body: Gender in the Making of Modern Science*. Beacon Press, Boston.

Schiebinger, Londa (2003) Philosopher's Beard Women and Gender in science. In Roy Porter (ed.) *Science in the Eighteenth Century*, Vol. 4, Cambridge History of Science series, Cambridge, Cambridge University Press, pp. 184–210.

Selby C.C. (ed.) (1999). *Women in Science and Engineering: Choices for Success*. New York Academy of Sciences, New York.

Shearer, Barbara S and Shearer, Benjamin F (1997) *Notable Women in the Physical Sciences: A Biographical Dictionary*. Greenwood Press, Westport, CT.

She Figures (2003) *Women and Science Statistics and Indicators*. European Commission/ Directorate-General for Research.

She Figures (2006) *Women and Science Statistics and Indicators* (Luxembourg: Office for Official Publications of the European Communities, European Commission).

She Figures (2009) *Women and Science Statistics and Indicators* (Luxembourg: Office for Official Publications of the European Communities, European Commission).

Shteir, Ann B. (1996). *Cultivating Women, Cultivating Science: Flora's Daughters and Botany in England, 1760 to 1860*. The Johns Hopkins University Press, Baltimore.

Shulman, Bonnie Jean. (1994) Implications of Feminist Critiques of Science for the Teaching of Mathematics and Science. *Journal of Women and Minorities in Science and Engineering*, 1(1), pp. 1–15.

Sindermann, Carl and Yentsch, Clarice, M (1992) *The Woman Scientist Meeting the Challenges for a Successful Career*. Plenum Press, New York.

Skelton, Christine, Becky Francis, and Lisa Smulyan (eds.) (2006) *The Sage Handbook of Gender and Education*, Sage, Thousand Oaks, CA.

Sonnert, G. and Holton, G. (1995). *Gender Differences in Science Careers*. Rutgers University Press, New Brunswick.

Sonnert, G. and Holton, G. (1996) Career Patterns of Women and Men in the Sciences. *American Scientist*, 84: 63–71.

Sonnert, Gerhard and Gerald Holton (1995) *Who Succeeds In Science? The Gender Dimension*, Rutgers University Press, New Brunswick.

Spelke, E. S. (2005) Sex differences in intrinsic aptitude for mathematics and science? A critical review. *American Psychologist*, 60(9): 950–958.

Spelke, E. S., and Grace, A. D. (2006) Sex, mathematics, and science. In S. J. Ceci and W. M. Williams (eds.), *Why aren't more women in science? Eminent researchers debate the evidence on a key controversy of our time*, pp. 57–68. American Psychological Association, Washington, DC.

Spencer, S. J., Steele, C. M., and Quinn, D. M. (1999) Stereotype threat and women's math performance. *Journal of Experimental Social Psychology*, 35(1): 4–28.

Spivak, Gayatri Chakravorty (1983) Displacement and the Discourse of Woman in Mark Krupnik (ed.) *Displacement: Derrida and After*, Indiana University Press, Bloomington, pp. 169–95.

Srinivas, M.N (1998) *Village, Caste, Gender and Method: Essays in Indian Social Anthropology*, Oxford University Press, New Delhi.

Steele, C. M. (1997) A threat in the air: How stereotypes shape intellectual identity and performance. *American Psychologist*, 52(6): 613–629.

Steinke, Jocelyn (2005) Cultural Representations of Gender and Science: Portrayals of Female Scientists and Engineers in Popular Films. *Science Communication*, 27(1): 27–63.

Stephen J. Ceci, Wendy M. Williams, and Susan M. Barnett (2009) Women's Underrepresentation in Science: Sociocultural and Biological Considerations, *Psychological Bulletin*, 135(2): 218–261

Steinke, J. (1997) A portrait of a woman as a scientist: Breaking down barriers created by gender role stereotypes. *Public Understanding of Science* 6 (4): 409–28.

Steinke, Jocelyn (2005) Cultural Representations of Gender and Science: Portrayals of Female Scientists and Engineers in Popular Films Science Communication 2005 27: 27–63 *Science Communication*, 27(1): 27–63.

Stephen Kulis (2002) More Than a Pipeline Problem: Labor Supply Constraints and gender Stratification across Academic Science Disciplines, *Research in Higher Education*, 43(6): 657–691.

Subrahmanyan, L (1998) *Women scientists in the third world: the Indian Experience*, New Delhi, Sage Publications.

Summers, L. H. (2005) Remarks on diversifying the science and engineering workforce. Cambridge, MA, National Bureau of Economic Research.

Thomas, Kim. (1988) Gender and the Arts/Science Divide in Higher Education. *Studies in Higher Education* 13(2): 123–137.

Tobias, Sheila. (1992) Women in Science-Women and Science. *Journal of College Science Teaching* 21(5): 276–278.

Toren, N. (1993) The Temporal Dimensions of Gender Inequality in Academia. *Higher Education*, 25(4): 439–55.

Trankina, M.L. (1993) Gender Differences in Attitudes Toward Science. *Psychological Reports* 73 (1): 123–30.

Traweek, Sharon (1988) 'Feminist Perspectives on Science Studies', *Science, Technology and Human Values*, 13(no 3–4): 250–253.

Traweek, Sharon (1988) *Beamtimes and Lifetimes: The World of High Energy Physics*, Harvard University Press, Cambridge.

Traweek, S. (1988) *Beamtimes and lifetimes: The World of High Energy Physicists.* Harvard University Press, Cambridge, MA.

Tuana, Nancy, (ed.) (1989) *Feminism and Science.* Indiana University Press: Bloomington.

Uberoi, Chanchal. (1993) India's Long History of Women in Science, *Physics Today* 46(4): p.92.

UNESCO Institute for Statistics. (November 2006). Women in Science: Under-represented and Under-measured. *UIS Bulletin on Science and Technology Statistics* Issue No. 3, November 2006.

Valian, Virginia (1998) Why so Slow? *The Advancement of Women:* MIT Press, Cambridge, Mass.

Vetter, Betty M (1976) Women in the Natural Sciences. *Signs: Journal of Women in Culture and Society,* 1(3): 713–720

Vetter, Betty M. (1988) Participation of Women and Minorities in Science on Decline. *Science,* 239(4840): 653–654

Wajcman, Judy (1991): *Feminism confronts Technology,* Polity Press, Cambridge.

Wajcman, Judy (2000) 'Reflections on Gender and Technology Studies: In What State is the Art?' *Social Studies of Science* 30(3): 447–64.

Wajcman, Judy (2004) *Technofeminism.* Polity Press, London.

Warner, Deborah. (1978) Science Education for Women in Nineteenth-Century America. *ISIS* Vol. 69, (1): 58–67

Wasserman Elga (2000) The Door in the Dream: Conversations with Eminent Women in Science, Joseph Henry Press.

Wazir, Rekha (ed.) (2000) The Gender Gap in Basic Education: NGOs as Change Agents. Sage Publications, New Delhi.

Weasel, Lisa H (2004) Feminist Intersections in Science: Race, Gender and Sexuality through the Microscope, *Hypatia,* 19(1): 183–193

Wenneras, C and Wold, A (1997), 'Nepotism and sexism in peer-review', *Nature,* 387, (6631): 341–343.

Wertheim, Margaret (1995) Pythagoras' Trousers: God, Physics, And The Gender Wars. Times Books, New York.

Women in Science (1992) (thematic issue). *Science* 255: 1333, 1365–1386.

Women in Science: A Man's World (thematic issue). (1975). *Impact of Science on Society,* 25.

Women in Science (1994) Comparisons across Cultures (thematic issue). *Science,* 263: 1389–93, 1458–1459.

Women, Science, and Society (1978) (thematic issue). *Signs: Journal of Women in Culture and Society.*

Women's Health Research (thematic section). (1995) *Science,* 269: 765–801.

Women, Technology and Innovation (1981). (thematic issue). *Women's Studies International Quarterly* Ed. by Joan Rothschild.

Wylie, Alison (1997) The Engendering of Archeology: Refiguring Feminist Science Studies, *Osiris Osiris*, Vol. 12, pp. 80–99

Xie, Yu and Shauman, K.A. 1998): Sex Differences in Research Productivity: New Evidence about an Old Puzzle. *American Sociological Review*, 63(6): 847–870.

Xie, Y and Shauman, K.A. (1999) Gender differences in research productivity, *The Scientist*,, 13(19): p. 10.

Xie, Yu and. Shauman, K.A. (2003) *Women in Science: Career Processes and Outcomes*, Harvard University Press, Cambridge.

Zuckerman, H., J. R. Cole, and J. T. Bruer, (eds.) (1991) *The Outer Circle: Women in the Scientific Community*. W. W. Norton and Company, New York, NY.

Index